AMERICAN POPULAR MUSIC BUSINESS IN THE 20TH CENTURY

Russell Sanjek
David Sanjek

New York Oxford
OXFORD UNIVERSITY PRESS
1991

Oxford University Press

Oxford New York Toronto
Delhi Bombay Calcutta Madras Karachi
Petaling Jaya Singapore Hong Kong Tokyo
Nairobi Dar es Salaam Cape Town
Melbourne Auckland

and associated companies in
Berlin Ibadan

This volume is an abridged and expanded version of
Volume 3 of *American Popular Music and Its Business: From 1900 to 1984*,
copyright © 1988 by Elizabeth Sanjek

Published by Oxford University Press, Inc.,
200 Madison Avenue, New York, New York 10016

Library of Congress Cataloging-in-Publication Data
Sanjek, Russell.
American popular music business in the 20th century
/ Russell Sanjek, David Sanjek.
p. cm.
Includes bibliographical references and index.
ISBN 0-19-505828-3
1. Popular music—United States—History and criticism.
2. Music—United States—20th century—History and criticism.
3. Music trade—United States.
I. Sanjek, David. II. Title.
ML200.S263 1991 90-47745

9 8 7 6 5 4 3 2 1

Printed in the United States of America
on acid-free paper

Preface

THIS book is an abridgment of the third volume of *American Popular Music and Its Business—The First Four Hundred Years* by Russell Sanjek, my late father. It covers the years 1900 to 1984, a rich and provocative period in the history of American entertainment, one marked by persistent technological innovation, an expansion of markets, the refinement of techniques of commercial exploitation, and the ongoing democratization of American culture. The invention of motion picture film, radio, and television as well as the current range of reproductive technologies, from audio cassettes to the compact disc and digital audio technology, has broadened the avenues through which popular music can reach the public. Publishers and record company executives in turn have utilized each of these media to market their products while the body of writers and performers increasingly has represented a broader range of American society. The creation of such diverse musical forms as blues, jazz, country and western, latin and salsa, and rap music reflects the diversity of our cultural landscape. These many developments have also necessitated the rewriting of the Copyright Act, as new rights and privileges of composers and performers have arisen in response to expanding markets and novel technologies. Today American culture is one of this nation's major growth industries and perhaps our principal export to the world at large.

To allow this rich and complex history to reach the widest possible audience in the clearest possible manner, I have prepared this abridgment and updating of my late father's work. This has required reducing the original volume by half, thereby streamlining the narrative without, I hope, sacrificing clarity or readability. The original thirty-seven chapters have been telescoped into thirteen by joining together common subjects and reordering the temporal arrangement of others to carry out the overall narrative. In addition, I have included in this Preface information germane to the narrative from the second volume, *From 1790 to 1909*, of *American Popular Music and Its Business*. While the present work concerns the twentieth century, events from the first decade of this period as well as noteworthy material from the nineteenth century discussed in the second volume have been summarized so as to provide an appropriate context for the main body of the text. Finally, I have added a conclusion that addresses the salient developments in the American music industry since 1984, the point at which my father ended his original manuscript. While I have been forced to sacrifice much fascinating anecdotal material from the longer work, I hope I have not thereby diminished any of my father's passionate fascination with history's rich pageant. The three volumes that comprise *American Popular Music and Its Business—The First Four Hundred Years* were a labor of love written during a long and eventually losing battle with cancer. Truly, the work prolonged his life as it allowed him to tell the story of what fascinated him most: the diverse forms in which the American consciousness has expressed itself musically.

Several themes guide the anedotal record this book provides and—as much as the many individuals, inventions, recordings, and legal wrangles—drive its narrative. They are, first, the many, many technological innovations that have permitted the transmission of music through a wide range of media. Corporate, creative, and scientific interests have worked together to create, market, and utilize those media to bring music to the public's attention. Second, each of these inventions has also been utilized by commercial interests to profit from the work of music artists. The progress of the twentieth century has shown that the control of the entertainment marketplace increasingly rests in a small body of conglomerates. It has also illustrated that on occasion less than legal means have been used to permit music to reach the public through manipulation of the marketplace. Third, the course of the twentieth century has seen the forms of popular music broaden and, in effect, become ever more democratic. The musical canvas has become richer and more diverse as the work of marginalized members of American society has entered the mainstream. Without this diversity, American culture would constitute a narrow tradition. Finally, technological evolution has in turn necessitated that the laws respecting copyright undergo a similar transformation, although it is inescapable that the law will always lag behind the laboratory. Each time an invention increases the variety of means whereby composers or performers might earn royalties for their work, the law has had to determine how these royalties might be calculated and how that technology affects the concept of intellectual property.

In turning to the material germane to the central discussion of this book, we must first explore the principal technological innovations of the latter half of the nineteenth century that had an important impact upon the music industry: the phonograph and the pianola, otherwise known as the player piano. Both were a boon to music publishers and composers as well as the public. They made it possible not to have to be physically present at a performance to sample many forms of music and opened new avenues of profit.

Thomas Alva Edison lays claim to the invention of the phonograph although several individuals helped to perfect it. Edison began experimenting with a speaking machine in 1877 in an attempt to capitalize upon Alexander Graham Bell's telephone, which had been publicly unveiled the year before. It was his hope that a talking machine might be marketed as a stenographic aid and a benefit to the deaf. He had stumbled upon the principle behind the phonograph when an embossing needle he was using to capture vibrations on paraffin paper accidentally pricked his finger. No one knows what epithet or expletive Edison might have uttered, but soon thereafter his agent was marketing a turntable and cylinder recording apparatus as a "talking machine," one of whose possible uses was illustrated when Edison recorded himself singing "Mary Had a Little Lamb."

In 1878 the Edison Speaking Phonograph Company was founded. Initial public response was great, and some exhibitors made as much as $1800 a week demonstrating the invention by making on-the-spot recordings. Interest waned, however, when the technology proved too frail. Into the breach stepped Alexander Graham Bell, who with two partners—Chichester A. Bell, an English-born cousin, and Charles Sumner Tainter, an American acoustic engineer—sought to create a more dependable recording system than Edison's fragile tin foil-covered cylinder. In 1884 they came up with a wax-coated cardboard cylinder, an improved diaphragm, a fluctuating stylus guided by the recording grooves, and an electric battery-powered motor instead of the original, unsteady hand-cranked device. Edison, recognizing the superiority of Bell's innovations, sought to pool patents and establish a joint enterprise, but Bell struck out on his own. Bell demonstrated the new "graphophone" in 1887 and established the American Graphophone Company.

To regain a competitive edge, Edison attracted public attention by bringing musical performers to his Menlo Park, New Jersey, laboratories to make recordings. While the recordings were initially for publicity's sake alone, their popularity led to the manufacture of coin-operated machines on which disks could be played. By 1889 Edison started to offer his recordings for sale, although he complained that other entrepreneurs were stealing his patents. He was unable to exert a monopoly, however, and other companies began to spring up, including the Columbia Phonograph Company, which was founded by some of Graphophone's original investors. They and others advertised their cylinders for home use or on coin-operated machines in record parlors. These predecessors to the present-day jukebox soon caught the public's attention and emptied its pockets. One dealer reported that in only six months five of his machines brought in over $3,000.

Despite such receipts, Edison stubbornly resisted commercial recordings, while Columbia, soon to be the largest and most successful holder of an Edison patent release, concentrated on the coin-operated automatic players. A market clearly existed for a home phonograph machine, but a disk was needed to replace the cumbersome cylinders. It was invented by the German-born Emile Berliner. In 1888, after having created vital transmitters for Bell Telephone, he conceived of engraving sound vibrations on a metal disk by a chemical action and using a hand-cranked machine on which to play them. His business, however, the Gramophone Company, remained a modest endeavor, and his technological innovation lacked adequate marketing, particularly as a result of the uncertainty of turntable speed produced by manual operation. Edison re-entered the scene and formed the National Phonograph Company. He soon developed a hand-wound spring-driven home phonograph as well as a system to mass produce records, five masters at a time. Columbia at the same time still ruled the cylinder market and had begun to sign well-known performers to exclusive recording contracts, the first said to be the U.S. Marine Band under the baton of the celebrated composer John Philip Sousa.

The hand-cranked machine perfected by Berliner remained inefficient but was the only available technology until 1896, when mechanical genius Eldridge Reeve Johnson, in conjunction with Berliner's company, invented the Victrola. It had a motor-powered drive and an improved sound box. Its new seven-inch recordings were made of celluloid, rather than vulcanized rubber, and their louder sound was an advantage over cylinders. Disgruntled competitors attempted to remove Berliner's firm as a serious competitor by campaigning against Johnson's innovations. His company withstood the opposition, expanded its operations internationally, and opened the first record factory in Hanover, Germany, in 1898. Soon thereafter Berliner and Johnson began to quarrel over patents, and Johnson separated from his exployer. In 1900 Johnson set up his own firm, the Consolidated Talking Machine Company, and started to sell records and a player under the name Gramophone. Columbia sued Johnson and sought to have him forbidden to use the Gramophone name. The courts agreed with Columbia's demand, and Johnson replaced the name with one of his own: Victor. Desirous of regaining his association with Johnson, Berliner horse-traded a set of agreements with the inventor, and they formed the Victor Talking Machine Company in 1901.

While debates over patent rights continued, a structured phonographic marketplace was established. Three giants dominated the commercial arena with varied products: Edison, cylinder only; Victor, disks only; and Columbia, both. Production of recorded music multiplied many times in a decade, from 3,750,000 to 27,500,000 in 1909, with a cylinder-to-disk ratio of two to one at the start, and one to nine in 1914, several years after Columbia adopted a disks-only policy. The number of cylinder and record players grew in proportion, from an estimated near million produced in 1904. Victor alone produced 250,000 talking machines in the next five years.

Distribution of both recordings and phonographs was initially handled by mail order and outlet chains, but soon they could be purchased anywhere, from bicycle shops to department stores. Victor and Columbia fixed prices and offered a 15 percent discount on all merchandise. Edison, on the other hand, catered to a mid-American market of rural and small-town consumers. Victor soon dominated the market and benefited from international marketing agreements, particularly those with British Gramophone, who contributed the now famous "Little Nipper" logo, the dog listening to "his master's voice," which began to appear on all Victor records in 1902. British Gramophone that year also inaugurated Victor's higher priced, culturally auspicious Red Seal line, which catered to sophisticated tastes by making some of the world's greatest concert artists available. The Real Seal line began with recordings of Russia's Imperial Opera and soon added the work of opera's Enrico Caruso to their roster. Caruso would become one of Victor's biggest sellers and help the label to dominate the concert field for years.

Victor also initiated royalty payments to artists for recordings. Caruso's earnings illustrate not only how quickly the recording industry had become a major commercial market but also the potential earnings that performers stood to make. When Caruso signed his Victor contract in 1904, he was paid $4,000 for the first ten sides and forty cents per disk, plus an advance on royalties of $10,000 for the coming ten years. His income from recordings over his lifetime was reckoned to be from two to five million dollars.

The other major technological innovation of the second half of the nineteenth century was the pianola, or mechanical piano, first demonstrated at the Philadelphia Exposition in 1876. Its substantial popularity must be examined in the context of the American public's fascination with keyboards and home entertainment. It is estimated that by 1887 some 800,000 pianos had been purchased and more than half a million students studied the instrument. In 1902, another survey found that 92,000 musicians and music teachers were employed in this country, twice as many as in England, a society supposedly more cultured than that of the rough and tumble United States.

In 1850 piano mechanic John McTammany, whose tombstone reads "Inventor of the Player Piano," was granted a series of patents on the mechanical piano. It operated when one pumped a small bellows, which provided air to activate organ reeds and simultaneously worked a perforated paper roll. While McTammany had the wherewithal to design the pianola, he lacked the means to market it, and in 1888 he sold his patents to William B. Tremaine, who merchandised them as he did the products of the Aeolian Organ Company, which he owned as well. Later in 1899 Tremaine introduced the "Aeriola," a self-playing piano perfected by inventor Edwin Scott Votey. It was a pedal-operated machine which, when attached to a piano, enabled one to play the instrument mechanically while using a piano roll. Eventually 75,000 player pianos and a million music rolls were sold. Companies other than Tremaine's

Aeolian benefited from consumer interest. In 1898, the Wurlitzer Company of Cincinnati, Ohio, marketed the first successful coin-operated player piano, and piano parlors sprang up around the country. Enthusiasm for the mechanical keyboard peaked in the early 1920s by which time it had become a $10 million business. In 1921, of the 341,652 keyboards sold, 208,541 were mechanically powered.

The popular music of the day played on victrolas or placed upon piano rolls largely came from the commercial publishers and composers we have come to identify with Tin Pan Alley. That name came into being when the publisher M. Witmark and Sons relocated their offices at 49–51 West 28th Street. It is said, perhaps apocryphally, to have been coined by writer Monroe H. Rosenfeld to describe the firm's piano, which had been fixed to produce the tinkling syncopation of a current popular form of music by interweaving paper strips between its strings.

While American music publishing has a long and illustrious history, it was advances in the marketing and publicizing of catalogues that typified its activities in the second half of the nineteenth century. Practices were developed that allowed material to reach the broadest range of consumers through the widest possible means. Stress was put upon using popular performers to advertise the publishers' wares in return for various forms of financial remuneration—what came to be identified as "pay for play" and later as "payola." "Advance" or "professional" copies of sheet music were given to performers, often featuring their pictures on the cover. This might also involve giving them the "exclusive" rights to perform a particular composition. Often a popular artist would interpolate such an "exclusive" piece into a performance, stopping the program and thereby giving the publisher publicity. Another strategy was to have performance instigated by a "plant" in the audience. Publishers also employed individuals who came to be known as "song pluggers," often the writers of the very material they promoted, who demonstrated compositions before performers or theatrical managers or music sellers in order to entice them to feature the material. Orchestra and dance band leaders were provided with sheet music free of charge, often accompanied by large-sized folders to hold musicians' music on the back of which was displayed the publisher's name in three-inch letters. Innovative forms of marketing were employed as new technologies opened up novel agencies for publicity. One was the projection of stereoptican slides with illustrated song lyrics at theaters, which led to the birth of the "sing along." This was first used to promote Robert Lowry's sentimental classic "Where Is My Wandering Boy Tonight?," and soon songs were being written specifically for stereoptican projection.

Costs for promotion rose quickly and began to eat into publisher's profits. By the middle of the 1890s it was estimated to cost $1300 to launch a song: $250 for 10,000 professional copies; $50 to print a star's picture on the cover of regular copies; $500 for advertising in the trade papers; an initial payment of $500 to a performer guaranteeing to feature the song professionally. Understandably, the odds against success rose equally with the potential profit margin. Only one out of 200 songs made a substantial profit. Fewer than half made back the initial $1300 promotion investment.

At the turn of the century publishers' interest focused particularly upon "production music"—that which was written for musical comedies and four-act plays. It was from this repertoire that most orchestras in large hotels, plush restaurants, and vaudeville houses drew their selections. The attention paid to this branch of entertainment indicated that as much as the publishers of Tin Pan Alley controlled most popular music, a small body of theatrical producers monopolized what appeared on the stages not only of Broadway but also across the country. By 1906 the American theater was a $200-million enterprise employing over 100,000 people, principally under the control of the Klaw-Erlanger organization and newcomers like the Shubert brothers. Klaw-Erlanger, together with several other booking agents, formed a theatrical trust, "the Syndicate," in 1896 which controlled more than 700 theaters nation-wide. Other producers and theater managers trusted their judgment, but whether they did or not "the Syndicate" forced them to accept what it offered, regardless of quality. Its virtual stranglehold over popular entertainment held fast until 1916.

The most illustrious popular composers of the late nineteenth and early twentieth century were Victor Herbert, John Philip Sousa, and George M. Cohan. Herbert was perhaps the first popular American composer to realize the potential monetary gains *writers* might make from their work. In the past composers too often sold their material outright or failed to maximize the possible avenues of exploitation. Herbert insisted that no less than 5 percent of box office receipts be paid to him and his collaborators; that no songs or music by other writers be interpolated into performances of his work; and that he must control the final libretto. His string of hits—35 successful musicals over 30 years—illustrates the degree to which his music appealed to the public. Herbert was also among the first to institutionalize the collection of royalties through his participation in the founding of the American Society of Composers and Publishers (ASCAP).

John Philip Sousa, like Herbert, hit a responsive chord with his many marches and, although a less effective businessman than Herbert, managed to become one of the highest paid composers of his day. Originally intending to be, like Herbert, an operetta composer, Sousa instead became the bandmaster of the U.S. Marine Band, and it was the marches he composed especially for that ensemble that gained him international fame. He resigned that post in 1892 and soon thereafter signed a contract with the John Church Company, of Cincinnati, giving him royalties that eventually rose to 15 percent of the retail price from sheet music sales. In the period July to September 1894 three marches alone brought in $6,588.94. Later he would also profit from the new technology of phonograph recording as performances by his ensemble were among the top sellers of the day.

George M. Cohan proved very popular on the stage with musicals that featured down-to-earth vernacular dialogue and show-stopping tunes at a time when most stage musicals were European-influenced and British musical comedies predominated. The multi-talented Cohen, who wrote the book, music, and lyrics in addition to starring in his shows, appealed to the patriotic spirit of the American public. Such classics as "Yankee Doodle Dandy," "You're a Grand Old Flag," and "Over There" helped to earn him a Congressional medal, the first such honor awarded to an American songwriter. In 1910, when the prolific writer/performer had six hit shows running on Broadway and concurrently on tour, his friend Marc Klaw of the theatrical firm Klaw-Erlanger said Cohan represented "the spirit and energy of the twentieth century—a concentrated essence, four-cylinder power—a protest against and apology for the elimination of the palmy days of the drama . . . This youngster struck a universal chord in his songs and plays, and that is why we know and love him."

While individuals like Herbert, Sousa, and Cohan represented the mainstream tastes of the period, American popular music was beginning to be diversified if not in substance at least in the ranks of those who produced and profited from it. While the Civil War and the period of Reconstruction had modified the legal status of black Americans, racism was far from eradicated even if avenues for black advancement, including in the music business, did begin to appear. Admittedly, the genre to which they were connected had the unappetizing appelation of "coon music," a none too substantial advancement over the prior "darky music." The term applied to all songs sung in black dialect or in some way involving a black person—although it must be remembered that most "coon songs" were performed by whites. Nonetheless, black performers did gain fame, one of the earliest of whom was Ernest Hogan, author of "All Coons Look Alike to Me." This song was one of the era's greatest hits, both as a vocal performance and in a band arrangement by white musician Max Hoffman. Hogan went on to write and appear in a number of musicals and, up to his death in 1907, was said to be the greatest performer ever seen in the American black theater. When pressed about the racial nature of the song that made him famous, Hogan stated he felt it was composed at a time when music needed a new direction, and its success made the path easier for other black artists to follow. Those included Irving Jones, author of "Ballin' the Jack" (1911); Gussie L. Davis, best known for waltzes; Bob Cole, and the brothers James Weldon and J. Rosamund Johnson, who, in addition to several successful musical comedies and comic operas, penned "Lift Every Voice and Sing," which became the National Association for the Advancement of Colored People's official song. Most successful, perhaps, was the stage performer Bert Williams, who, along with his partner George Walker, introduced the cakewalk to the American dance tradition and starred in several popular musicals. Upon Walker's death in 1911 Williams went on to even greater fame as the only black performer in Florenz Ziegfeld's *Follies* in 1910 and he appeared in each subsequent edition until his death. Also, Williams had started recording for the Victor Talking Machine Company in 1901 and was the first black artist to have a sustained career in the recording business.

The musical landscape was most memorably transformed by black writers during this period through the introduction of ragtime, which took embellished or syncopated melody lines and played them off against the "slow drag" of parade music. One of the predecessors of American jazz, ragtime first appeared on the scene in 1896 when Max Hoffman, who served the same function for Ernest Hogan, transcribed performer Ben Harney's "Mr. Johnson Turn Me Loose." The rhythmic eccentricities of Harney's style began the ragtime vogue and produced a number of million sellers in the form of sheet music, recordings, and piano rolls. The greatest ragtime composer certainly was Scott Joplin. His best-selling composition "Maple Leaf Rag" (1906) was recorded by the U.S. Marine Band and sold over a million copies in sheet music. Between 1899 and his death in 1917, Joplin published over fifty pieces, many of which are among the classic ragtime compositions. Joplin, however, had higher aims, although his publisher, John Stark, did not care for some of his more inventive works. This caused Joplin to abandon the popular music business and devote himself to the completion of his opera *Treemonisha,* a 230-page vocal-piano score with twenty-seven interpolated songs, first printed in 1911. During Joplin's lifetime it was performed but once, and under rough-hewn conditions, in 1915 to an invited audience, with Joplin on piano as the sole accompanist. Long after his death the innovative nature of the score was recognized and *Treemonisha* received a full-scale production.

For all ragtime's success, however, its acceptance was far from complete as the racial origin of its principal composers generated virulent prejudice. At its national meeting in 1901 the American Federation of Musicians sought to ban ragtime and all other manifestations of "the negro school." Battle cries were mounted against the purported debasement of musical taste and threat to public morality that black music was said to represent. Such efforts indicated that while the business of American music was being democratized and the body of composers now represented a wider range of the populace, the taint of prejudice was no less evident in the entertainment world than in society at large.

If the music industry did not regulate the racism in American society, it did cooperate with efforts to legislate protection of foreign authors' and composers' rights through the Copyright Act of 1891 and its subsequent revisions in 1897 and 1909. Before 1891, there was no regulation of the works of foreigners, and publishers were free to issue whatever works by foreigners they wished with no legal liabilities or payments to the writer. Creators pressed for some form of protection. The European artistic community had already enacted a binding system of protection. It had been initiated by France's Société des Autuers, Compositeurs et Editeurs de Musique (SACEM), founded in 1851, which was an impetus to other European countries to convene the Berne Convention in 1886. This agreement did not account for mechanical reproduction of music but did assert that any writer who complied with the copyright law of his own country enjoyed full protection against pirated editions or, in the case of printed literature, unauthorized translations in all signatory nations, which, at that time, did not include the United States. When publishing works by foreigners, publishers in the United States were operating under a gentlemen's agreement known to members of the U.S. Board of Trade as "courtesy of the trade." It implied that once a publisher publicly staked out an initial claim to a particular foreign work, it could be expected that others would not initiate an edition of their own.

The supervision of the registration of American copyrights had been entrusted to the Library of Congress and its librarian in 1860. Five years later, a penalty of twenty-five dollars for each failure to make a proper deposit was enacted. A recording fee of fifty cents for each transaction was instituted in 1870, as was a penalty on one dollar for every sheet of printed material found that had been manufactured without the copyright owner's written permission, half of the collected sum to go "to the use of the United States." These regulations doubled the fines incorporated in the first federal Copyright Act signed in 1790. The policing and collection function for the 1865 and 1870 regulations were left to the copyright owner, who had to apply to civil courts for redress.

However, despite these changes, it was recognized that the system was makeshift and not altogether dependable. New calls were issued for further reforms and led to the foundation of the American Publishers' Copyright League in 1887. Their efforts met with success in 1891 when the first major revision since 1790 of the U.S. Copyright Act was made. Known at the Chace Act, its provisions extended copyright protection to twenty-eight years for both resident and nonresident authors if reciprocal copyright relations existed between the two nations involved; books, chromos, and lithographs had to be manufactured in the United States to obtain copyright; foreign copyrighted books could be imported, subject to duty.

Despite the substantial changes wrought by the 1891 Chace Act, music publishing only fully came of age, insofar as the legislative process involving copyright was concerned, in 1895 with the formation of the new Music Publishers Association of the United States. The seventeen members aimed to elevate the tone and character of their business by correcting evident abuses and most of all by achieving adequate revision of the copyright system so that their enterprise was more fully protected. Their efforts resulted in the 1897 revision of the Copyright

Act. It inaugurated a government-supported Copyright Office, increased penalties, and strengthened law against piracy. Most important, it added the words *"and musical"* to the statute enacted in 1856 extending protection to dramatists against unlicensed public performance of their work. This change specifically covered the kind of dramatic performances—operas, farces, extravaganzas, and other forms of musical theater—popular in the period.

Technological developments led in 1906 to a further revision of the Copyright Act, starting with legislative discussion in the Senate and House Patent Committees. The question at hand was whether or not mechanical reproduction of music in the form of piano rolls constituted an infringement. The government argued that the rolls were in effect "writing" and within the scope of the present law. Yet it was two additional provisions unconnected to the piano roll debate that would have a lasting effect on the future of music publishing: protection against unlicensed public performance for profit and compulsory licensing with a two-cent royalty fee on all sheet music. The final hearings in 1908 took into account a recent Supreme Court decision in *White, Smith, vs Apollo* that copyright protection must be extended to cover mechanically recorded music, a provision that, with the growth of the record industry, was to have enormous influence. So too would the final version of the new Copyright Act, signed by President Theodore Roosevelt before he left office in 1909. It included a provision that guaranteed the *exclusive* right "to perform the copyrighted work publicly for profit if it be a musical composition and for the purpose of public performance for profit." Although few in the music industry paid much heed to this new legislation, not too far ahead income from this right would provide the major portion of financial returns from copyrighted music.

Larchmont, New York
February 1991

D. S.

Acknowledgments

COMPLETION of this manuscript over the course of the last three years was achieved with the help of a number of people. First, I thank my family, for suggesting that I take on the project and for believing in my capacity to do justice to my father's work. Second, Sheldon Meyer of Oxford University for confirming my family's confidence in my abilities and Leona Capeless, Karen Wolny, and Stephanie Sakson-Ford for their editorial assistance. Third, the Computer Center of New York University, whose staff helped a technological troglodyte enter the computer age, for I could not have done what I did without the benefit of word processing. Fourth, my many friends who asked how my father's work was coming and put up with my frequent absence from their lives.

Finally, my mother, Betty. As she was for my father, she has been the true guiding spirit who saw me through the lengthy process this abridgment represents. Without her help, it could not have been done. I dedicate it to her, for words cannot express my debt.

Contents

Abbreviations

ACA	American Composers Alliance
AFM	American Federation of Musicians
AGAC	American Guild of Authors and Composers
AIR	American Independent Radio
AMP	Associated Music Publishers
AOR	Album Oriented Rock
ARMADA	American Record Manufacturers and Distributors Association
ASCAP	American Society of Composers and Publishers
BMI	Broadcast Music Incorporated
CATV	Cable Television
CGA	Composers Guild of America
CHR	Contemporary Hits Radio
CISAC	Confédération Internationale des Sociétés d'Auteurs et Compositeurs
CLGA	Composers and Lyricists Guild
CMA	Country Music Association
CMDJA	Country Music Disk Jockey Association
CRT	Copyright Royalty Tribunal
CWC	Current Writers Committee
EMI	Electrical & Musical Industries
ERPI	Electrical Research Products Incorporated
HMV	His Majesty's Voice
IFPMP	International Federation of Popular Music Publishers
IRNA	Independent Radio Network Associates
KAO	Keith-Albee-Orpheum
MCA	Music Corporation of America
MDS	Music Dealers Service
MGA	Musicians Guild of America
MICC	Music Industries Chamber of Commerce

MOA	Music Operators of America
MPA	Music Publishers Association
MPCE	Music Publishers Contact Employees
MPHC	Music Publishers Holding Corporation
MPPA	Music Publishers Protective Association
MPPC	Motion Picture Patents Company
NAB	National Association of Broadcasters
NAPA	National Association of Performing Artists
NARAS	National Academy of Recording Arts & Sciences
NARM	National Association of Record Manufacturers
NARTB	National Association of Radio and Television Broadcasters
NCRA	National Commission for the Recording Arts
NMPA	National Music Publishers Association
NVA	National Vaudeville Artists
PMA	Publishers and Managers Association
PMRC	Parents Music Resource Council
PRS	British Performing Rights Society
RIAA	Recording Industry Association of America
RPF	Radio Program Foundation
SACEM	Société des Auteurs, Compositeurs et Editeurs de Musique
SARA	Society of American Recording Artists
SCA	Screen Composers Association
SESAC	Society of European Stage Authors and Composers
SOA	Songwriters of America
SPA	Songwriters Protective Association
UBO	United Booking Office
UCC	Universal Copyright Convention
VMA	Vaudeville Managers Association
VMPA	Vaudeville Managers Protective Association

American Popular Music Business in the 20th Century

1 The Birth of the Movies and the Decline of Vaudeville

DURING the mid-1880s, Thomas Edison assigned William K. L. Dickson to construct an apparatus that would permit viewing through a magnifying lens of continuous pictures. Yet once his patent application for the Kinetoscope, or "moving views," was filed, in 1888, Edison abandoned the cylinder design. He had seen the work of George Eastman's new Kodak camera, and he called on the inventor for assistance. Eastman obliged with a fifty-foot strip of improved thin blank film. In 1891, Edison's new Apparatus for Exhibiting Photographs of Moving Objects was ready for a patent application.

Edison believed the $150 for foreign patent applications too high and left the Kinetoscope free for unlicensed development in England and on the Continent. In 1894, Edison's Kinetoscope Company, sole vendor of the device and holder of territorial exhibition rights, purchased the first ten viewing machines to be manufactured and installed them in a Kinetoscope Parlor in New York. Other exhibition rooms followed.

As Edison believed his invention was a money-making novelty whose appeal would soon end, others improved the technology. In 1895, Thomas Armat developed an improved projector, the Vitascope, which had a "beater-mechanism" to provide smoother feeding of film, and took his machine to Edison. The famous inventor found little new except the feeding process, but he was persuaded to manufacture it under his own name and thus be the first to market a practical projection machine.

The Kinetoscope's first important competitor came from France. B. F. Keith imported an early Cinematographe, and in 1896 two New York Keith theaters were showing moving pictures. The Edison monopoly prevailed until the American Biograph Company moved into vaudeville houses with projectors and movies of its own manufacture, to which only projectionists in its employ had access. The industry reached a plateau of development and an exhibition policy that prevailed for several years. Neither Edison nor Biograph sold their cameras, in order to maintain control of film making. As competition for audiences grew hotter, leading to the presentation of a greater variety of expensive live talent, and the movies tended to bore, because they showed the same old subjects, managers used moving pictures only in order to clear their houses between live presentations.

Film took a major step forward in 1903 with the first American "story picture," *The Great Train Robbery*. Made in the wilds of New Jersey, its exciting action and simple story won Americans over to the new medium. The first true movie theater opened in 1905 in McKeesport, Pennsylvania, where two exhibitor entrepreneurs remodeled a street-level store, installed a piano, and began to show twenty minutes of movies continuously from eight in the morning until midnight. They called it a "nickelodeon." Within two years, 5,000 similar places were thriving throughout the nation. The programs consisted of three to five reels and were changed almost daily.

It was inevitable that those who owned and monopolized patents, production, and distribution of motion pictures would form a trust. For almost a decade, Edison's legal staff had been engaged in litigation with rival companies over patent misappropriation. From this chaos, in 1908, the Motion Picture Patents Company emerged, a combination of Edison, American Biograph, Kalem, Lubin, Selig, the Essaney Company, and two foreign concerns, Pathe Freres and George Melies et Cie of Paris. The MPPC trust licensed motion-picture theaters, taxed each two dollars per week to use officially approved equipment, and rented films at a cost of $15 to $125 per day. The trust also waged a vicious war on unlicensed exhibitors, raided theaters, smashed projectors, and destroyed "outlaw" films made by independent companies. By 1910, some 5,281 of the country's 9,490 screen theaters had been brought into line. The MPPC next formed the General Film Exchange, a syndicate of more than half the 100 suppliers of moving pictures.

Outlaw independents began to defy the MPPC by making movies of more than one reel, including the furrier Adolph Zukor, the one-time cloth sponger William Fox, and the nickelodeon and exchange owners Harry and Jack Warner. William Fox, the most recalcitrant of the independents and the owner of the only rental company that failed to join the General Film Exchange, began a prolonged legal battle with the movie trust in 1912. His license to deal in any MPPC product was canceled, compelling him to obtain an injunction. Simultaneously he made use of his strong contacts with New York's powerful Tammany Hall political organization to get the government to institute antitrust action. The combine was forced to do business with him, though the case dragged through the lower courts before it finally was heard by the Supreme Court in 1915.

Adolph Zukor became a sudden new power in the independent movement in 1912, when he purchased the American rights to a French-made two-reeler starring Sarah Bernhardt, bypassed the MPPC, and showed *Queen Elizabeth* in midsummer of 1912 in a legitimate New York theater. Admission was an uncalled-for one dollar. Territorial rights were auctioned independently, producing an immediate profit of 300 percent. This was surpassed in 1913 by George Kleine, of Kalem, with the eight-reel Italian feature *Quo Vadis*. It played in a Broadway theater at a dollar a head for twenty-two weeks, and then went on the road, bringing even greater grosses.

Such impressive returns attracted the legitimate theater syndicates. Klaw & Erlanger, the earliest of these monopolies, arranged for film distribution through Biograph. The company filmed several features, but exhibitors found their fifty-dollar-a-day rental fee too steep, and abbreviated versions were offered at standard prices instead. The Shubert brothers were more fortunate in their moving-picture ventures. One of the leading independent movie makers, Carl Laemmle, found himself with a ten-reel movie on prostitution. With $5,700 already invested, he was loath to risk more, so he sold a one-third interest and all exploitation rights to the Shuberts for $33,000. The film showed in twenty-eight Shubert theaters in the New York area and grossed $440,000. Some of this profit went into World Special Films, to produce films of Broadway plays with the original casts. This in time became part of Zukor's Famous Players-Paramount.

The advent of the first two-dollar movie, D. W. Griffith's *The Birth of a Nation*, in 1915, ended the desperate hope held by vaudeville troupers that the enormous popularity of moving pictures would wane. The movie was seen by more people than had read any book in the history of man other than the Bible. It reaped a profit of $50 million in less than forty years.

Using the legitimate theater's multiple-road presentation method, Zukor's Paramount distribution wing sent out twelve copies of the film. Each was supplied with a 151-page piano score and all orchestral parts of the first important American score of accompanying film music. Working with an unimportant composer of art and operatic music, Joseph Carl Breil, Griffith chose appropriate selections from Wagner, Grieg, Beethoven, Schubert, Rossini, and others. Most of their music had not yet been heard by the majority of Americans, and thus what we today regard as classical war-horses was a great novelty.

In 1894, when the Edison Kinetoscope Company began the first mass production of short moving pictures, it also offered accompanying cylinder recordings. A Kineto-Phone phonograph was available for $350 to provide a sort of synchronization of music and filmed action. The Camerophone Company introduced its record player coupled with three separate projectors in 1909. A number of leading Broadway singing actresses were filmed and disk recordings of their singing made, but the unresolved technical problem was how to achieve volume sufficient to fill a large theater.

Pianos were first installed in nickelodeons in order to cover the whirring of the projection machines, and it was left to the pianist to select appropriate music. In the mid-1900s, the Edison Company first programmed suitable instrumental music to be used with its releases. In 1909, the Vitagraph Company of Brooklyn introduced arranged piano scores for its new films. The following year, Vitagraph began to issue "music cue" sheets for piano and the three-piece orchestras. A Cleveland music publisher, Sam Fox, was the first regularly to issue collections of selected "mood" music, beginning in 1913.

The first original musical score written for a moving picture was composed in 1907 by Camille Saint-Saëns for a French production, *L'Assassinat du Duc de Guise*. The famous film composer Max Steiner believes himself to be the first to conduct a large orchestra in an American screen theater, playing music written by him for a feature movie, *The Bondman,* shown in 1915. Steiner suggested that something special was needed to entice ticket buyers, and produced an original score for a hundred-piece orchestra, to be played behind the feature film.

Sixteen million people daily were going to the movies in 1914, and they sparked a building boom that brought the total of U.S. screen theaters to 30,000, 15 percent of them seating over 600. For the large houses, admission prices rose to $1.50, the result of inflation as well as staggering increases in production costs. These had escalated ten times for the average feature film.

The independents' war with the MPPC was wreaking its toll. Exorbitant salary offers had lured major Broadway and vaudeville luminaries to California, but, finding the work tedious, most of them left. Hitherto unknown performers, such as Mary Pickford and Charlie Chaplin, became screen idols almost overnight by flitting from company to company, responding to bidding wars for their services. The attraction of Southern California's benign weather, nonexistent tax structure, and semiofficial attitude against union labor drew the industry west. Seventy production companies were soon in business, and their annual budgets approached $25 million. This changing economic balance led to the inevitable demise of the once all-powerful movie trust, the Motion Picture Patents Company.

The years after 1916 were golden ones for the movie business. New production companies sprouted up overnight. Thirty-five million Americans were going to screen theaters at least once a week, and it appeared to be impossible to lose money on a new feature. Wall Street brokers and other speculators begged to share in the profits. The international banking house of Kuhn, Loab underwrote $10 million of an issue off Famous Players-Paramount preferred stock and then cleared $21 million. Moviemakers with a record of success had to fight off investors.

It remained for Thomas Ince to restructure movie making into the production-line system. In 1914, Zukor, Lasky, and others still thought of their function as that of a theatrical producer, responsible for financing, choice of story and actors, and hiring a director and placing responsibility for the finished product on him. To keep affiliated exhibitors supplied with a new feature movie each week was difficult. Ince soon acquired the reputation of a man who made "box-office" movies rather than artistic triumphs. In 1915, he, along with D. W. Griffith and Mack Sennett, formed the new Triangle Film Corporation. While Griffith was at work on his epic *Intolerance* and Sennett created Keystone comedies, Ince assumed general control of all production. Unrealistic shooting schedules and mounting costs were restrained by Ince's appointment of supervisors over all aspects of individual features.

The power of the MPPC abated. William Fox's lawsuit was finally settled by a Supreme Court decision that warned defendants to desist from further "unlawful acts" or suffer the wrath of a government dedicated to fighting monopoly. A patent-infringement suit brought by the movie trust, in hopes of sustaining the combine's structure, concluded with a decision that negated many of its patents.

Prosperity now began to grow at the expense of independent exhibitors, whose rental fees threatened to rise because of the block-booking policies introduced by Adolph Zukor. In order to get a star vehicle, it was necessary to buy Zukor's entire output for a season. A second rebellion ensued, directed by Thomas Talley, the fiercely independent westerner who had opened America's first movie house fifteen years before. Feeling that exhibitors could bypass the producers and deal directly with stars, Talley formed the First National Exhibitors Circuit in 1917, a combination of 27 theater owners, which soon hired Charlie Chaplin, Mary Pickford, and Thomas Ince. The new company became serious competition for Zukor and part of the Big Three that dominated Hollywood until the advent of talking pictures. The third element, Metro Pictures Corporation (later Metro-Goldyn-Mayer), became part of the group in early 1919 after Marcus Loew purchased the floundering company for $3 million.

First National's success made Zukor build a chain of company-owned first-run screen theaters. A new moving-picture monopoly was now in place; it controlled most of the world's total movie production, the distribution systems, and a substantial portion of America's theaters. The antiquated Motion Picture Patents combine had been replaced by the Big Three. The only enemy challenging their hold on American popular entertainment, E. F. Albee's vaudeville monopoly, was about to join in battle for supreme control.

Even before vaudeville came of age, the variety stage had been a principal ally of popular-music publishers, by promoting their sheet music. Paying for play had begun in the early 1880s when some American publishers paid variety artists and concert-room singers to "plug" their songs. The sheet-music business prospered, growing to forty-five major companies, capitalized at $1.6 million and with a value of $2.2 million in 1904.

The fees paid to vaudevillians to "plug" new songs had become important even to the best-paid variety artists, whether they appeared in "big-time" or "small-time" vaudeville. The former was staged twice a day, whereas the latter offered three to six performances daily, usually in "pic-vaude" houses, where moving pictures and live variety had become the standard bill of fare by 1912. Neither form of vaudeville offered exorbitant compensation to headline artists, who were, moreover, subservient to a monopoly that became the fiefdom of a single man: E. F. Albee, who assumed control of the Keith-Albee circuit in 1917.

Born in 1857, Albee at seventeen began to work at B. F. Keith's Boston variety theater. Soon partners, Keith and Albee opened the Bijou Theater and the first continuous vaudeville show on July 6, 1885. Some 700 Keith theaters soon followed, whose permanent slogan was "Cleanliness, courtesy and comfort." Vaudeville had become a big business, with its own monopolistic apparatus controlled by Keith in the East and by Martin Beck's Orpheum from Chicago to the Pacific.

During the 1890s, most variety artists secured their bookings through a dozen agents located in New York and Chicago, who received 10 percent for working on the performers' behalf. In 1900 some 100 vaudeville managers organized the Vaudeville Managers Association. To centralize all bookings and control salaries, a satellite hiring organization, the United Booking Office (UBO), was formed and given the power to manage performers willing to cooperate with VMA.

The new organization was fought by means of a strike by the White Rats ("star" spelled backward), a group of the most militant variety artists. Keith and Albee took control of VMA, whose powers grew and profits increased, but only for cooperative theater managers. In 1906, a theatrical monopoly owned by the producers Marc Klaw and Abe Erlanger joined their most important legitimate-theater competitors, the Shubert brothers, to build and operate a competing national vaudeville circuit. Once it appeared that this venture might succeed, secret negotiations conducted by Albee scuttled it, in return for one million dollars in cash. Klaw and Erlanger and the Shuberts agreed that they would not go into vaudeville for ten years. Albee went back to swallowing up all competitors, and successfully accommodated the Keith-Albee circuit to the challenge of Edison's moving pictures.

The White Rats, whose membership had fallen off, attained their first victory over Albee's monopoly in 1910. A new law in New York State which put a ceiling of 5 percent on an agent's commission roused many performers against the UBO, which demanded 10. By the end of that year, a new actors' organization, with 11,000 members, became affiliated with the American Federation of Labor under the name Rat Actors Union of America.

The following year, New York's state commissioner of licenses reported that 104,000 vaudeville contracts had been approved by his office. The only source of employment was some 1,000 theaters giving bit-time, or "class," entertainment on a two-a-day basis, and about 4,000 small-time houses, many of them pic-vaude theaters. Some performers attempted to evade the power of the UBO and the VMA by playing the latter theaters during the regular twelve-week layoff period between bookings. If caught, however, they faced barring for life from any VMA theater.

Abuses against artists by UBO, VMA, and Keith's were manifold. Because of New York State's 5 percent limitation of booking commissions, the fiction of "personal representatives," who collected an additional 5 percent, was created. At least half of their commissions were kicked back to a Keith collection agency. Variety artists also paid for all traveling expenses, any stage costs, and all musical arrangements. The growing inequities were responsible in great measure for the rising price of "pay for play," about which many music publishers began to complain openly.

The trade paper *Variety* was their chief sounding-board in connection with the gratuity system. Founded in 1905 by Sime Silverman, the publication became temporarily the chief advocate for the White Rats and the vaudevillians' cause. Performers who advertised in it were refused UBO bookings. Publishers who advertised new songs there found their entire catalogues barred from VMA theaters. Silverman persevered, however, and when poor times in 1913 brought industrywide booking cuts, he opened his pages to the White Rats.

B. F. Keith died in 1914. Albee appointed John J. Murdock general manager and plotted to destroy all opposition from vaudeville performers. Under Murdock's command, a blacklist of the White Rats was prepared. Backed with Keith's money, a new union was formed, the National Vaudeville Artists. Pat Casey was made head of the new Vaudeville Managers Protective Association, whose task it was to enforce the blacklist. On Murdock's orders, a new clause was inserted into standard UBO contracts: it warranted that each performer seeking to do business with the UBO was a member of NVA. If earlier White Rats affiliation was ever discovered, it would mean expulsion for life.

A White Rat strike broke out in the Midwest in 1916. It was handled with such dispatch by Edwin Claude Mills, the manager of Oklahoma City vaudeville houses, that he was summoned to the VMPA's New York office to work with Casey and Murdock and succeeded in breaking other White Rat strikes. The White Rats went into bankruptcy, and their headquarters in New York was bought by Albee.

The price of victory was an action by the Federal Trade Commission, in May 1918, charging that the Vaudeville Managers Protective Association was an illegal combination operating in restraint of trade, collecting excessive fees from performers, and using a blacklist to punish union members. Also named in the complaint were the National Vaudeville

Artists, United Booking Office, and the Keith Vaudeville Collection Agency, to which professional representatives paid back a share of their commissions. In all of these, Murdock and Casey were officers.

Pat Casey was the principal Albee spokesman during hearings held by the FTC. He claimed that the association benefited performers and managers. Inaccurately asserting that the VMPA controlled fewer than half of all vaudeville houses, he said that in the course of a year Keith's circuit presented some 6,000 acts, giving employment to 8,000 or 9,000 actors. He omitted to mention that they represented only a third of available performers. Yet the VMPA agreed to remove the clause warranting membership in the NVA to get bookings.

Given a clean bill of health, the VMPA and other Albee interests prospered. In 1918, Albee changed the name of his empire to the B. F. Keith Exchange. His profits remained over $10 million annually through 1921. Albee also joined the fight by legitimate theater managers in 1919 to smash the Actors' Equity Association. He served during its four-week-long walkout as second vice president of the United Managers Protective Association, and when the actors won became an even more ruthless antilabor man. At the same time, bad times drastically changed vaudeville. One-quarter of all theaters dropped variety completely or replaced it with one-reel movies. Those who vied for the 5,000 jobs available took whatever Albee and others offered.

Once the government's antitrust action against the vaudeville managers terminated, Albee and general manager John Murdock reorganized the Keith-Albee empire. They purchased additional theaters and increased prices to one dollar, two dollars at their flagship arena, the Palace Theater.

Observing the successful solidification of Albee's control of two-thirds of all big-time variety, the Shubert brothers, whose agreement not to enter vaudeville for a decade was nearing its end, prepared a new Advanced Vaudeville Circuit, which opened in 1922. Some of Albee's biggest attractions were lured away with promises of roles in forthcoming Shubert Broadway productions and higher salaries. Albee, determined to weather the desertions and have his revenge, offered, for the first time, long-term contracts and special considerations to loyalists. When the Shubert incursion came to an end, with a $1.5-million loss, Albee welcomed back the unfaithful only after they confessed their transgressions with paid advertising in the pages of *Variety*.

However, radio soon became a potentially profitable medium, with one million sets in use by mid-1922, and while Keith-Albee added a clause to all contracts banning any appearance on wireless broadcasts, nothing could thwart Albee's efforts to keep his chief attractions off the air. Big-time vaudeville was dying when Albee declared 1926 its one-hundredth anniversary and let loose his underlings to celebrate that occasion by going on a five-a-day schedule. It was hardly the time for cheering. Some 97 percent of the country's 21,000 theaters were by then film houses. The remainder was made up of 500 legitimate theaters, many burlesque houses, and a dozen two-a-day variety-only palaces. The movie industry had also made serious inroads into theater real estate despite the Federal Trade Commission's injunction that the Big Five divest themselves of such holdings.

By now one of the world's most hated men, Albee was blissfully unaware of the approaching end of his power and was basking in the glory of *Variety*'s report that he was the seventh most prosperous man in show business, with holdings of $25 million. In the face of life-threatening competition, Albee was persuaded to approve long-term contracts with an annual salary increase for the 300 headline acts that his United Booking Office forced on many of the country's independent theaters. General manager Murdock's experience in the picture business dated back to his days as head of America's first color-movie company, and he now negotiated the purchase by Keith-Albee of an important interest in Pathé-De Mille Pictures. His well-known friendship with Joseph Kennedy was felt to effect a merger of FBO and Pathé. But both had other fish to fry. The long-secret business collusion between Keith-Albee and the dominant western Orpheum circuit was made public by the official merger that created Keith-Albee-Orpheum.

In 1928, a meeting was held by Kennedy and Albee, during which an offer to buy 200,000 shares of KAO stock for $4.2 million was accepted. The syndicate of Wall Street houses for which Kennedy acted gave him a bonus of $150,000, elected him chairman of the board, and approved an option to buy 75,000 shares of KAO. Immediately, a purge of Albee's favorites was initiated; only Pat Casey and Murdock emerged apparently unscathed and still loyal.

Murdock and Kennedy announced a few months later that Pathé and FBO had signed contracts with RCA Photophone. The latter would install its equipment in 200 KAO theaters. FBO and KAO merged with RCA Photophone to create RKO, from whose title all reference to Albee was deliberately omitted. Kennedy was named chairman, with Murdock as leading associate.

Vaudeville's decline and the ineptitude of the new management only raised the organization's deficit, bringing in hard-eyed AT&T men to run the corporation. Kennedy resigned, as did Murdock. When RKO stock reached fifty dollars, Kennedy exercised his option and realized a profit of some two million dollars. The following March Edward F. Albee was dead of a heart attack.

No one could save vaudeville. Only four theaters in the country offered variety entertainment exclusively. With its holdings in picture and sound equipment and the licensing business at stake, RKO offered units of four acts on a fifty-week basis to theaters. The Vaudeville Managers Protective Association, once so powerful it manipulated music publishers and songwriters, was bankrupt. The White Rats now bowed to the inevitable and surrendered the American Federation of Labor charter granted twenty years earlier. Soon the only true vaudeville troupe in the old tradition being sent out by RKO was its Varieties unit, whose biggest-drawing attractions generally were people from the movies.

As the movies thrived and vaudeville declined, the rights and privileges of American publishers, composers, authors, and songwriters were undergoing a profound transformation. The principal agent of that change was the revised Copyright Act of 1909. It dealt with the mechanical reproduction of music, mandating a royalty of two cents for each cylinder, recording, or paper music roll manufactured. This compromise provision had its origin in 1907, when President Theodore Roosevelt learned that a "giant music monopoly" granted exclusive piano-roll recording rights to a single manufacturer for a period of thirty-five years. James F. Bowers, president of the Music Publishers Association, had been instrumental in creating that arrangement with the Aeolian Company whose Pianola and Victrola were changing the way Americans heard music. Bowers persuaded eighty-seven publisher members of the association to sign exclusive agreements with Aeolian to cut piano rolls of their copyrighted music in return for a royalty of 10 percent of retail price.

Exposure of the scheme brought Roosevelt's attention and the cry of "Monopoly!" from Aeolian's competitors. In spite of efforts by a number of important composers and publishers, the unfavorable (to them) legislation was passed. For the first time in U.S. history, the peacetime bargaining process between supplier and user was to be regulated by the government. The fixed two-cent royalty and a compulsory-licensing provision were to guard against any future music-copyright monopoly.

With popular sheet music selling for from twenty-five to sixty cents a copy, the wholesale value of printed music more than tripled between 1890 and 1909. In 1909, more than 27 million phonograph records and cylinders were manufactured, having a wholesale value of nearly $12 million. More than 25,000 songs were entered for copyright registration, complementing the parallel growth of piano sales to a new high of 365,000 instruments.

Prior to World War I, the major companies devoted a principal portion of their production to vernacular music and novelty recordings, but put major emphasis in their advertising on music that was good for their general image and attracted customers of the "better kind." The Victor Talking Machine Company's $1.5-million advertising campaign in 1912 concentrated on its Red Seal line—one-side-only disks, three and a half minutes long at most, of grand-opera selections, familiar old songs, and classical instrumental music. Victor's best-seller was Enrico Caruso, the first performer to be signed to an exclusive contract and royalty arrangement by an American firm. The 36 recordings Caruso cut before 1912 became the cornerstone of that year's Red Seal catalogue. With an ambitious but inferior grand-opera catalogue, the Columbia Phonograph Company was a poor second to Victor.

Even the most cautious recording executives became convinced of the musical theater's power to create popular hit songs after the success in 1907 of Franz Lehar's *Merry Widow*. Public demand became so great that F. W. Woolworth restored a sheet-music counter to his growing chain of five-and-ten-cent stores. The economic consequences reshaped the structure of music publishing.

After 1911 a dance craze overwhelmed upper-class America and then the world. Dances like the two-step, cakewalk, and turkey trot inspired by Afro-American theatrical and sporting figures freed ragtime couples from the strait-jacket of the nineteenth century's figure and set dances. Hits like Irving Berlin's "Alexander's Ragtime Band" set feet to moving, even in the world's finest social circles.

This musical liberation was embodied by Vernon and Irene Castle. Vernon, an Englishman, married the American-born Irene in 1911 and took her to Paris, where they showed off the Texas Tommy, the turkey trot, and other dances. The music of Irving Berlin was ideal for these steps, and the Castles introduced both to Europe and the world. The new dances they popularized after their return to New York in 1912 were speedy, simple, and rhythmic, but they were performed by the Castles with a grace that few of their pupils could emulate.

The Syncopated Society Orchestra of black and Puerto Rican musicians, which usually accompanied the Castles, was directed by James Reese Europe. He dedicated himself to presenting Afro-American musicians as capable of playing for any audience and as creators of a vital music that blended their own past with that of white Americans. He conducted the music for all the Castles' public appearances, served as musical director for their string of dancing salons, and wrote the music for most of their dances. When Europe persuaded Vernon Castle to slow down his dancing, to move to tempos that stressed the back beat, as in black music, he introduced the dancer to the one-step.

Record companies were eager to obtain the Castles as phonograph artists. In frantic bidding, Victor won and began a line of dance records that remained profitable until singing stars displaced them after World War II. The first black band to be signed by a recording company, James Europe's fourteen-piece Syncopated Society Orchestra, was added to the roster.

World War I ended Victor's partnership with the Castles. Vernon joined the Royal Flying Corps and died in a plane crash in 1918. James Europe enlisted in the New York Negro Regiment, for which he formed a military band of first-class black and Puerto Rican players. The group, known as the "Hellfighters," was sent to France and upon its return to this country was about to embark on a national tour when Europe was killed by a member of his orchestra.

In a continuing profitable position as the world's largest record company, Victor maintained its lead over its two major competitors. In 1914, the company produced a bit more than half of the total of 27 million recorded units. The Edison company, which continued to make cylinders until the late 1920s, issued almost all of the four million produced that year. As the original talking-machine patents began to expire, around 1914, Columbia found itself sharing sales with new companies formed to take advantage of a fading technological monopoly of pooled patents. The most prominent was Aeolian-Vocalian, formed by the player-piano company, and Brunswick-Balke-Collender, an Iowa piano manufacturing company. Concentrating on the Midwest market, Brunswick made both popular and classical music "personality" records.

Victor still dominated record sales in 1916 and doubled production to 31 million records three years later. When the war curtailed all recording of classical music in Europe, Calvin G. Child, director of Victor's Artists Bureau, looked to America for talent for the Red Seal line. Red Seal sales symbolized Victor's prestige, but they never showed a sales ratio of more than 20 percent of the popular black-label product. Sales went from 2.1 million in 1914 to 9.5 million six years later, when the total value of American phonograph-record production jumped to 106 million.

The superior Victor product, along with Aeolian's player-piano rolls, enjoyed the majority of all retail sales. However, the introduction, after 1910, of coin-operated automatic pianos quickly displaced nickel-in-the-slot phonographs and peep-show machines and significantly increased Aeolian's hold on the family market. This grew larger in 1916, with the first real innovation in the player piano's history—the "word-roll," on whose margin the lyrics of popular songs were printed. The two-cent royalty mandated in the 1909 Copyright Act was regularly paid, but publishers held that the use of lyrics was not mechanical reproduction, only duplication of protected material not covered by federal law and therefore not subject to the two-cent royalty requirement. That position was twice upheld by the Supreme Court, and royalties from word-rolls soon rose to twelve and sixteen cents. The six leading popular-music houses—Irving Berlin; Leo Feist; T. B. Harms with Francis, Day & Hunter; Shapiro, Bernstein; Waterson, Berlin & Snyder; M. Witmark & Sons—formed the Consolidated Music Corporation to handle their piano-roll licensing.

Because the participants controlled a majority of the best-selling songs, Consolidated could insist on a minimum twelve-cent word-roll royalty in addition to the two-cent mechanical rate. It also demanded the production of two additional music rolls from each partner's catalogue before permission would be given to use one of his major hit songs. The first recorded antitrust action involving music publishers was instituted in 1920 by the United States government, against Consolidated and its owners. This was also followed by quick dissolution of the alleged trust. Licensing of piano rolls was eventually assigned to the Music Publishers Protective Association.

The Victor Talking Machine Company was not the first to record a jazz band, but it did make the first great profits from the new American musical form. Several authentic jazz bands had appeared in New York venues, where they won favorable press notices and audience plaudits, but little attention from record companies. The Original Dixie Land Jazz Band, however, sold a million records for the Victor Company. This group of five white musicians from New Orleans scored an instant triumph when it played at a New York restaurant in 1917. A Victor session in February 1917 produced one of the company's earliest million-seller Black Label records: "The Livery Stable Blues," coupled with "Original Dixie Land One-Step." Jazz mania was wedded to the dancing craze. Calvin Child brought Paul Whiteman's nine-piece Ambassador Orchestra into the studio. Their rendition of "Whispering" backed by "Japanese Sandman" was responsible for the sale of millions of disks and for Whiteman's climb to the throne of jazz, where he ruled as a press agent-appointed king.

The fast-dance mania and new jazz music were shaped by Afro-American culture, a fact completely disregarded by those who capitalized on their union. For years, the record industry had shown little interest in the potential market of 11 million black Americans, who represented nearly one-tenth of the entire population. The purchasing power of this large block was discovered by pure chance in 1920, when the first recording of music written by a black songwriter and performed by black talent was consumed by black record-buyers. The "blues" first came to the attention of most whites with William C. Handy's "Memphis Blues" (1912) and "St. Louis Blues" (1914). Both were "vaudeville blues" and not in the traditional three-line AAB form.

Handy did make records for Columbia in 1917, but he failed to include either of his two blues hits. The first authentic blues was recorded for the same label in 1916 by a white man. "Nigger Blues," written by a black minstrel-show performer in 1912 and published before either of Handy's pieces, was recorded for Columbia four years later by a Washington attorney, lobbyist, and gifted mimic George O'Connor.

Like many competing whites, most black professional songwriters preferred the work in the more elastic vaudeville blues form, not as strict as the traditional blues. Among those who used both forms was Perry Bradford, a successful black vaudeville performer, agent, songwriter, and manager. When he played some new songs for Fred Hager, musical director for OKeh Records, a session was arranged for Mamie Smith, who Bradford managed, to record two of Bradford's vaudeville blues. Response was favorable to the record, released in 1920 with no indication of the singer's color.

Whether or not out of deference to those black leaders of the postwar period who saw themselves and their brothers and sisters as "the Race," the Mamie Smith recordings and other blues records were listed as "race records." It was by this name that they continued to be known throughout the 1920s, when black Americans bought as many as 10 million blues and black gospel records in a single year, and the 1930s, when those sales slumped with the failing economy. In the more socially aware time that followed, the designation "rhythm and blues" became general.

2 The Formation of ASCAP and the Diversification of the Radio and Recording Industries

IN 1910, as sheet music sold 30 million copies, Tin Pan Alley implemented novel principles of salesmanship. To increase familiarity with their product in potential customers, song pluggers publicized material through constant repetition. The best song pluggers often proved to be the songwriters themselves, and Tin Pan Alley hired them to serve both functions. Under the writer-for-hire arrangement, songs became the exclusive property of the employer.

Live vaudeville performers were crucial to familiarizing the public with new songs and the music trade's chief means of exploitation. Music publishers paid for the musical arrangements that performers were required to supply as well as providing copies of new songs, exclusive performing rights, food and drink, and, finally, hard cash. When Keith-Albee control of vaudeville extended to the extraction of sizable portions of a week's salary, the music publishers opened branch offices in important cities to make life easier for performers who were financially dependent on them.

Ten-cent popular-song sheet music was selling in greater quantities than ever, but profits decreased when the chief distributor, the five-and-ten-cent stores such as those of Frank W. Woolworth, received a 40 percent discount. Woolworth realized the power popular music had to attract customers. They poured in to buy merchandise bearing an image of McGinty, for example, who went "down to the bottom of the sea" in a best-selling song. Woolworth therefore started selling sheet music, and one dime became the standard price for all popular music other than production and high-class music.

When George M. Cohan's "Over There" sold two million copies in sheet music and one million records, it was purchased for $25,000 by Leo Feist, last of the major publishers to join in the pay-for-play practice. Learning that jobbers and retailers ordered music only after a song was successful, he spent nearly $100,000 annually for advertising to validate his company's slogan asserted, "You can't go wrong with any Feist song!"

In this period when ten-cent sheet music ruled, only Irving Berlin's work maintained the former thirty-cent price. In 1909, Berlin went to work for Ted Snyder as song plugger and lyric writer. In its first month, his "Alexander's Ragtime Band" reportedly sold half a million copies. To capitalize on this commercial talent, Henry Waterson, Synder's manager, organized the firm of Waterson, Berlin & Snyder. A transposing piano was given to the songwriter, who never learned to play in more than one key. His long string of hits over the next half-century eventually was controlled by Irving Berlin, Inc., and its successors.

Many songs during this period became hits due to their subsidized exposure by vaudeville performers. Only a concerted industrywide effort to change this system could save music publishing. The single business association in a position to marshal such cooperation was the Music Publishers Association. However, when the MPA primarily protected high-class standard music to the exclusion of popular songs, a group of leading Tin Pan Alley houses left it in 1907, only to resume membership when Witmark, Stern, and Marks, Feist, and others acquired standard-music catalogues. However, in 1914, they again found their interests diverging over the advancement of standard and educational music and the extinction of a gratuity system affecting their popular catalogues.

This situation soon transformed. When his business fell off due to Keith's boycott of songs publicized in *Variety*, Sime Silverman began to report that leading publishers sought to end payments to singers and raise wholesale prices of sheet music to ten cents. In 1913, a group of production songwriters and authors met to discuss the formation of an American songwriters' and publishers' society.

The 1909 Copyright Act provided an exclusive right to license performance of a copyrighted work "publicly for profit if it be a musical composition and for the purpose of a public performance for profit." Recovery damages for infringement were fixed at not less than $250 or more than $5,000. In 1897, the words *and musical* had been added to statutes adopted earlier requiring the permission of the owner of a stage work in order to give a public performance. The 1909 provision mentioned both dramatic and musical performing rights, but no effort was made to license the public performance for profit of music in America until 1911.

That year, the world's first performing rights society, the French Société des Auteurs, Compositeurs et Editeurs de Musique (SACEM), opened offices in New York to collect 5 percent of gross receipts when French compositions registered in the United States were played.

Nathan Burkan, attorney for the Witmark company, offered SACEM's Articles of Association to six writers and their publishers who gathered in October 1913. Those articles included a stipulation that performance income be divided equally among the writer of the music, of the words, and their publisher. Four months later, the American Society of Composers, Authors, and Publishers was formed for a term of ninety-nine years. George Maxwell was elected president and stated that ASCAP would prevent the playing of all copyrighted music at any public function unless a royalty was paid.

With Nathan Burkan as acting counsel, in 1914, fourteen New York publishers formed a new Music Publishers' Board of Trade and approved a fine of $5,000 for direct or indirect payment to performers for song-boosting. Leo Feist did not participate, saying that he had been the last to fall in line and pay singers and that he would be the last to cease.

ASCAP's rates were more modest, ranging from five dollars to fifteen dollars monthly, depending on the type of music used, the size of the establishment, and its gross receipts. Some infringers successfully were brought to court. The most productive of these suits, filed in 1915, charged that Victor Herbert, his collaborators, and their publisher, G. Schirmer (not an ASCAP member), had "heard" an unlicensed performance of their song "Sweethearts" in a restaurant. The action was based on the dramatic-work licensing provision of the copyright law.

Variety's revenues from music publishers' advertisements were declining in 1916, and the new Board of Trade appeared to be falling apart. Silverman turned the matter over to John J. O'Connor, head of advertising. To replace the impotent board, O'Connor appealed to the New York publishing fraternity to curb the subsidization of performers. Three publishers showed no interest—Feist, Remick, and Max Dreyfus, of T. B. Harms.

O'Connor then moved to get the vaudeville monopoly's support and the full backing of both the VMA and the UBO to enforce a ban on music owned by Feist and Remick. Having effectively destroyed the White Rats union, Murdock saw an opportunity to extend control over vaudeville artists. Murdoch demanded that publishers change the system or their songs would be banned from all UBO and VMPA theaters. They in turn joined the new Music Publishers Protective Association in 1917. Leo Feist and all other holdouts were persuaded to join after their music was boycotted in Keith houses. Meetings of the MPPA were held in VMPA's headquarters office, and Maurice Goodman, attorney for Keith and his satellite groups, headed a three-men executive committee on which Pat Casey and Claude Mills but no music men served. The gratuity system was thus brought under control, as publishers were punished through VMPA pressure. ASCAP progress was less satisfactory. Rates for motion-picture houses were fixed in 1917 at an average of ten cents a seat, and opposition immediately arose. Injunctions against ASCAP collections were obtained, and heavy use was made of the tax-free music.

ASCAP's victory in the Supreme Court indicated to Albee and Murdock that eventually vaudeville too would be taxed. With the MPPA under control, the financially beleaguered society was ready for seduction. In 1917 the vaudeville magnates offered ASCAP their services to ensure regular collections from music users and to obtain licenses from the pic-vaude houses and all non-VMPA-affiliated theaters. In return, the VMPA members could use ASCAP music without payment, and 50 percent of all gross receipts collected by Casey in the first two years would be passed on to Murdock, one-third in the following three. The terms were agreed to, but when the government's antitrust suit against the vaudeville combine was filed in 1918, an unsigned document between the parties became a vague "gentlemen's understanding," as ASCAP failed to collect from the Albee interests.

Other elements of the music trade, chiefly those dealing with "high-class" music and manufacturers or vendors of musical instruments, banded together in 1918 to form the Music Industries Chamber of Commerce. During the next decade of hearings in Washington on copyright revision, the MICC usually appeared in opposition to the MPPA and ASCAP. The printed music trade's only representative in the MICC was the Music Publishers Protective Association, whose sole interest lay in fixing and maintaining a higher price for standard and art music.

Variety stated in 1919 that a prospering music industry owed much to the MPPA. New members were regularly approved for admissions, the payment system appeared to be only a memory, and mechanical royalties improved. The annual production of printed music had increased to 30,000 titles. Although the ten-cent and syndicate stores' stranglehold on wholesale prices continued, these were growing as royalty income from mechanical reproduction on paper rolls and recordings rose. The MPPA drafted a new standard contract that raised the ceiling on piano-roll royalties to twelve and a half cents, with seven and a half the minimum.

In early 1920, Leo Feist, Charles K. Harris, Louis Bernstein, and Isidore Witmark, music publishers affiliated with both MPPA and ASCAP—rather than Pat Casey, who represented Keith and Murdock—served as heralds of a new day. Casey remained in charge of the MPPA. E. Claude Mills was brought in as his assistant and chairman of the executive committee. Some publishers indicated to ASCAP that Mills had brought the MPPA to maturity with a uniform word and music-roll licensing contract, a bureau to educate dealers, and a censoring service to guard against blue lyrics. The MPPA paid his $20,000 salary and insisted that without Mills there could be no effective merger to present a united front on the question of collecting license fees from music users.

Meanwhile, a major problem had arisen that ASCAP officials wished to conceal. The standard form of agreement between composers, authors, and publishers conveyed all rights to the publisher, but it gave the writers no power to grant public-performance rights to ASCAP. Many nonmember publishers who controlled important catalogues were unable to enter ASCAP. The softening of ASCAP officials on behalf of Mills and the acceptance of important tax-free catalogues owned by MPPA members led to an agreement. Each organization would have half of the members of a new twenty-four-man board of directors. The publishers insisted on distribution of ASCAP income, half to them, the balance to authors and composers. This new plan passed in 1920 with a special clause written into the updated Articles of Association: a two-thirds majority was required in order to affirm any action.

By 1921, the record business reached peak sales at retail of $106.5 million ($47.8 million wholesale). *Variety* added a monthly best-selling records chart. Two of the industry's original three giants, Edison and Victor, were in good financial shape in 1921. Victor profited from the death in mid-1921 of Enrico Caruso; it stimulated sales of $2.5 million of his Red Seal recordings. Furthermore, their purchase of a half-interest in Victor's British associate, the Gramophone Company-His Master's Voice, allowed access to the entire world's phonograph market. Columbia, on the other hand, was not faring well. Poor management by Wall Street speculators, overproduction of disks and phonograph machines, and the depression of 1921 speeded the company's involuntary bankruptcy and receivership.

The industry's fundamental economics in 1921 were easy to comprehend. A single disk cost twenty cents to make. The sale of 5,000 recordings effectively wiped out all production costs and made further sales mostly profit. Get-rich-quick operators spurred the formation of 150 companies to fill growing markets. This increased supply of cheaper labels brought about abandonment of the traditional release-date policy. The release of a new copyrighted piece had always been withheld until the publisher felt that sheet-music sales were exhausted. Competition now led record makers to release recordings whenever they chose.

The improvement of records and phonograph machines further upset the normal balance of power. In 1922, Columbia introduced a silent record surface developed by technicians in its British branch, which was about to be sold to Louis Sterling in the interests of building an American cash reserve. Sterling was readier to accept the new technology than his debt-ridden U.S. counterparts. Meanwhile, as sales dropped, Victor improved its own surfaces and brought out the first double-faced Red Seals.

David Sarnoff unsuccessfully had attempted to persuade Victor officials of radio's potential. Sarnoff therefore turned to the new second-largest American record company, Brunswick, and sold $1.5 million worth of receiving equipment for installation in a new line of Radiola models, merchandised by RCA. Victor relented, meeting the challenge with a new console model into which a radio set could be placed.

At the same time, forms of recorded music diversified, beginning the process of the "democratization" of American music. Columbia began to issue blues recordings and other minority-interest music on their sub-label OKeh. This venture, begun in 1920, had been surprisingly successful, and it had brought about the introduction of "race records" by most other labels.

An economic system governing race recordings solidified. Whether written by the artist or the talent scout-producer, race songs initially were assigned to a producer-owned publishing company. Talent was paid by the piece, which most artists preferred. The production of "hillbilly" music, which accounted for as much as a quarter of all popular sales by 1930, began as an accommodation to an Atlanta OKeh record distributor, Polk Brockman, who believed in a large southern market for old-time music, suggested that OKeh record local artist Fiddlin' John Carson, and offered to buy a large quantity of the pressings. Ralph Peer, assistant to Fred Hager, chief of OKeh production, set up a 1923 recording session whose creditable sales justified Brockman's faith. OKeh subsequently embarked on assembling the industry's first hillbilly record catalogue, for which music was often recorded by Ralph Peer.

Peer evolved a royalty and artist-payment system which became standard for both hillbilly and race music and later spilled over into popular songs. He never recorded an established selection or artists who could not write their own music and created what became one of two standard contracts offered to recording artists for original music. It assigned all rights to OKeh for a twenty-five-dollar talent fee per side and gave a guaranteed royalty to the composer-artist of 0.005 cents per side sold.

Hoping to increase sales and stem their losses, the new Big Three—Victor, Brunswick, and Columbia—rushed to make large offers to successful dance-band leaders and vocalists. However, a glut of new releases inundated the market. Victor failed to persuade new investors that the industry could again prosper.

The introduction of improved technology, heralded by the installation of electric recording microphones and equipment in the Victor and Columbia studios, again proved to be the phonograph and record makers' salvation. Experimentation during World War I brought about the development of quality amplification devices and microphones to improve wireless telephony. David Sarnoff, principal force behind the technology's development, insisted upon incorporating a radio and a talking machine in a single enclosure based on experiments at AT&T. Great strides were made in the production of an electromagnetic recording head and a reproducing machine that could be amplified to any desired volume with vacuum tubes. The result was more lifelike sound, with increased frequency response.

Sarnoff's plan to establish the world's largest manufacturing and broadcasting empire called for the acquisition of Victor's furniture-making facilities, although Victor was not ready to talk of merger or sale. Representatives of Western Electric in 1924 demonstrated the new electric recording process for uninterested Victor executives and then entered into negotiations with American Columbia. In London,

Louis Sterling, impressed by test pressings, in 1925 reached an agreement with Western Electric, purchased American Columbia, and incorporated the business into his European operation. A nonexclusive contract was negotiated with Western Electric, and a similar arrangement was made by Victor. Secrecy was the keyword at Camden, so the public would not learn that machines and disks would be rendered obsolete. Nationwide cut-rate sales were organized to dispose of stock on hand, and in the autumn of 1925 Columbia and Victor put out their first electrically recorded disks. At the same time, Victor introduced a full line of new Orthophonic Victrolas. Victor's superior manufacturing and merchandising caused profits to rise sharply as sales of Orthophonic recordings increased.

Recording techniques were changing rapidly, but industry practices lagged behind. Records were no longer a byproduct in the manufacture of phonographs, but executives still relied on music publishers to work on recommended advance copies or lead sheets of music. Average sales of individual releases tumbled, and a scatter-shot policy of picking material became the custom. Believing singing talent would increase sales, record companies spent six million dollars on a talent hunt. At the same time, successful artists were signed to long-term contracts, which only helped them to gain more live performances, their chief source of income. Soon, however, too many unknown performers recorded too many unknown songs.

Victor demonstrated the power of promotion over radio in 1925 after making an arrangement with AT&T to waive the customary advertising fees in return for an hour of the "world's greatest music" on which John McCormack performed his Victor release of Irving Berlin's "All Alone." Within a month, orders poured in for a quarter-million records, a million copies of sheet music, and 160,000 player-piano rolls.

The record business returned to the prosperity of the postwar period. Victor was in the forefront, with profits stimulated by the improving quality of both players and disks. In 1926, twice as many homes owned phonographs as owned radios. The radio rapidly changed in favor of broadcasting when the price of sets powered by regular house current fell.

Wall Street introduced David Sarnoff to a challenger to his ambition to own the Victor company: the combination of the Speyer and J. & W. Seligman investment houses. Negotiations were concluded in 1926, and Gene Austin justified the transaction when his 1927 recording of "My Blue Heaven" sold in the millions.

Ralph Peer was now associated with Victor, in charge of race and hillbilly recording. In lieu of salary he was given control of any music he recorded. Fearful that he would destroy the company's reputation for paying artists the highest amounts, Victor executives insisted that Peer double his twenty-five-dollar standard fee. Nonetheless, in the first year under this understanding, Victor paid Peer $250,000 in a single quarter.

for his publisher's 75 percent of mechanical royalties. The most important contributor to his income was a tubercular white ex-railroad man from Mississippi, Jimmie Rodgers. Peer first recorded him in 1927, and slightly more than 100 songs were recorded by the Singing Brakeman in the next three years, and published by Peer.

Alarmed by rumors of a merger of Victor with the RCA interests, Sterling made a bold move to save his American Columbia holdings. United Independent Broadcasters, an improvised network of about a dozen stations, could provide a ready-made outlet for advertising and promoting Columbia's product. Sterling offered to pay a bonus of $165,000 to a reorganized Columbia Phonograph Broadcasting Company and would receive in return ten hours a week on the air. Shortly before the contract was signed, in 1927, the Federal Radio Commission, formed to regulate broadcasting, ruled that failure to inform listeners when music mechanically reproduced on piano rolls and phonograph records was used constituted a fraud. This raised some question about the use of radio for effective promotion, but the Sterling deal was made, and Columbia unsuccessfully began to hunt for advertisers on its nightly broadcasts. Sterling recognized that his investment was ill-advised and the operation was purchased by a group that included William Paley. The network name was changed to the Columbia Broadcasting System, eliminating any connection with the phonograph business. In the twenty-seven-year-old Paley, broadcasting got a new and daring innovator, whose contribution would match that of David Sarnoff.

More than one-third of all nonclassical music recorded by Victor in 1928 was controlled by Ralph Peer. In order to conceal his steadily rising royalty income, Peer formed a number of music firms, the most important of which was Southern Music Publishing Company. Despite the precaution, Peer's returns increased, and in 1928 Peer turned over the operation of Southern Music to the record company, although he still shared control of all material already copyrighted and could continue to add to his catalogue.

Sarnoff's master plan for RCA approached realization in 1929 when the board of RCA, of which he was now vice president, approved his proposal to acquire a first-class furniture factory to make cabinets for RCA receivers and facilities for the manufacture of RCA sets, tubes, and allied products. The Victor Talking Machine Company, with total assets of about $69 million, was purchased by the transfer of $150 million worth of RCA common stock in exchange for a seven-eighths interest in the record company, which was divided among RCA, General Electric, and Westinghouse. It took months of corporate fighting and intrigue for Sarnoff to get control, then reorganize and consolidate the combined RCA Victor company, of which he became president.

Other developments of a lesser consequence were taking place at Columbia, Brunswick, and the American Record Company. The last was chief manufacturer of three-for-one-dollar records. Sterling temporarily revitalized his American Columbia company by capturing Paul Whiteman from Victor. When RCA Victor first reduced Red Seal production, Sterling filled the gap by importing European masters. Yet in spite of the major commitment of its facilities to the manufacture of RCA radio phonographs, Camden beat Sterling and all others, producing 34.5 million of the 65 million disks sold in 1929.

Variety announced the stock market collapse of 1929 with the headline "Wall Street Lays an Egg." Two days later, the Edison company terminated all manufacture of all recordings, retaining only the production of radios and dictating machines. RCA stock fell from 114 to 20 in the next few weeks. With the Great Depression, record sales fell by 39 percent in 1930 alone.

The race market was the first to evaporate; only the dance bands briefly managed to hold on, principally due to large demand by college students. Hillbilly records enjoyed a 25 percent share of all popular-music sales. But as the Depression deepened, hillbilly record sales, too, began to slide.

Victor's Red Seal series was slowly phased out. Most artists' contracts were not renewed and were replaced by recordings by small ensembles or soloists. Special sales emphasis was put on albums of classically influenced and artistically arranged versions of popular songs and instrumental music.

In 1930, Sarnoff was celebrating his first month as president of the Radio Corporation of America when word came that the Justice Department had filed an antitrust action to dissolve the complex of pooled patents, manufacturing facilities, and ownership of broadcasting, vaudeville, motion picture, and electronic properties, which Sarnoff had so laboriously assembled. But it was radio's fate for which he feared the most.

Radio began in America in 1877 when music was broadcast over wire lines during a demonstration of his telephone by Elisha Gray, a rival to Alexander Graham Bell, but the development of broadcasting proceeded faster abroad. In Italy Guglielmo Marconi constructed a wireless device that could send electric signals over distances. First demonstrated in 1901, Marconi's wireless sent the prearranged letter S from England to an experimental station in Newfoundland.

Many experimenters were simultaneously working in the United States. One of the most innovative was Lee De Forest, who introduced the vacuum "Audion" tube in 1906. This key element in the advancement of broadcasting was duly patented and unsuccessfully merchandised by the De Forest Radio Telephone Company. De Forest sought relief from the near failure of his company and the panic of 1907 in regular experimental broadcasts from the Eiffel Tower.

Broadcasting history waited for tragedy at sea. It was announced by twenty-one-year-old David Sarnoff, wireless operator of Wanamaker's New

York store's experimental broadcasting station. In 1912, a series of dots and dashes spelled out the news of the sinking of the S.S. *Titanic.* The fame Sarnoff won for the seventy-two hours, without sleep, of reporting news of the disaster brought him the first of a series of responsible positions with Marconi.

Wireless telephone advanced quickly. Soon known as "radio," short for "radiotelephony," it was a plaything of tinkerers. Yet when the American Telephone & Telegraph Company completed installation of the first giant radio station in the United States, Sarnoff sent a memorandum to his superior: "I have in mind a plan of development which would make radio a 'household utility,' in the same sense as the piano or phonograph. The idea is to bring music into the home by wireless."

Sarnoff envisioned a single "Radio Music Box," supplied with amplifying tubes and a loudspeaker telephone, estimated to cost around seventy-five dollars if made in large quantities. Its acceptance by a mere 7 percent of all American families would mean a gross business of about $75 million.

The Marconi Company disregarded Sarnoff's visionary 1916 memorandum. Instead, it concentrated on the development of high-powered equipment for military use. When America declared war, the 6,000 professional and amateur radio stations were closed down or turned over to the navy. Radio patents were expropriated, and experimentation was subject to government control.

As wartime restrictions were lifted, business in radio components boomed. Had it not been for Washington's prevention of the largest wartime supplier, the British-owned and -operated Marconi Company, from controlling international wireless broadcasting, the electric companies might never have developed Sarnoff's Radio Music Box. First with private support and then with semiofficial government assurance against antitrust complications, the General Electric Company formed the Radio Corporation of America in 1919. GE equipment would be sold and operated by RCA. All patents were mutually cross-licensed. American Marconi was dissolved, and its stations acquired by GE. With them came David Sarnoff, in an upper-management position. Soon the leading operator of ship-to-shore wireless installations, the United Fruit Company, joined AT&T, General Electric, Westinghouse, and others in a vast patent-pooling and cross-licensing arrangement, in which the telephone interests had major control.

Licensed transmitters were slow to return to the air. The best known, KDKA, operated in the Pittsburgh area and on November 2, 1920, sent out what is regarded as the first scheduled news-and-music program. Returns in the 1920 presidential election were read between selections of recorded music. The response indicated to Westinghouse that a market existed for a home radio receiver. Sales mounted, and other manufacturers followed Westinghouse into the marketplace. Only RCA evinced little interest in the manufacture of receivers. When Sarnoff resurrected his 1916 memorandum, the company reluctantly appropriated $2,000 to develop a receiver, the Radiola.

Thirty-one stations were granted licenses in 1921, and 576 in the next year, when between 500,000 and a million sets were in use. One tenth of these were assembled by the "hams," who spent $50 million for parts, and $60 million was spent for factory-assembled receivers.

Most license holders viewed station ownership as an indirect source of monetary returns. Almost all programming was musical, most of what was regarded as the "best of standard good music" provided without charge. License-holders generally did not conceive of the special use for commercial advertising. There was general opposition to AT&T's 1922 announcement to engage in "toll" broadcasting, having won the sole right for direct commercial purposes. AT&T soon was broadcasting over WEAF, in New York. The first toll-paying "sponsor" was heard in 1922. Only a few advertisers took advantage at first, but by mid-1923 their number had grown sufficiently to bring the broadcasters into a legal confrontation with a most unlikely foe—the music business.

As early as 1922, daily "music by wireless" caused fear of diminishing phonograph sales. Performers were broadcasting without payment, succumbing to the lure of free advertising. Dealers were unable to keep up with the rising demand for radio parts; orders accumulated for half a year and stock was exhausted as soon as it hit the shelves. ASCAP meet with representatives of the leading radio interests to advise them of the federal laws regarding public performance of copyrighted music. Negotiations continued through 1922, after ASCAP had granted permission to use its catalogue to stations owned by the large electric companies. The only stipulation was the announcement that the music was performed through the courtesy of the society and the copyright owners.

A conference at ASCAP's New York office, arranged by Claude Mills, brought together representatives of the large electric companies, the Department of Commerce, some unaffiliated publishers, and other interested parties. Mills stated that ASCAP controlled 90 percent of all music and proposed a minimum fee of five dollars a day for its use from radio stations. Those representing radio rejected the suggestion on the ground that no station except WEAF intended to ask for money for the use of its facilities.

Mills, Nathan Burkan, and J. C. Rosenthal, ASCAP's general manager, immediately instituted a legal action to confirm that a radio performance was a performance for profit and thus to establish ASCAP's right to collect fees. Shortly afterward, applications for ASCAP licenses were sent out. In November, *Variety* floated the rumor that the MPPA planned to install its own broadcasting station. Sarnoff, meanwhile, installed two 100-foot-high towers to send out signals through RCA's new property, station WJZ. Once on the air, Sarnoff had the world's most powerful radio station and became a serious rival to AT&T.

Restricted by the government's ban on "direct" advertising and AT&T's exclusive rights to use its transmitters and long-distance lines for commercial purposes, many stations used direct action or subterfuge to get along with the telephone company's facilities.

The radio manufacturers considered any cost of operation to be a general public-relations and advertising expense and not connected with "direct" advertising. Their regular transmissions created an audience of over 20 million persons in 1923–24 and were responsible for the sale of two million receiving sets. Operating expenses for the average station were less than $50,000 in 1924, due in great measure to airing only free talent, although a small number were beginning to pay performers.

Two court decisions affected the future of broadcasting: the ASCAP victory in an action against the operators of a Newark station, which sustained the society's claim that the least vestige of advertising on the air infringed the public-performance right; and the settlement in an action against Marcus Loew and his station WHN by AT&T, which opened the way for commercial broadcasting. Before New Year's Day, 1925, several hundred stations were selling time for "indirect advertising," which provided only discreet identification of the sponsor. ASCAP income from 199 radio licenses was $130,000, up from the previous year's $35,000. The continuing fight divided the music business when publishers joined the tax-free (non-ASCAP) movement in order to get time on the air for their music. Another dimension occurred when the MPPA initiated its own campaign to remove the two-cent compulsory license fee for mechanically reproduced music from the copyright law.

The secondary issue created an unexpected union of many of ASCAP's traditional music-user antagonists: the motion-picture industry and exhibitors, hotel and other entertainment people. Broadcasters attempted to get special legislation favoring their causes, but compulsory licensing continued, and all radio-industry-sponsored bills failed. In the course of the fighting, a new lobbying entity was born—the National Association of Broadcasters.

The radio business was divided in its approach to a solution of the music-licensing problem. Some station owners insisted on government action to curb ASCAP's "monopolistic practices." Others proposed the creation of a strong, industry-supported alternate source of music. These two courses were joined in 1923 by the activation of the NAB. Paul B. Klugh, founder of the Music Industries Chamber of Commerce, was hired as executive secretary. He suggested that Claude Mills be hired away from the MPPA and put in charge of the organizations Tax-Free Bureau. Mills insisted that the NAB get industrywide support before he would accept the position. It was slow in coming, and so were lost the services of Mills, who became one of broadcasting's most commanding antagonists.

Members of Chicago's Associated Independent (non-ASCAP) Music Publishers provided broadcasters with many popular songs. They joined the broadcasters in the hope of having their songs used in motion pictures, following the admission of the Motion Picture Theater Owners Association to the NAB. Equally important was the MPA's decision to use their standard and concert music for broadcasting free of charge. America's most important serious music was added to the ASCAP repertory only after its publishers received four places on the society's board as well as free licenses to major concert-music auditoriums.

The defection of Henry Waterson and his important Waterson, Berlin & Snyder catalogue from ASCAP and an antitrust suit filed by that firm in April 1924, asking for restraint of the society's influence with radio performances of his music, provided broadcasters with new arguments. Waterson openly opposed ASCAP's plan to license radio use and announced that he would exploit it by granting temporary permission to use specific songs without charge. He filed his lawsuit only after ASCAP refused to release him from a membership agreement. Forming two subsidiary non-ASCAP houses, he offered 70 copyrighted pieces to all broadcasters.

The heavy use of music on broadcasts over a number of northeastern stations at ten dollars a minute threw a new light on broadcasters' pleas for free music. Music programs used professional musicians who were friendly with the song pluggers. RCA's introduction of a new and much-improved "superheterodyne" receiver in 1924 created an affluent audience which demanded more professional entertainment. Victor's new series of programs to sell records of high-class popular and standard music was already showing the importance of radio performances in pushing sales.

The Washington hearings that began in 1925 for a new copyright law immediately polarized music users. Its main features extended copyright protection for the lifetime of the author plus fifty years, retained public-performance licensing (removing only the "for profit" language), and abandoned the two-cent royalty limit on phonograph records and piano rolls. For the first time, AT&T and RCA cooperated with the NAB to fight the proposed changes. Nathan Burkan and Claude Mills repeated their statesmanlike presentations on ASCAP's behalf in Washington. Gene Buck, ASCAP's new president, introduced songwriters whose income had been reduced by 40 percent by what he said was broadcasters' use of their works. It soon became evident that radio would have to pay for music, and any copyright revision temporarily was dead.

Many radio men conceded that those who owned a commodity that occupied between 60 and 70 percent of all air time should be compensated. That judgment gained greater advocacy after the Supreme Court upheld several decisions in ASCAP's favor. Having concluded that the public would pay for its programs by means of commercial advertising, both groups were brought together at the annual national Radio Conference in 1925. Claude Mills introduced a new phrase—"blanket licensing"—which gave total access to the ASCAP repertoire but allowed the society to restrict specified selections on written notice.

Radio was still in the experimental stage it had been in when ASCAP abandoned free licenses and let loose regional representatives, who retained one-third of collections made within fifty miles of their offices and half from those beyond. The majority of stations continued to refuse to pay for music and performers. While 200 stations had signed contracts with ASCAP, no standard by which to determine fees existed. Among them was RCA, which had a special license calling for payment of twenty dollars an hour. Thus a single five-times-a-week program now cost in excess of the society's top fee of $5,000. Many advertisers were now paying AT&T $2,600 for an hour on its basic thirteen-station network, producing a total of $75,000 from time sales in 1925.

Still a victim of AT&T's monopoly, RCA maintained an indirect advertising policy for the seven stations that made up its network. Because AT&T or sponsors paid musicians and artists, Sarnoff found himself obliged to do the same, and his board of directors considered alternate financing. Sarnoff constructed the first experimental 50,000-watt transmitter, which played an important role in the RCA board's conclusions.

The drawn-out, complex, and secret struggle between the broadcast and the telephone components of the industry was formally resolved in 1926 by AT&T's abandonment of broadcasting. Its station licenses were dropped, and WEAF sold to RCA. A ten-year contract gave access to AT&T's long lines to the new National Broadcasting Company, in which RCA held a half-interest, GE 30 percent, and Westinghouse the balance. In the midst of these negotiations, Mills announced on behalf of ASCAP that because broadcasters were now "willing within their limitations to pay fair and reasonable amounts for the right to use music," peace was made with the radio trust.

National network radio was introduced on November 15, 1926, with an estimated 12 million listeners. The National Broadcasting Company spent $50,000 on the program, but, it pointed out, advertising would henceforth pay for the programming. Thus, NBC was engaged in the direct-advertising business.

The path to that decision had been smoothed by the successful outcome of a suit challenging the Secretary of Commerce's right to assign wavelengths and power, and to regulate the industry. In the absence of a new regulatory and licensing agency, without government approval NBC put together two networks. From New York, WEAF fed the fifteen-station Red Network, which sponsored programs of conservative music and entertainment, while WJZ fed the ten-station Blue Network with educational talks and cultural presentations.

Those who attended the NAB's fourth annual convention in 1926 heatedly discussed looming federal regulation and the changes inherent in RCA's proposed paid-advertising network scheme. Only a few agreed to support a new radio-program service whose offerings would rely on tax-free music. Arthur Judson, a concert-artist booker and manager of the Philadelphia Orchestra, was appointed director of the proposed Judson Radio Program Corporation, and soon met with Sarnoff to offer the bureau's services in signing talent and building programs. Sarnoff rejected the offer, as RCA had opened its own talent bureau.

The United Independent Broadcasters network was formed by Judson soon after, with the third-ranking New York area station, WOR, as flagship. With talent assembled by Judson, the network went on the air in 1927 and lost $100,000 the first month. After a series of temporary injections of venture capital, the network was taken over and became the Columbia Broadcasting System.

The Radio Act of 1927 made only a few changes in the 1912 legislation. It mentioned neither network broadcasting nor paid radio advertising, nor did it create a new regulatory commission. Under these circumstances, there was little action except for a temporary halt to the proliferation of stations. The number of licensed facilities was then 732, only ten of which made any profit in 1927. With no expense except that for a radio set and having available the best free entertainment, the American public flooded Washington with protests every time any change in the broadcasting system was proposed. Nourished by regulatory inaction on the part of an underfunded and understaffed commission, commercial broadcasting thrived with a 24 percent saturation of receivers in American homes.

At the end of 1928, regular coast-to-coast broadcasting began with the hookup of sixty-nine NBC affiliates. It could now be heard on over 80 percent of the 11 million receivers in 9.6 million homes. In September the Lucky Strike Company sponsored the first experimental national program series away from the usual nighttime dramatic series or classical and semiclassical music.

With the promise in 1928 of increased commercial support, NBC expanded its radio and electric patents-connected interests. Sarnoff already engineered deals through Joseph Kennedy for control of the Keith-Albee vaudeville holdings, leading to the creation of RKO and the introduction of RCA Photophone into motion-picture studios and theaters. Now the RCA board approved his plan to consolidate the manufacture of all radio sets and equipment into a single corporate entity. Sarnoff was elected president of the unified RCA Victor Company, and for 6.5 million shares of new RCA common stock, GE and Westinghouse turned over all manufacturing rights for radio equipment, phonographs, sound-picture technology, and many white goods.

The stock market crash occurred just as final contracts were being approved, yet RCA continued to carry out Sarnoff's plan to dominate every phase of the entertainment business. By the end of 1929, E. Claude Mills, who resigned from the MPPA and ASCAP, announced that his new employer, the Radio Corporation of America, was forming an additional subsidiary, Radio Music Company, which purchased Leo Feist and Carl Fischer. Mills allayed fears that RCA was attempting to reduce ASCAP's monopoly and promised that there would be no boycotting of music published by competing firms.

Unconvinced, ASCAP pointed out that its contracts with the Feist and Fischer songwriters had been extended through 1935. NBC renegotiated its yearly blanket licenses, and in 1930 the five stations from which network programs originated agreed to pay more. Fees for non-ASCAP music were negotiated with the publisher, and then charged to the sponsor. Mills also became involved in bargaining for NBC's music licenses and played a role in increasing ASCAP's income from broadcasters.

The listening audience continued to grow in 1930, but only one out of every seven stations made money, and profits of the three networks fell. Advertising agencies played a leading role in this program development. An hour of peak broadcasting time, between 7 and 11 p.m., cost approximately $10,000, out of which the network rebated a 15 percent commission to the advertising agency. The networks began to look to advertisers and their agencies to provide and pay for all entertainment. For this, the agencies collected an additional 15 percent commission.

When they could not afford network time rates or wished to tap the Pacific Coast audience during the evening, advertisers used the "electrical transcription." This medium had emerged from the sixteen-inch recordings and large 33⅓-speed turntable developed in connection with ERPI's sound-on-disk talking-picture process, which was refined for radio use. In 1930 a number of advertisers improvised temporary networks by shipping recorded programs to stations around the country. Transcriptions also were accepted by all three networks for "spot" advertising; they were played during the regular hour and half-hour breaks that were originally intended only for station identification.

Nonetheless, during 1930 commercial radio spent $250,000 more on operating expenses than was taken in from advertising. Millions were spent on nonpaying public service, religious, and political broadcasts, as well as for land-wire leases. RCA was AT&T's best customer, paying three million dollars to connect its two chains of affiliated stations. In the antitrust action, charging industrywide restraint of trade, filed in 1930, both user and supplier and other subsidiaries were named as codefendants. The government sought a total reorganization of the manufacture of radio sets and equipment. RCA was deemed to control the business through exclusive ownership of significant patents. The Justice Department asked that GE and Westinghouse dispose of their holdings in RCA. This and restoration of free access to patents would create vigorous competition with the industry giant, whose profits in 1930 were more than $5.5 million.

In December 1931, AT&T agreed to cancel all cross-licensing arrangements with RCA. The final disposition of other charges awaited the conclusion of negotiations looking toward a consent decree and settlement.

Of more concern to broadcasting's smaller fry was ASCAP's announced intention to raise license fees, which represented less than 2 percent of their 1930 gross receipts. Profits were, in the main, nonexistent.

3 Hollywood and Movie Music

THE value of a song in promoting a moving picture was demonstrated by the popularity of "Mickey (Pretty Mickey)," written by Charles N. Daniels for the 1918 movie of the same name. Only ASCAP's attempt to collect an anticipated $1.5 million annually from movie exhibitors prevented the use of Tin Pan Alley hit songs to attract people to box offices. As ASCAP filed actions against screen-theater owners for the use of copyrighted music, they began to use nontaxable material. Production companies and some non-ASCAP standard-music publishers also printed collections of thematic music appropriate for use in movie theaters.

Hollywood producers recognized that music improved even an inferior picture. Classically trained composers provided music available in full score or in parts and transcriptions. Principal among them was Erno Rapee, director of the first symphonic orchestra in a motion-picture house (New York's Rivoli) and then musical director at Radio City Music Hall from its opening until his death in 1945. Owing to a large thumb-indexed volume, *Motion Picture Moods for Pianists and Organists*, Rapee became the most performed composer for the screen and the chief reference for movie-house musicians.

The sudden withdrawal of major catalogues of unlicensed music following ASCAP's reorganization immediately affected motion-picture exhibitors, who faced a stepped-up drive to collect fees from them. Lists of tax-free music prepared by the new Motion Picture Theater Owners Association were rendered obsolete overnight. The destruction of ASCAP became their immediate objective.

Active fighters against organized music publishers and songwriters were prosecuted by the society and generally found guilty of violating the Copyright Act. Theater owners began a well-financed attack on copyright. A major setback occurred when the Federal Trade Commission in 1923 found no actionable grounds against ASCAP.

Representatives of the theater owners and the new radio business became frequent witnesses in Washington during anti-ASCAP licensing hearings held from 1923 to 1926. Both groups tried, unsuccessfully, to survive with tax-free music offered in 1923 by the defection from AS-

CAP of several major publishers, and by all members of the Music Publishers Association. Theater owners became active supporters of the new National Association of Broadcasters, formed in 1923 to fight ASCAP by creating a plentiful supply of tax-free music. Peace came in 1926 as 11,000 exhibitors became ASCAP licensees and paid $525,000, which represented more than half ASCAP's income that year.

The Motion Picture Owners Association could no longer compete effectively with the theater chains, owned by Wall Street men who also owned major Hollywood interests. These powers were apathetic to any notion of making pictures speak. The talking picture was nothing new, however. It went back to Edison and Lee De Forest, who as early as 1906 had experimented with sound on film. His interest was revived in the early 1920s, after RCA purchased most rights to his vacuum tube. De Forest acquired a partial interest in the sound-on-film process owned by Theodore E. Case, which he improved, and then formed the Phonofilm Corporation in 1925.

At the same time, the Western Electric division of AT&T had secretly been at work on similar experiments. A 1924 test picture elicited little enthusiasm. Western Electric sold the process to Wall Street broker Walter Rich in 1925. He came across the Warner brothers, who bought a half-interest from Rich, and the Vitaphone Company was formed. Production began of short popular- and classical-music and comedy movies to accompany full-length silent Warner features, starting with the 1925 movie *Don Juan,* starring John Barrymore, which featured music and no dialogue.

With rare exceptions, motion-picture "theme songs" usually failed. The astonishing success in 1926 of Erno Rapee's theme song for *What Price Glory?*, "Charmaine," revived multimillion-copy sales of sheet music, and Tin Pan Alley's interest in the production of short musical two-reelers.

William Fox, producer of *What Price Glory?* and *Seventh Heaven,* which featured Erno Rapee's second million-selling hit, "Diane," at first opposed sound films but now saw them as a way of reducing the expense of large orchestras for his movie-house chain. Fox secured a sublicense from Vitaphone and obtained rights for Theodore Case's sound-on-film process. Sensing the revival of interest, Western Electric recovered its bargain-price exploitation rights from the Warners and Case in 1927 and created Electrical Research Products, Inc., to license and service the process to which it held all basic patents.

More than 100 first-run movie houses around the country were now wired by Warners. In 1927, *The Jazz Singer,* a silent feature with songs chosen and recorded by Al Jolson, opened to great acclaim. Reluctant Paramount executives committed themselves to the GE-RCA Photophone process for a single picture, the aviation epic *Wings,* in 1927. The movie was a box-office success, but ERPI and Vitaphone had outwitted their competitors by concluding contracts with seventy-five first-run screen theaters and all five major production companies.

Once committed to Western Electric's sound-on-record system, the Warners believed that they were entitled to the compulsory mechanical-licensing provision of copyright law and its two-cents-per-side royalty. The MPPA pointed out that commercial recordings were made for home use only and threatened to pull out of ASCAP. In such an event, every theater showing a Warner product that featured interpolated songs would violate the law. The Warners therefore agreed in 1926 to pay a minimum $100,000 annual royalty against recorded or synchronized music rights.

After Western Electric pressed the Warner-Rich Vitaphone Company to return its initial license, and ERPI was formed in 1928, the MPPA forged a new agreement with it, again licensing synchronization rights at a rate higher than that called for by the law. These were passed along to other film makers under their licenses from ERPI. Acting as an agent for MPPA members and as chairman of their board, Claude Mills concluded an arrangement for sixty-three music houses. The income was based on a sliding-scale fee of from two and a half to five cents a seat annually from a maximum of two million theaters in which ERPI equipment was installed, in addition to the usual ten-cent-per-seat payment to ASCAP.

When ERPI and its licensees realized in 1929 that the agreement did not include world rights for talking pictures, the contract was modified. Higher new fees were set for the use of each individual song. With the MPPA in a temporarily moribund state, publishers received the payment directly.

Talking pictures became the principal means for plugging popular music. Al Jolson's recording of "Sonny Boy," backed with "Rainbow Round My Shoulder," both featured in his second Warner feature, *The Singing Fool*, sold a million copies of sheet music in less than a year. Warner Brothers' assets rose from $18 million in 1928 to $239 million a year later. It was obvious that the film industry must control a sufficient number of music copyrights to make it totally independent of any combine of music publishers, songwriters, or copyright owners. The ERPI contracts with the MPPA were soon to expire and in order to raise the price of their product by freeing exhibitors from the obligation to pay ASCAP and the publishers, a giant tax-free library of music must be assembled.

Warner bought M. Witmark & Songs and the Dreyfus music interests, providing the company with the exclusive services of the pick of the country's best songwriters. Considerable profits were anticipated from the Dreyfus operation, the Chappell-Harms firm alone accounting for an annual surplus of $500,000. With Warner owning a majority interest in Remick (in which Dreyfus had a 50 percent interest), Harms and

T. B. Harms, New World Music, DeSylvia, Brown & Henderson, and several other music houses, they could control a major portion of the music publishers' vote in ASCAP policy making and the distribution of its revenues.

The Depression, however, eroded all entertainment profits. Sound was now installed in 83 percent of all theaters, but box-office receipts fell. Hollywood cut expenses and eased songwriters out of their contracts and sent them back to New York.

These songwriters returned to a tempestuous music publishing world. Fights between publishers and songwriters arose over profit sharing. When a song could not be bought outright, publishers offered contracts with a maximum "2 and 33" royalty: a two-cent royalty of sheet-music sales and up to one-third of phonograph record royalties. When numerous collaborators shared in royalties, some were acknowledged on the sheet-music cover, while others got a percentage for "cooperative contributions." These ranged from a gratuity in return for plugging the song on stage or radio to paying performers and bandleaders in order to obtain performances, but all such practices radically reduced the writer's share. Publishers complained that the major users of their music were responsible for their declining profits. Yet record and piano-roll royalties averaged about $2 million a year; sales of printed music remained around $15 million annually; and income from licensing public performance of music through ASCAP rose from about $250,000 in 1921 to nearly $2 million in 1930.

The 1920 compromise that ended the production-music writers' control of ASCAP brought tax-free music houses back as licensees and strengthened the society as a collection agency. The writers planned to assign all performing rights to ASCAP for ninety-nine years, but publishers voted instead to keep assignments on a five-year, renewable basis. Most songwriters regarded its Claude Mills as a watchdog on behalf of the few important music publishers. This judgment was validated when a group of small publishers proved that the large firms were spending between $1,000 and $2,000 in cash inducements to performers and Mills took no action.

The music publishers were wary of enforcing the right of public performance for profit when it threatened their sources of exploitation. They argued that vaudeville was covered by the dramatic rights they controlled and was not responsible to ASCAP. Nearly two years passed before the thirty-eight MPPA houses agreed to print the warning "All Rights Reserved Including Public Performance for Profit" on all printed music.

As ASCAP income rose, pressure mounted to have it classify authors and composers and allot performance royalties on that basis. In fixing the status of a writer, the "number, nature and character" of his catalogue, the "popularity and vogue" for his work, and the length of his association with ASCAP should be taken into account. The reorganization was possible only after the new publisher members insisted that provision be made for two-thirds of their directors to determine the share to be paid to the music companies. This created a self-perpetuating board.

A serious examination of ASCAP's business practices in 1923 persuaded the board to appoint a strong central management. A three-man advisory committee was named: Nathan Burkan's protégé J. C. Rosenthal, who remained as general manager in charge of legal affairs and daily operations; founding member Silvio Hein, a writer, named as traveling secretary; and E. Claude Mills, the publisher's nominee.

In 1923, Gene Buck was elected ASCAP's unpaid president. At the end of his first year, Buck announced the largest quarterly distribution in ASCAP's history to date: $80,000, from an annual income of $600,000, a 25 percent increase over the previous year. He convinced the MPPA to remove the music its members published from the tax-free category. This brought in fifty-three "name" standard-music composers, whose publishers immediately were elected to the Class A group, the highest classification, and paid in four figures.

Burkan, Buck, and Mills successfully protected ASCAP interests. Mills acted as official spokesman for the music publishers and served their needs for the society. While they ignored mechanical royalties, printed-music sales, and steadily rising ASCAP income, Mills and the music men he represented hammered on radio's excessive use of popular songs and the resulting decline in royalties. Meanwhile, the most aggressive houses rededicated their efforts to securing on-the-air play.

In 1924, *Variety* suggested that the day of the million-copy seller was gone. With retail prices of twenty-five and thirty cents for popular songs, financial returns were greater even if overhead had grown. Modern promotion techniques permitted a song to be made into a hit, but only with a tremendous outlay of money and attention.

Radio changed the profile of the average sheet-music buyer, who now dialed around the radio frequencies looking for a new song to learn. Tin Pan Alley's writers of production songs enriched the coffers of the most prosperous firms, in particular those belonging to the Dreyfus brothers, Max and Louis, who purchased the most valuable music business, the Harms firm, in 1906. In the early 1920s, they purchased the British Chappell's American outlet and formed Chappell-Harms, as a repository for non-production music.

Key to their success was the work of Jerome Kern. In 1904, Kern became a major owner, in association with the Dreyfus brothers, and the firm devoted itself principally to publication of his music. A parallel company, Harms, Inc., owned entirely by the Dreyfus family, was formed. Kern's success illustrated Max Dreyfus's instinct for discerning genuine talent. Among the many writers he promoted are Rudolf Friml, George Gershwin, Vincent Youmans, and the team of Rodgers and Hart.

Important changes occurred on Broadway. The Ziegfeld production of *Sunny*, starring dancer Marilyn Miller, marked the influence of Ruth St. Dennis and Isadora Duncan on modern dancing as an art form and a principal feature of contemporary stage musicals. The musical-comedy composer now had to compose and orchestrate complex dance music.

At Dreyfus's insistence, ASCAP, in 1925, invoked a clause in its radio contracts allowing for the restriction of specific music on written notice. Broadway producers had been delighted when stations broadcast part or all of their productions, but they complained when these cut down ticket sales. Arthur Hammerstein, head of the Producers and Managers Association, instituted a drive to obtain a larger share of publisher royalties and a part of the ASCAP fees. The PMA threatened to take over all rights and ask for a half-share in the money broadcasters paid ASCAP. This was avoided when Mills signed letters from ASCAP to all stations announcing the restriction of six songs from *Rose Marie*. The restriction policy was difficult to enforce, but music publishers, Harms in particular, invoked it often.

In 1925 Shapiro, Bernstein shipped more than one million copies each of several pieces, including "The Prisoner's Song." The industry equated sales of 400,000 copies with a hit, which this hillbilly song exceeded. Written by Guy Massey, it was recorded for Victor by his cousin Vernon Dalhart. With little or no promotion, "The Prisoner's Song" became one of the best-selling vocal records and songs of all time. Dalhart collected more than $85,000 from Shapiro, Bernstein, 5 percent of which he paid Massey until his death.

No music house had a windfall to equal "The Prisoner's Song," but business turned a corner with Victor's introduction of the Orthophonic Victrola line and Brunswick's better-sounding new Panatrope. Direct-advertising radio stations looked to music from the old masters and the standard houses. Popular syncopating orchestras and their singing entertainers plugged hit songs over the air after 11 p.m., a time when the "right" audience was not expected to be listening.

The Feist business fine-tuned radio plugging. They concentrated on important radio bands, and when demand arose, sheet music was rushed to stores around the country. Other music companies had almost succumbed to such changing ways. M. Witmark averted a financial collapse because of Sigmund Romberg's loyalty and financial support. With their credit restored, the Witmarks bought the Arthur W. Tams Music Library. The combined Tams-Witmark Library became the largest source of musical-comedy and operatic music for amateur productions.

Variety music editor Abel Green's *Inside Story on How to Write Popular Songs* frankly dealt with the prevailing state of the music business in 1927 and emphasized that a connection with any of the famous baton wielders was imperative. Their "plans in editing and suggestions" together with live performance were "compensated with royalty interests" of one-half, one-third, or one-quarter and also credited on the title page.

All was far from serene in the music business and ASCAP in 1927.

In order to raise the ASCAP ratings of his supporters at the MPPA, Mills instituted a special four-week survey of radio performances. On the basis of these findings, a new class of writers, AA, was formed to supercede the A group, and its members were awarded an additional $100,000 for performances. Publishers argued that this practice could not recognize and report on old-time songs and standard music, which formed the bulk of radio and motion-picture-house use. Half of the society's income was coming from the latter source.

An anti-Mills faction had existed. Abel Green wrote that the MPPA had lost sight of its purpose—the elimination of payment—and worked only for the benefit of a "favored few." Some leading MPPA houses had set up dummy corporations, Green alleged, in order to evade the organization's rules against paying singers and song plugging on the radio. Mills used his position, Green claimed, only to increase his ASCAP-MPPA salaries and by his unilateral actions effectively displaced J. C. Rosenthal as ASCAP's general manager. Unperturbed, Mills continued negotiations for picture-synchronization rights with Western Electric's ERPI and concluded a five-year agreement that guaranteed the MPPA an expected grand total of one million dollars. ASCAP, the sole custodian of performing rights in screen theaters, had not been party to the negotiations, but Mills indicated that authors and composers would reap additional income through separate agreements with their publishers.

Mills's often autocratic actions at ASCAP brought about a public confrontation with young songwriters, dissident new writers, and Class B and C publishers in 1928. His planned elevation of Berlin, Feist, Remick, and Shapiro, Bernstein, plus Harms, to the AA category was opposed by those publishers who would be set back one or two classifications.

In 1930, demands were presented, at the first general business meeting of ASCAP in ten years, for specific information about operations that represented 20 percent of income, distribution, the classification process, and for an itemized list of expenditures. Questions were raised about Mills's negotiations with Vitaphone and RCA that would bring more than $100,000 to the publishers, rather than to ASCAP.

Serious discussion of the society took place. A survey of licensees revealed that the music houses with national branch offices dominated. It was argued that Marks, Belwin Music, Sam Fox, Witmark, Robbins, Rischer, and Schirmer should be ranked with the five class A firms. Newly successful songwriter-publishers believed that the old school of music men deserved special consideration, but that the young writers should have an active role in operations as well as a larger allowance of total income. Many reforms were proposed, among them open annual meetings, a standard songwriters' contract, the end of the self-perpetuating board, and a permanent Class B category for authors and composers who had been members for more than ten years.

The sudden economic impact of talking pictures was an equal factor in the temporary suspension of ancient antagonisms. Hollywood recognized the awesome power of the talking picture to catapult songs into success. One could print dance arrangements of a new movie song then sit back and wait for the screen to work, with little other expense or effort.

William Fox was the first in Hollywood to make an alliance with a leading Tin Pan Alley house, De Sylva, Brown & Henderson, to be the outlet for his picture music. The writers had an unexpected hit in a second *Singing Fool* song, "Sonny Boy." Coupled with "There's a Rainbow Round my Shoulder," Jolson's recording of "Sonny Boy" for Brunswick sold over a million copies. Within six months, "Sonny Boy" sold over a million sheet-music copies. In 1929, four of their musical comedies—*Follow Through, Good News, Three Cheers, Hold Everything,* and a musical revue, *George White's Scandals*—were running simultaneously on Broadway. Fox took the team to Hollywood in 1929 where they wrote four hits for one of the earliest musical-comedy talkies, *Sunny Side Up.*

The first original movie musical, *Broadway Melody,* was released by MGM in 1929, with songs that went into Robbins Music's catalogue. A 51 percent interest in the business had been picked up from Jack Robbins and the catalogue offered MGM a medium for exploiting new screen musicals, as well as a hedge against successful commercial sound-and-picture broadcasting. Robbins took advantage of the MGM chain of theaters by putting uninformed pages in the lobbies to sell sheet music and recordings of songs from *Broadway Melody.* Under his MGM contract, he retained all copyrights in his own name.

The music division of Warner Brothers anticipated a problem within three years, when ASCAP's per-seat contract with movie houses and MPPA's agreement with ERPI would expire simultaneously. To avoid these dilemmas, Warner decided to control all music copyrights used in its talking pictures. In 1929, Warner copyrights were assigned to the film company and put in Witmark's catalogue. *Variety* viewed this as leading to the inevitable disintegration of ASCAP. Once a catalogue of a sufficient number of copyrights made Warners independent of existing music-licensing organizations, the company would grant performing rights to exhibitors without charge, and force competitors to pay whatever the traffic would bear for synchronization licenses. Anticipating such a strategy, ASCAP persuaded songwriters to extend their membership for an additional two years, to coincide with the expiration of the original MPPA-ERPI agreement.

Warner Brothers' purchase of the Witmark business and the impending deal to buy all of the Dreyfus music holdings clearly indicated that they intended to pull the MPPA down. Harms's books indicated that Max Dreyfus got five times his ASCAP income by selling synchronization rights on his own, at rates higher than the MPPA's.

Warner took over all assets and copyrights in 1929 of Harms and

Chappell-Harms, De Sylva, Brown & Henderson, Remick Music, Green & Stept, Famous Music, T. B. Harms, and George Gershwin's New World Music. A new corporate entity was formed to handle the merger, Music Publishers Holding Company.

Hearing rumors of a vast takeover of music companies by RCA, Mills journeyed to Los Angeles in April 1929—publicly, to plead with ASCAP songwriters for mutual action against the moving-picture companies that threatened to topple ASCAP; privately, with the hope of negotiating new synchronization fees with those enemies. The movie men directed Mills to meet with their legal counsel in New York for a discussion of already "exorbitant" fees.

Bitter antagonisms within the ASCAP ranks resurfaced when Mills urged that public-performance rights be reserved in any proposed contract. However, of more concern to the writers was the discrimination by Class A members of the East Coast committees against West Coast writers in lower classes.

An ultimately more pressing problem for ASCAP was the displacement of live musicians by mechanical record players. The coin-in-the-slot phonograph industry had not used disks until 1908, when the first automatic coin machines went on sale. Though the mechanical player disappeared from circulation in 1920, it returned in 1927, when the Automatic Musical Instrument Company presented the Selective Phonograph, which offered twenty recorded selections. In three years, 12,000 of them were in use.

Hollywood's compensation to song-writers gave those who stayed in the East a new fear that their ASCAP ratings would fall if their western counterparts controlled their society. There had been unsuccessful attempts to form a New York-based counterpart to the MPPA. The most recent, the Songwriters, had been dormant for several years when its secretary, Leo Woods, met with 200 writers and composers in 1929. A new standard contract was proposed with a minimum guaranteed salary, improved accounting methods, elimination of all cut-ins, and the obligation of the publisher to work on a new song within six months of buying it or return it to the writer. Little was accomplished, however, while the film industry reconsidered the value of popular songwriters.

That confusion was reflected by the record companies. When Hollywood created a glut of material, this led to a superabundance of recorded movie music, which sold pictures but not recordings. A Hollywood-connected publisher could no longer assure record makers that a specific film song would be plugged, cut off after a few bars, or completely eliminated.

Louis Dreyfus meanwhile purchased a controlling interest in the British firm Chappell & Co. and thereby became an important figure in negotiations with the British Performing Rights Society, which led to the first payment made from England to ASCAP.

Finally succumbing to the series of offers from ASCAP-licensed music users, Claude Mills resigned from both ASCAP and the MPPA to become head of RCA-NBC's Radio Music Corporation. He then negotiated for the purchase of several music catalogues while being the chief target for the Warner companies' complaints that NBC boycotted their catalogues in favor of those controlled by Radio Music. NBC's reported suggestion to radio artists that they favor Radio Music copyrights brought threats from Warner to restrict 3,000 of their songs if the order was not rescinded.

Three important changes in the Copyright Act were introduced in early 1930, but had little support from music-business figures. As a result, passage was doomed for American membership in the Berne Convention, which would have brought harmony to international copyright protection through elimination of the compulsary licensing clause and the two-cent royalty and an amendment to permit divisibility of rights so that individual copyright privileges could be assigned separately.

Radio broadcasts now were the most potent force yet for promoting popular music. Compensated radio artists and bandleaders unfailingly identified their material. Network officials were unable to restrain the fast-talking bandleaders for accepting cash for air play. A cut-in contract appeared to be the only way to do business. Publishers complained about having to pay for performance, but they approved a new MPPA regulation reducing the mandatory $1,500 fine for it to $250.

The stock market crash caught up with the popular-music business. Wholesale orders were down 75 percent. The record business suffered, too, a victim of overproduction and tumbling sales. In 1930, John G. Paine, the successor to Claude Mills at the MPPA, called a meeting of the publishers. The only outsider present, Abel Green, urged that ASCAP control and regulate radio performances by restricting songs. One thing was learned, Green concluded: "the suspicion that radio, through its own publishing interests, was endeavoring to create its own library of songs in order to be rid of royalty obligations, is immediately disproved." Not enough sufficiently worthy material could possibly be turned out to fill radio's needs. The MPPA instituted a fee of fifty cents per song per station when electrically transcribed commercial programs were broadcast, which was to be paid to the association for distribution to the appropriate publishers.

Hollywood reduced synchronized music and songs and no longer dominated New York. Finding that its new music holdings were not making profits, Warner permitted their Harms and Witmark firms to operate in the traditional flamboyant Tin Pan Alley manner. The songwriters' hegira to Hollywood was slowly reversing. Hollywood entered a period of box-office decline and production cost-cutting that continued until 1934. Warner, which had spent nine million dollars to buy a major share of Tin Pan Alley, turned operation of its music holdings over to the former owners. Studio music staffs were cut down to a single person. Most films now were accompanied by bits and snatches from old releases or music written by the few trained composers still on salary.

Despite a drop from the 143 songwriters under contract to all companies during the musical talkies' short-lived heyday to fewer than 20 in 1931, several hundred songs were used, generally acquired directly from music publishers. The business had come half-circle since 1929, when music reigned and a film was rarely released without a title song and original popular music. The elaborate musical picture was no longer affordable.

The industry turned to the currently most popular entertainment medium—network radio—for entertainers with guaranteed appeal. A number of studios again planned musicals. Warner went again to the Broadway and Tin Pan Alley songwriters for its songs. Harry Warren and Al Dubin wrote the score for *42nd Street,* a low-budget film. Production costs mounted due to elaborate production numbers staged by choreographer Busby Berkeley, yet profits from these spectaculars permitted excesses long after the industry had turned to the personality musical, featuring stage and radio entertainers and slim opera divas.

The first of these was RKO's 1933 *Flying Down to Rio,* with Fred Astaire and Ginger Rogers. The elegance of their dancing raised screen dance and music to a level of perfection and public appeal. In 1934, film music was first recognized by the Motion Picture Academy, with Oscars for "The Continental," from Astaire's *The Gay Divorcée* and the score for *One Night of Love,* starring Grace Moore, a Metropolitan Opera star. Quality screen songs also sold sheet music: those from *42nd Street* and the *Gold Diggers* series had a combined sale of 400,000 copies.

Herman Starr, Warner Brothers' vice president of music operations, regularly shifted all rights in movie scores to less active firms in order to improve their ASCAP ratings, which, owing to recent changes in classification standards, were based on network-radio performances. Determined to enforce Jack Warner's directive that his music holdings must provide at least a million dollars annually by 1936, Starr prepared to withdraw from ASCAP unless performance royalties increased significantly. Throughout 1935, company employees concentrated on radio plugs to ensure that Warner copyrights represented at least 40 percent of all the music used by network radio. The competition for higher ASCAP payments was intense in a year when most of the ten most-performed songs came from movies.

A new sense of music's value permeated Hollywood. There was an increase in the creation of original music for all films, and experienced film composers were hunted with fervor. Dr. Erich Wolfgang Korngold, a former Viennese child prodigy and composer of the opera *Die Tote Stadt,* joined the staff at Warners. After years of experience in man-of-all-work assignments, Max Steiner, Alfred Newman, and others found themselves installed in music departments that had doubled in size almost overnight.

General wisdom had it that a successful film song added significantly to box-office returns and that radio play was an essential factor, a relationship Herman Starr preferred to ignore. Believing that the Warner copyrights were vital to the sponsors who used musical programs to attract audiences, he asserted his musical rights were worth more than he could ever receive through continued affiliation with ASCAP. On January 1, 1936, the entire Warner music catalogue was pulled out of the society and off the air. Overtures unsuccessfully had been made to the networks and individual sponsors for separate music licenses. Forced to plead for the best deal possible, in the summer of 1936, Warner Brothers' music interests returned to ASCAP.

Sixty-three of New York's top songwriters moved west during the Warner boycott. With the largest song-writing unit on any lot, MGM purchased the Feist music-publishing business. Starr was caught up in the splurge of film musicals and began to negotiate for the remaining Chappell properties.

Studio heads, after reading *Variety*'s list of the twenty-five songs with most airplay, berated their East Coast music men, who drove their staff to get more film music on that list. During 1936 and for the next half-dozen years film songs were generally at the top of the charts. Irving Berlin questioned the validity of the promotion represented by the *Variety* list and whether popular music really added to the box-office grosses of movies. Jack Robbins disagreed. Responsible for MGM's acquisition of the Feist business as well as the recent purchase of Miller Music, he believed that a company should be prepared to pour out a quarter of a million dollars a year for advertising and promotion of music. Hollywood was not prepared for that, and music men continued to bury the sums spent for subsidies. But overplugging was driving the independent music houses out of business by reducing opportunities for airplay.

Because ASCAP distributions were based mainly on network play, the independents' income was reduced. They turned to Washington. Confronted with an investigation by the Federal Trade Commission, the movie-affiliated publishers resorted to a corrective code of trade regulations. Investigation revealed that paying for plugs was still industrywide. Film-company music houses continued to dominate *Variety*'s weekly list, by the use of "money, gift, bonus, refund, rebate, royalty, service, favor, and other things or acts of value," as the new code catalogued the practices. Subscription to the new rules by a majority of MPPA members was sufficient to call off further FTC investigation, but in 1937, the thirteen music houses owned by or connected with Hollywood interests continued to share 65 percent of the ASCAP publisher distributions.

The 1937–38 period was the greatest film-musical season in history, with plans for $750,000 and million-dollar productions. Stories more smoothly integrated music and songs. Professional managers and company heads were moved to California to coordinate the selection, spotting, and promotion of all new songs. Many expensive film musicals were probably saved by the inclusion of songs recognized by the new liaison men as "commercial."

No one in the film industry yet dreamt of the astonishing sums that would accrue with television, but as early as 1935 some did concern themselves with future royalties from a medium still in its experimental stages. That year also, ASCAP mailed a new five-year extension agreement to all its members, which included television rights. The earliest American development of television was undertaken in 1910 by Vladimir Zvorykin, employed at the Westinghouse laboratories. Zvorykin conducted the first public demonstration of a television set in 1927. Radio stations began to apply to the Federal Radio Commission for experimental television licenses, and Paramount's participation in CBS in 1929 was predicated on the belief in the inevitability of picture-casting. The soundness of that judgment was confirmed in 1935 when David Sarnoff announced that RCA-NBC would spend a million dollars on over-the-air-demonstrations. Zvorykin had persuaded Sarnoff that only $100,000 lay between experimental and full-development stages. Six years and much money later, a working television station was erected atop the Empire State Building.

The movie business had witnessed the growth of broadcasting into a multimillion-dollar business, much of which depended on the use of their music, licensed for a fee regarded as completely unrealistic. The broadcasters did not intend to perpetuate that inequity. Hollywood-affiliated ASCAP publishers held that any use of music by telecasting would be similar to synchronization of music on film and therefore outside the society's scope; a separate license would be required, which only the publisher could offer. The six major ASCAP houses refused to sign any ASCAP document that referred to television, and a revised version omitting the word was distributed, preparing for future battles over television performing rights.

The MPPA's most recent code of practice was two years old in 1939, and five Hollywood-New York music-house combines—Warner Harms, MGM, Robbins/Feist Miller, 20th Century-Fox/Movietone/Berlin, Paramount Famous, and RKO Chappell/Gershwin—collected the lion's share of all ASCAP distributions. The MPPA itself protested to the Federal Trade Commission that, in addition to the "fabulous sums," paid for airing their music, the companies lured bandleaders with promises for short commercial films of their bands in actual performances. Over the protests of picture-music publishers, ASCAP cast about for months before the current means of paying publishers was changed. A notable decline in film songs on the most-played lists followed.

Meanwhile the audience appeal of romantic musical dramas and comedies with high quality songs had seen its day. The story became the thing with which to win a music-conscious public. People turned out for *Gone With the Wind,* which had a memorable musical main theme, the three-million-dollar Technicolor production *The Wizard of Oz,* Walt Disney's *Pinocchio,* a revival of Victor Herbert's *Sweethearts,* and screen biographies of Herbert, George M. Cohan, and other popular music figures.

Government figures released in late 1940 put an end to the decade-long assertion that the picture industry was America's fifth-largest business. With $2.5 billion invested in theaters, studios, and distribution, and another billion overseas, it was, in fact, the thirty-fourth. In only one year, 1937, did gross income exceed expenses, but 1940 promised to be a profitable year. Attendance at the country's 19,000 theaters had flattened out at 80 million. Admission prices had peaked at an average of thirty cents in 1929, then fell to twenty cents in 1933–34, and were now rising to twenty-five cents. Optimism was leavened by government imposition of an addition 10 percent tax in the second half of 1940 and the possibility of involvement in the European war, which had drastically cut income from foreign sources. There was also the uncertainty caused by the consent-decree settlement in the antitrust suit brought against Paramount, Loew's-MGM, RKO, Warner Brothers, and 20th Century-Fox by the Department of Justice. Block booking of a season's entire production was outlawed, and exhibitors could see new releases in advance. The possibility also loomed that the major studios would have to divest themselves of the 20 percent of all theaters they owned, which provided 54 percent of their total revenues.

Equally important for the future was the prospect of another kind of war at home, one between ASCAP and the broadcasters, who had, surprisingly, united against the new contract with increased rates being offered to all stations. The lesson of 1936, the Warner war with radio, was forgotten, with consequences that were to shake every aspect of popular entertainment.

4 The Fall and Rise of the Record Business

IN 1925 phonograph recording tumbled by almost half from the 1921 high of $106 million. The Victor Talking Machine Company failed to pay dividends for the first time since 1901. Sales continued to slip, eventually dropping by 60 percent in a single year, to $16.9 million in 1931.

The new Los Angeles RCA facility remained mostly unused. RCA Victor merged its distribution system and product-merchandising operation, while the Camden factory concentrated on the Radiola. Recording was cut to a single wax per selection. Expensive artists were dropped by Victor, as well as Columbia and Brunswick, its two major competitors.

Among the few artists retained at Victor was Jimmie Rodgers, whose record sales had slipped dramatically. Ralph Peer, instrumental in retaining Rodgers, made fellow recording supervisors jealous of his partnership with Victor through his music house, Southern Music Publishing Company. Eli Oberstein, a special enemy, had been brought by Peer to Victor's sales accounting group to protect his interests. Despite rumors of various crimes, Peer clung on tenaciously and continued to build his catalog.

Conditions in the British record business resembled those in America. The two major manufacturers, the Gramophone Company-His Master's Voice (HMV) and Columbia, suffered a joint decline of 90 percent in profits during 1930–31 and were forced to merge. Under Louis Sterling's guidance, Columbia controlled European machine and recording manufacture. In 1925, he purchased factories in a number of foreign countries, which were consolidated into a holding company, Columbia International.

In 1931, Sterling's international phonograph combine joined Gramophone-HMV to form Electrical & Musical Industries. David Sarnoff represented RCA on the EMI board until 1935, when RCA Victor surrendered all interests in EMI, leaving the latter without any American affiliation, Sterling having disposed of Columbia several weeks after EMI became operative. Grigsby-Grunow, manufacturers of radios and household appliances, become proprietors of Columbia Records.

Impoverished Americans more and more turned to radio for their entertainment. The increasing use of quarter-hour radio programs recorded on electrical transcriptions persuaded Victor executives that consumers might purchase fifteen-minute recordings. Victor engineers and furniture designers produced but did not market a fifteen-minute disk and handsomely encased equipment with which to play it. This provided the Erwin-Wasey advertising agency with an opportunity in 1931 to be first to market a double-faced recording bearing five minutes of the newest hit songs on each side.

Goaded into action, Victor introduced longer-playing records and doubled the number of grooves on a new compound called Victrolac, which cut down playing speed to 33 ⅓ revolutions per minute. Columbia entered the race, with long-playing disks and compatible machines. However, enthusiasm for long-playing records waned quickly due to the high price of two-speed machines.

In 1932 sales declined 40 percent. Warner Brothers disposed of its Brunswick phonograph and record holdings to American Record, the major purveyor of twenty-five cent and three-for-a-dollar records. Sears, Roebuck, the most important of the latter, was catering chiefly to the middle-American market for hillbilly music, which had tripled since 1930.

Now second to American Record, Victor cut down on all field recording and instead brought authentic hillbillies to metropolitan studios. Now in full charge of Victor's race and hillbilly catalogues, Oberstein undermined Peer. Hostilities between the two former friends concluded when Sarnoff learned the terms of RCA Victor's deal with Peer. Justice Department attorneys were seeking anything that might discredit the corporate giant. Determined to avoid any such possibility, Sarnoff gave orders to dispose of Southern Music to Peer.

RCA Victor's negotiations with the F. W. Woolworth Company to provide a line of cheap records nearly foundered when the latter insisted they be priced no higher than twenty cents. In 1932 RCA introduced eight-inch popular-dance-music records made specifically for the chain, to sell for a dime, followed by regular ten-inch Victor Electradisks at the same price. When demand warranted and distribution of the fifteen-cent Hit of the Week releases was terminated, the price was raised to the agreed-upon twenty cents.

Recorded music now was heard frequently over the approximately 600 independent radio stations. Many of them featured live performers and electrical transcriptions. Live talent was paid as little as a few dollars. Electrical transcriptions cost between $40 and $150 a week. As a consequence, broadcasters began to buy records at list price, or get them free in return for publicity.

By request of the Music Publishers Protective Association, in 1933 RCA Victor, Columbia, and Brunswick printed "Not Licensed for Radio

Broadcast" on every new pressing. Despite the broadcaster concerns, the warning meant nothing, for property rights ended once a record was sold, a legal opinion vigorously opposed by manufacturers and many leading artists. Victor artists Fred Waring and the "Pennsylvanians" negotiated for a new contract that would reserve the use of Waring disks by radio. Other bandleaders insisted that meager royalties from record sales failed to compensate for the depreciation of their value to network broadcasting. Music publishers pointed out that they could not join in legal action against broadcasters because an ASCAP license gave stations the right to perform music in any form without restriction.

Sales of twenty-five cent popular hits had fallen from 50,000 to 5,000, so expenses were cut further. Bandleaders were hired to record four numbers in single takes and paid union scale and little or no royalty share. Production costs for Victor's new Bluebird records were the same as those for the more expensive black labels, and sales were less than half. All three major companies seriously considered cutting prices of their most popular products. Edward "Ted" Wallerstein, Victor's new record division manager, encouraged the revival of the record-buying habit by promoting the RCA radio line, which had enjoyed a comfortable share of the $300 million spent in 1933 on receivers.

With Ralph Peer's departure, Fred Erdman, in charge of all recording for Victor, found another music man with broad experience in both race music and popular-song-and-dance bands: Irving Mills, a talent scout and producer for cheap-label manufacturers. Mills, who was instrumental in establishing Edward "Duke" Ellington, built up the RCA Victor talent roster by directing his press agent to encourage newspapers and magazines to review new Victor popular records, which would be provided without charge. This sort of promotion became the company's single means of general press relations and exploitation for more than a decade.

The unexpected success in 1933 of "The Last Roundup," with sales of 100,000 copies on all labels, signaled a small upturn in industry income, the first since 1929. The new technical recording techniques responsible for that upturn included an effort to meet the challenge of sharp and clean-cut master disks from England and a daring new course being considered by Jack Kapp, general manager of Brunswick. RCA Victor had issued a number of imported British HMV masters with markedly superior sound qualities due to their oversized studios. Victor, Columbia, and Brunswick searched for sufficient space in which to reproduce the British sound. Brunswick was the first to do so by recording in a Los Angeles motion-picture studio, acquired from a bankrupt customer by Herbert Yates, the majority stockholder in Brunswick's new parent company, American Record.

Yates intended to use British financing to purchase Columbia Records and its licensed Western Electric recording equipment from the creditors of the bankrupt Grigsby-Grunow Company. The British Decca

Record Company had prospered under founder E. R. "Ted" Lewis. As the British and European distributor of the subsidiary Brunswick and Melotone labels, it had also become the chief source of profits for American Record. Ready to buy a 50 percent interest in Yates's option to purchase Columbia Records from the Grigsby-Grunow creditors, Lewis went to the United States in 1934.

The urbane Lewis and the freewheeling Hollywood buccaneer Yates made a strained partnership, which soon dissolved when Yates planned to scrap Columbia's Bridgeport plant and remove the pressing equipment to his plant in Scranton, Pennsylvania. The decision was opposed by both Lewis and Kapp. The Englishman offered to buy Yates out of Brunswick instead, but found the $750,000 asking price too steep. Yates accepted his general manager's resignation.

One of the period's most influential and successful record men, Jack Kapp, was hired by Brunswick in 1925 to run its race-music division out of Chicago, where he and younger brother David operated a record and mail-order business. His insistence in 1928 that Brunswick release Al Jolson's versions of "Sonny Boy" backed by "There's a Rainbow Round My Shoulder" and its subsequent million-copy sale were responsible for his supervising the entire operation. He quickly became known as a "man of no taste, so corny he's good," but he built a stable of best-selling artists, including Bing Crosby, the Mills Brothers, and the Dorsey brothers.

A series of complicated business transactions took place as Kapp assembled a complete recording, merchandising, and distributing operation based on the investment of $250,000 by Lewis. He ensured the services of nearly the entire Brunswick roster. Warner Brothers was given a one-fifth interest in the new venture through one of its wholly owned subsidiaries, the Brunswick Radio Company, and sole use of the facilities and all recording equipment of yet another Warner company, United Research. Lewis was named chairman and Kapp president of American Decca Records. The price for Decca disks was actually fixed at thirty-five cents each or three for a dollar.

Meanwhile Yates merged Brunswick and Columbia and closed the Bridgeport plant. He sent all pressing and processing work to his Scranton factory, notorious for second-rate work. Stripped of almost the entire artist line-up when Kapp moved to Decca, Brunswick's executives hired Irving Mills to procure artists. To guard against future mass departure of talent, Yates and American Record instituted a new artist-acquisition policy, under which all contracts were made directly with the recording company.

In 1934, RCA Victor announced a new and inexpensive record-player attachment, its Duo Junior. Competing inexpensive record attachments soon attracted a new and chiefly youthful market for dance records. This audience, responsible for 40 percent of sales, was drawn by a new kind of energetic dancing, known as "swing."

The vast majority of radio stations not affiliated with the networks and dependent on regional or local sponsors for revenue began to use more recorded music. They played as much as eighteen hours a day of canned dance and vocal music but neglected to identify the source of entertainment, in spite of a Federal Radio Commission requirement. More prosperous operators also used electrical transcriptions to supplement live or record shows, and thereby incurred the wrath of the American Federation of Musicians. Correctly anticipating that canned music would finally eliminate the employment of live musicians, AFM leaders, chiefly James C. Petrillo of the Chicago local, forced the radio industry to employ union musicians even to manipulate turntables and program commercial disks.

Just such a practice was being popularized by Martin Block of WNEW in New York, who ran the "Make Believe Ballroom" and purchased his own records. The program was an immediate success. WNEW raised Block's salary, while his sponsor was giving him 10 percent of all mail orders generated by his broadcasts. Soon Block was broadcasting twice a day and receiving a percentage of advertising revenue.

Actually, Al Jarvis and not Block was the first man to air recorded music on his L.A. program "The World's Largest Make Believe Ballroom." However, Block's style and the persuasive ad-libbed commercials made the profession of disk jockey a respectable one. Decca owed much of its initial survival to Block's regular introduction of new releases, although Jack Kapp adamantly opposed radio's use of his products. Contending that Victor, Columbia, and Brunswick were conspiring to keep his records off retail counters, Kapp instituted a million-dollar suit, charging restraint of trade. Despite pretrial postponements and technical objections, Kapp's action frustrated Brunswick's plan to open a price-cutting war with the regular release of twenty-five-cent disks.

Many of Kapp's artists helped form the National Association of Performing Artists, seeking to "curb promiscuous broadcasting" of their commercial recordings. However, a new form of bootleg disk containing unreleased material appeared: off-the-air transcriptions of broadcasts. Fred Waring became a prized target for transcription pirates. Waring had not entered a recording studio for three years and, as a result, his popular Ford Motor Company network show was frequently recorded. This motivated him to help organize NAPA, which not only worked with ASCAP to amend the copyright laws but also legally tested existing legislation.

Waring and NAPA received an immediate injunction against radio station WDAS, Philadelphia, for unauthorized use of Waring recordings. The WDAS lawyers argued that all Waring brought to the records other than his name were the special arrangements he purchased, but in early 1936, a Philadelphia court found for Waring, declaring that his "unique and individual interpretation of music compositions is important and increases the sale of recordings and compositions." The injunction was granted, but was stayed on appeal.

Another significant development was in libraries of popular dance-music selections, recorded for radio use only by World Broadcasting Service and Victor. Negotiations for the purchase of the Columbia label in 1938 were carried out to obtain its sixteen-inch transcription presses and equipment. By 1936, at least 350 stations made yearly contracts with one or more of the four services, each of which permitted only a single station in a market area to use their music in order to preserve exclusivity.

The transcription business added to Tin Pan Alley's income through licensing agreements, forged by the MPPA, that called for the payment by radio stations or advertisers of a fee for each performance of music copyrighted by its members. To simplify bookkeeping, the libraries worked out a new arrangement with MPPA: they paid fifteen dollars a year for each song recorded, regardless of the number of times it would be aired, to a maximum of 200 transcriptions manufactured.

Decca weathered its near-disastrous first year only after Kapp found his best customer to be the coin-operated music-machine business. He tailored his product to this market, which took 40 percent of all recordings for its 150,000 jukeboxes in 1936. Undercutting competition, he priced disks at twenty-one cents each.

In 1931, *Variety* reported that music boxes were doing a million-dollar-a-year business in a seven-state midwestern area. Three companies collected half a share of coins from five-, ten-, and twenty-five-cent machines. Now that the coin machines were consuming almost half of all production, price cutting became the rule. Decca cornered the market. By cutting corners, Kapp paid a publisher royalty of less than the now standard one and a half cents on thirty-five-cent disks, reported sales rather than the "*manufactured* and sold" number called for in the 1909 Copyright Act, and omitted reports on those delivered to coin-machine distributors. To curb payment schemes, several new requirements were introduced by the MPPA, among them a fifty-day limit on credit and quarterly rather than monthly reports. The MPPA board directed John Paine, its chief operating head, to create a separate mechanical-licensing office.

Unhappy with the way his artists were being handled by American Records, Irving Mills offered to join Decca in 1935 as supervisor of all dance recordings and to bring along all the bands and vocalists he controlled. After he lectured Decca on his expertise, the offer was declined, and Kapp refused to use any music or talent associated with Mills. Because Mills was an astute producer, however, the Brunswick, Okeh and Perfect recordings he directed sold well. His work now included the legend "For home use only," and all American Record contracts with him and his artists ceded only the right to release the disks for that purpose, reserving all other property rights to Mills.

Mills always expected to go into the record business, and in 1936 guaranteed $52,000 a year against royalties for distribution of his new products to Yates and American Records. Mills's new seventy-five-cent Master and thirty-five-cent Variety records appeared in 1937, and initially sold well. Mills desired international affiliation and negotiated with Boosey & Hawkes. This intrusion of competition into a controlled market was thwarted by EMI, to which American Record was committed. Mills failed, and he remained an executive and consultant, but all of his new releases appeared on either the Brunswick or Vocalian label.

RCA Victor, American Record-Brunswick, and Decca prepared to take one step further than Mills's plan to sell records for home use only to prevent their use by radio. Kapp complicated the issue of property rights in records in 1937 by arguing that manufacturers should issue performing licenses and collect license fees. Other firms concurred, hoping to make clear to NAPA that the recording companies created "the musical art which evolves in the form of records." With the approval of the AFM, manufacturers began to print on each record the legend "The use of this record has been licensed under specific patents. The resale of this record except for home use is prohibited."

On behalf of the music publishers, Harry Fox, general manager of the MPPA, informed the manufacturers that they were presuming on the copyright owners' privileges. He pointed out that the transcription companies paid fifteen dollars per copyright for 200 recordings, a larger royalty than the disk makers' usual one and a half and two cents, for which they received implicit permission to allow their products to be performed on radio. The MPPA did not intend to give an unfair competitive advantage to the makers of "for home use only" records. He added, however, that the publishers had no desire to frustrate any move by the manufacturers to protect their wares. American Record-Brunswick, Decca, and RCA Victor continued plans to license broadcasters and added an amplified restricted-use notice on all record envelopes.

The MPPA's response was the first probe of record-company account books, which showed that most manufacturers allowed unlimited return privileges to jukebox operators and suppliers and treated such dealings as rentals rather than outright sales. Special deals with recording companies and incomplete or inaccurate royalty accounting were also uncovered. The investigation's most tangible and lasting reform was the formation of a central MPPA mechanical-licensing bureau with standard and uniform contracts, supervised by Harry Fox.

Finding themselves victims of inaccurate weekly sales reports from the manufacturers, *Variety* abruptly dropped their recently restored weekly popularity charts in 1938. The corporate embroidery of sales figures was excused on ground that cutthroat competition in the jukebox market, where people relied on *Variety*'s figures when stocking their machines, necessitated the practice.

The big three companies were eager to capture the automatic-music-machine field which now accounted for 60 percent of all purchases, three-fourths of them Kapp's releases. RCA Victor in 1938 concentrated its Bluebird line on that market. On the basis of sales of 800,000 disks a month to jukeboxes, it was believed that Kapp cleared a profit of 1.8 cents from each thirty-five cent disk, causing Bluebird to slash their wholesale price to eighteen cents, a cut Kapp was not able to match.

The coin-machine trade grew at a rate of 25 percent above that of 1937. Informed sources put the number of automatic music machines at 225,000, and ASCAP and the MPPA remained convinced that nearly half a million were in use.

Despite the jukebox's influence the most effective and aggressive merchandising, directed by Victor's advertising vice president Thomas F. Joyce, was being done for classical recordings, where a profit of 10 to 20 percent now prevailed. On March 15, 1938, the Victor Record Society was launched; membership cost six dollars a year plus an initial purchase of eight dollars' worth of recordings. The society sent new subscribers a record-playing attachment that could be plugged into any radio receiver; *The Music America Loves Best*, an illustrated monthly magazine; *The Victor Society Preview;* and "correspondence privileges," giving them access to Victor's director of music.

The society's activities were supplemented by a scheme involving entrepreneur Nathan Hurwitz's Publishing Service Company, which gave away low-cost classical-music albums for $2.99 each. A no-frills record player was available for three dollars and a pledge to buy all ten advertised albums. In the winter of 1938–39 Hurwitz disposed of more than 300,000 albums manufactured by Victor from previously unreleased masters or ones cut from the active catalog.

While other labels released classical albums, RCA Victor monopolized that market which led to the acquisition of the American Record Company by CBS in late 1938, a move suggested by concert impresario Arthur Judson when CBS was ready to market electrical transcriptions. Judson had been largely responsible for the creation of the original CBS network and remained the second-largest stockholder. During the Depression, a coalition of independent concert managers, at Judson's suggestion, formed the Columbia Concerts Corporation, with William S. Paley as chairman, representing the majority stockholder, CBS, and Judson as president. The new coalition represented many important American conductors and distinguished instrumentalists. When many of them complained that the lack of exploitation implicit in a recording contract handicapped their securing international bookings, Judson suggested that CBS have exclusive access to its talent roster and become a major competitor of NBC and its Artists' Bureau.

Originally CBS intended only to retrieve the Columbia name by purchasing that record company. It soon succumbed to Yates's irresistibly modest asking price of $700,000 for American Record. With the company came the Bridgeport factory and Brunswick's new and remodeled studios. Edward Wallerstein was lured away from RCA to become president and developed a classical repertory at attractive prices.

The latest developments in recording technology were introduced into Columbia's studios, including Millertape, which was being used on an experimental basis. Its inventor, James A. Miller, perfected a new system for engraving sound impulses on a special film, producing the highest fidelity recording on tape prior to the 1950s. Cutting and editing of speech or music was easily accomplished on Millertape. With it, a radio show could be pieced together from different tapes.

Having won an outstanding victory in its fight to increase the employment of union musicians by broadcasters, the AFM resumed the war on canned music. New licensing agreements were approved by Victor, Columbia, and Decca in 1938. All recording sessions had to be registered with the local AFM chapter.

Working for almost a year through the symphony conductors he represented, Judson secured the services of several American orchestras for Columbia. To implement the "good music equals public service" programming used by the networks to win over the custodians of public taste and appease the Federal Communications Commission, CBS increased air time devoted to educational programming with emphasis on serious music. CBS broadcast the weekly Sunday afternoon concert by the New York Philharmonic. The Community Concerts division of Columbia Concerts Corporation set up a full session of classical-music concerts. All this activity exposed the classical-music lover, hence prospective record buyer, to more music than he had ever heard before.

The already hectic business faced the prospect of added turmoil with the addition of a new entrant: Eli Oberstein and his United States Record Corp. When Ralph Peer left Victor to concentrate on publishing, Oberstein supervised all race and hillbilly music. Never profligate with corporate funds, Victor paid him only $6,000 throughout the late 1930s. Oberstein's life-style demanded a much higher income, which allegedly came from a share in the cut-ins, under-the-desk deals, and bribery—all of which were regarded as normal practices at the time.

The success of Benny Goodman, Tommy Dorsey, Glenn Miller, and others he had brought to the company gave Oberstein a track record second only to Jack Kapp. In 1938, it came as a surprise when Frank Walker fired Oberstein, replacing him with Leonard Joy, an Oberstein assistant. Simultaneously, Ed Wallerstein left Columbia to head the entire Victor operation.

Oberstein's immediate announcement of a new recording company was reminiscent of Kapp's action of his departure from Brunswick in 1934. With an assured minimum order for one million disks a month from a syndicate of investors that controlled 150,000 jukeboxes, Oberstein overnight became a factor in the realignment of recording firms. But his United States Record Corporation was doomed. His declaration that Tommy Dorsey and other artists would also leave Victor proved inaccurate. Furthermore, Oberstein's original backers were found to have underworld connections. Oberstein then became embroiled in bitter legal actions with RCA charging misappropriation of funds. United States Record slid into bankruptcy.

The recording business was further complicated when in 1939 the Brunswick label was retired. Henceforth all its new releases would appear under a Columbia imprint and sell for fifty cents. This created an immediate problem for RCA Victor. Under its agreement with Benny Goodman that old releases could not be marketed at a price lower than that offered by any other manufacturer, it was impossible to move his records to the thirty-five-cent Bluebird line as planned.

Total Columbia sales in 1939 were up almost 600 percent over the previous year. With the best-selling single record of the year, "The Beer Barrel Polka," Victor reported total income up by a substantial 700 percent over the 1933 low mark. Decca's 90 percent share of all jukebox sales was responsible for total production of 13 million disks in 1939. It was the best year the record and music business had enjoyed since 1929; $750,000 was paid in mechanical royalties.

The progress was made without the promotion of new releases by radio stations. The still untangled web of property rights in commercial records had bogged down in the courts. Many broadcasters were reluctant to face the possibility of liability for payment to the manufacturers or to the artists or both. Broadcasters increased dependence on staff musicians and vocal performers or on the transcribed libraries. Their worst fears were confirmed in July 1939, when Judge Vincent L. Leibell, of the federal court in New York, found for Paul Whiteman in his action against radio station WNEW. The National Association of Performing Artists believed it increased artists' control over their mechanically reproduced music. Counsel for RCA, however, immediately claimed that NAPA was now effectively eliminated from the licensing picture by the ruling, which put that right firmly into the manufacturers' hands. After advising recording artists that it intended to share licensing revenue, Victor joined the Whiteman suit, in order to obtain a ruling that would exclude performers from licensing their own rights. Leibell ruled that Whiteman was privileged to reserve broadcast rights provided the company agreed, but that he could license use only in conjunction with RCA Victor.

Victor began to license broadcasters as did Columbia and Decca. An immediate outcry from broadcasters followed the announcement of a scale of fees for airplay. Stations in the top grade using the product of all three manufacturers were confronted with monthly fees in addition to their payments to ASCAP and the transcribed library service.

Many people believed that the record companies were really more interested in terminating all radio-station use of records than in collecting fees. Few stations signed licenses. More turned to the transcription companies, the quality of whose work exceeded that of commercial recordings or over-the-air sound of network programs. The transcription business demonstrated the commercial viability of a truer and more faithful reproduction of music and speech for home use. Improved amplification, speakers, and record-playing components were available in fully assembled but expensive units. Radio-phonographs assembled by small independent factories delivered high fidelity at a modest price, but most Americans were content with the comparatively antiquated modern record machine.

Columbia's price cut on all popular records had been responsible for renewed optimism. In 1940, Columbia invaded the classical market with a new Master Work line. Victor's long supremacy in that field soon suffered a serious sales blow, not only from Columbia but also from Decca. Price cutting escalated, and new seventy-five-cent and dollar Victor recordings went on sale. A general price reduction for the entire RCA Victor output came soon after the Columbia management dealt the final blow by reducing all Master Works disks to a dollar each.

As Wallerstein anticipated, every company benefited, and orders generally rose 200 to 1,000 percent, promising a 100-million record-sale year. After the reduction of its top-of-the-line records, Victor enjoyed an unprecedented buying spree; Columbia crowed about a general 400 percent increase in dollar volume over the period before the price war started.

For years, American opinion molders had praised radio when it offered a fairly limited classical-music fare, and condemned it for gross overuse of popular products. Judge Leibell appeared to have given the manufacturers power to curb that surfeit. Ruling on the appeal in *Waring v. WDAS*, the Pennsylvania Supreme Court affirmed the artist's right to prevent unauthorized performance of his music. In December 1940, the United States Supreme Court refused to review a subsequent reversal of the Pennsylvania decision. In effect, property rights ended with the sale of a record. Broadcasters could no longer be constrained from using music recorded "for home use only."

The golden age of the disk jockey was about to begin.

5 No Longer "For Home Use Only": The Battle Between Radio and Record Producers and the Creation of BMI

RADIO broadcasting was experiencing steady economic growth and increasing public acceptance throughout the 1930s. NBC and CBS enjoyed a combined gross of $35.7 million in 1931. However, the industry's total expenses exceeded total income by $237,000, and half of all stations operated in the red.

Because of the time difference, network programs were heard on the West Coast at 5 and 6 p.m., too early for the family audience. The prohibitive costs of two live broadcasts daily led to transcriptions being recorded for one-time use. Fewer than 100 stations were connected in the chain operations, and many therefore opted for the cost-saving broader market offered by transcriptions.

To remove the potential threat of government dissolution, General Electric and Westinghouse entered into a consent decree settlement of the antitrust action against them, AT&T, and RCA. They disposed of their RCA stock and voided all cross-licensing agreements. NBC and all its stations, the radio-connected manufacturing divisions, and other communication facilities were turned over to David Sarnoff's administration. Only RCA's control over the patents on radio tubes was found to violate antitrust laws. Now free to concentrate on making radio the principal medium for home entertainment, Sarnoff instituted West Coast NBC auxiliary networks.

The networks' prosperity created dissention within the industry, which was exacerbated by the latest resolution of radio's decade-long war with ASCAP. The music business was also plagued by declines in sheet-music profits and record royalties, and music publishers and songwriters depended on their ASCAP distributions. ASCAP had successfully thwarted passage of an amendment to the Copyright Act that would have revoked license fees and negotiated with the National Association of Broadcasters for a fixed fee on the use of music by individual stations on sustaining (commercial-free) programs. Several deadlines for ratification passed, and petitions were made to the government for dissolution of ASCAP.

Oscar Schuette, retained by the NAB to coordinate negotiations with ASCAP, advocated a waiting policy while a second committee negotiated the networks' contract. It was quickly approved by the NAB and ASCAP and signed by the chains and most of their affiliated stations. In 1933, the NAB began to undermine the music-trust monopoly in order to win a per-program system of payment to copyright owners. With Schuette as its president, the short-lived Radio Program Foundation lobbied for such a licensing process, as well as a source of tax-free music for radio, supplied by non-ASCAP songwriters and composers. One of the RPF's purposes was to "own stock in, lend money to, and otherwise assist" independent publishers.

President Roosevelt involved himself in the growing field of broadcasting by calling for a commission to regulate communication by telephone and broadcasting and to replace the Federal Radio Commission, temporarily formed in 1927. In 1934, the Federal Communications Commission was inaugurated and given a single restraint: the power to reject renewal applications after three years, on the basis of findings from public hearings. Broadcasters were already governed by two sets of regulations, their own NAB Code of Ethics and Practices, adopted in 1929, and the NRA Code, imposed by the federal government. Both outlawed manipulation of rate cards; monopolistic and discriminative practices; payment of gratuities for song plugging.

NBC and CBS's firm hold on chain broadcasting diminished in 1934 by the formation of several small networks. Multi-station hookups involving major NBC affiliates handled transcribed commercial programs and spot advertising on a cooperative basis; and MBS, the Mutual Broadcasting System, owned by the *Chicago Tribune,* was established. MBS by 1937 owned forty-five outlets and paid all stations their regular commercial rates, deducting only a small sales commission, advertising-agency rates, and wire charges. Programs were created by originating stations or by sponsors and their agencies, the network itself owning no studios or transmitting facilities.

Legislation to remedy the ASCAP situation continued. The NAB brought in Newton D. Baker, a powerful Democratic party figure, who helped instigate a Department of Justice antitrust suit against the society's officers, its members, and other music-business organizations. ASCAP filed an answer in late 1934. At the same time, complaints by NBC and CBS affiliates about their excessive share of ASCAP royalties and the meager compensation payments they received from the networks threatened to split the NAB. They were little mollified by announcements in 1935 of new affiliation rates. NBC abandoned any charge for sustaining programs and compensation to affiliates was boosted on the basis of a sliding-scale percentage of time rates.

As the arguments with ASCAP continued, rigid and homogenized formats were instigated by American Tobacco President George Washington Hill and practiced on Lucky Strike's "Your Hit Parade," first broadcast over NBC in 1935. His successful marketing of only the most tried-and-true easily accepted musical repertoire led to measurements of listener preference, foremost of which was that of statistician Archibald M. Crossley. He relied on listeners' memories to determine which programs they had heard the previous day. Experiments began in 1939 with the Nielsen Audimeter Survey—meters attached to radio sets that recorded the exact length of time various stations were tuned in.

The networks operated on the principle that radio coverage could be measured and based all time rates on 21 million sets. Audience size was reckoned to be the accumulated numbers of sets within range of each participating station's transmitter. The result was that advertisers spent three-fourths of their radio budgets on one-third of all stations.

At the 1935 NAB convention more than 400 broadcasters discussed the economic consequences of the networks' bombshell announcement of an extension of the 5 percent ASCAP rate for five years. The society insisted on collection from owned-and-operated network stations on the basis of card rates, but compromised when NBC accepted an increase in the sustaining-fee payments from flagship stations in New York beginning in 1936.

The government's antitrust suit against the major network broadcasters was adjourned. Network representatives argued that the suit had been inadequately prepared and would probably be lost, in which event far more onerous terms would be demanded. A complicating issue was Warner's 1935 resignation from ASCAP when the NAB negotiated for an improved Warner contract, asking for 40 percent of the ASCAP sustaining rate and 2 percent of gross receipts. The latter figure was grudgingly reduced, but at the new year between 20 and 40 percent of the total ASCAP repertory was not available to most American radio stations.

A more coincidental factor in Warner's defection was Jack Warner's hope to acquire control of a third national radio network. When his offer to purchase the Mutual Broadcasting System was rejected, plans accelerated for activation of Muzak-wired radio, to compete with network radio. The Muzak Corporation, founded in 1934, was located in ERPI-Warner Vitaphone's former New York studios. Muzak had an open-ended agreement with ASCAP, but depended for much of its music on Associated Music Publishers, a holding company for non-ASCAP music. In 1938, when the FCC would not allow Muzak to compete with the networks, it concentrated on special services designed for offices, factories, and homes, and a new transcribed-music library service for radio was instituted using the AMP label.

In 1937, network broadcasting enjoyed a $55-million year. NBC improved relations with affiliates by compensating for time used on chain broadcasts. A new, more equitable contract extended the term for cancellation privileges, guaranteed sixteen day and night hours a week for local commercial broadcasts, and continued to absorb line charges.

Anticipating the next confrontation with ASCAP, the NAB formed yet another tax-free catalogue, the Bureau of Copyrights, and recorded an initial 100 hours of tax-free music for sale to radio stations and prepared special arrangements of non-ASCAP music. The project foundered in 1937, owing to a recession and to the American Federation of Musicians' new demands for more jobs for members.

Local action in Chicago increased the number of standby union members. An ultimatum from the AFM, asking for the implementation of a similar standby practice by all stations and the guarantee of a minimum allocation of an additional $2.5 million for the employment of union studio musicians, came as a shock. A series of meetings between NBC and CBS and the AFM produced a tentative agreement that at least $1.5 million of this would come from the networks and their affiliates.

The firebrand of the AFM, James Caesar Petrillo, president of the Chicago local, lobbied for more radio money. He burst on the scene at the 1937 AFM convention and threatened a union walkout if broadcasters did not accept the group's new terms. The record manufacturers were persuaded to print a more stringent warning against use on the air of commercial disks and guaranteed the employment of union musicians and union-approved talent only.

Broadcasters again split into factions: one, representing the new National Independent Broadcasters group, ready for a fight to the finish; the other wanting to make the best deal possible. The second body, made up of network representatives and many of the most important affiliates, believed that a strike would erode profits and alienate sponsors. A new coalition—the Independent Radio Network Affiliates—was activated inside the NAB and assumed sole responsibility for dealing with the AFM. Network officials forced the union to postpone the strike until, in 1938, a final contract was ratified by a majority of stations. Terms for network affiliates were based on a percentage of time sales.

The NAB was further upset in 1937 when members asked for help in dealing with the Society of European Stage Authors and Composers, owned and operated by Paul Heinecke, which sold music licenses. SESAC had been formed in 1931 to handle music licensing for unaffiliated foreign publishers. SESAC's new contentiousness, impending failure of the Bureau of Copyrights and the NAB recorded music library, and the negotiations with the AFM disturbed many IRNA members. The networks and some more pragmatic broadcasters suggested the NAB elect new directors, staff, and a paid full-time president.

In 1938, attorney Neville Miller, former provost of Princeton University and mayor of Louisville, became the society's first paid president and immediately began to resolve the ASCAP problem. John Paine, ASCAP general manager, and Claude Mills, who served the society chiefly as a roving ambassador to broadcasters, reminded Miller that any new license would hinge on higher rates.

Nearly fifty bills and resolutions affecting broadcasting were presented during the 75th Congress, in 1937–38. Proposals called for more strenuous censorship of program content, an increase in public-service programs, more evidence of self-regulation, and amendments to the 1934 Communications Act that would curb the existing monopoly situation in national and regional chain radio. The FCC announced new hearings and began investigation of all networks. David Sarnoff's assertion that "self-regulation is the democratic way for American broadcasting" provided a battle-cry for the industry and forced the NAB to prepare a new code of government standards and sell it to sponsors, opinionmakers, legislators, and 722 broadcast licensees. Few paid attention to the earlier undemanding government codes or opposed the new NAB document.

Two technological developments shook the structure of the networks and all radio: television and frequency modulation (FM), the static-free transmission of sound to which radio is inextricably bound. RCA-NBC built the first modern television transmitting station in 1935 and began developing an experimental program service and manufacturing a limited number of RCA television sets. Sarnoff welcomed any competition since RCA was required to license its patents on a royalty basis. He inaugurated a limited "radiovision" service in the New York market, to begin in 1939.

Static-free reproduction of sound to accompany image, transmitted over ultrahigh frequencies, was imperative, and Sarnoff began negotiations with Edwin H. Armstrong, who in 1922 patented a new system of broader-range frequency-modulated signal transmission through upper frequencies. Experimental transmissions began in 1934, but RCA's concentration on television and Sarnoff's conviction that FM would terminate the present structure of broadcasting ended further collaboration with Armstrong. Armstrong persevered, eventually received permission from the FCC to build a 50,000-watt station, and began transmitting in 1939. In twelve months, the mass production of FM sets was in full swing, and the FCC was deluged with some 150 applications for FM stations.

Variety hailed the broadcasters' new code of practices, ratified at their June NAB convention, as historically significant. Other issues discussed at the meeting included the probability of a nationwide AFM strike; an adverse decision in a lawsuit challenging radio's right to play disks made "for home use"; and the need to call an emergency meeting of the entire industry to discuss the formation of a new music-licensing organization, if ASCAP did not cease its "run-around" tactics.

The radio business no longer wanted a per-use contract, having found it too expensive and time consuming. Except for NBC and CBS stations, payment at the network source was regarded as mandatory, royalties to be paid only for local programs on which ASCAP music was performed.

Neville Miller met with CBS lawyer Sydney Kaye, a leading authority in copyright. In 1935, when both NBC and CBS were considering a renewal of the favorable ASCAP contract, Kaye proposed the creation of a reservoir of non-ASCAP music. He also urged that inducements be offered to a few conspicuous American composers to let their ASCAP contracts terminate, allowing them to retain all publishing, stage-performance, and other property rights.

The Kaye blueprint emphasized that the proposed organization accrue no dividends, and he suggested initial broadcaster pledges of up to $1.75 million. Funding would be raised by an SEC-approved sale of 100,000 shares of capital stock of Broadcast Music Incorporated. Each investing broadcaster would receive stock in the amount nearest 25 percent of the sum paid to ASCAP in 1937, the only year in which a survey of total expenditure on music was undertaken. License fees would be equal to 40 percent of ASCAP payments. No provision was made for network payment at the source, and one-fifth of the BMI stock was allocated to NBC and CBS. Both agreed not to reconcile with ASCAP without their affiliates' approval. April 1, 1940, was set for the start of BMI operations. Only one-quarter of the million dollars pledged had come in by January 1940, and BMI opened an office in February, with a temporary staff of four. ASCAP responded to the plan with a new users' relations department to placate broadcasters and suggested it would do away with sustaining fees and apply the payment-at-the-source principle to all stations.

Armed with a new line, IRNA officials assured all that NBC was no longer wavering, and CBS was more anxious than ever to win the fight. Privately, both tended toward a settlement that involved a specific sum, which they would in turn allocate among their affiliates.

The threat of war in Europe plus the uncertain economy were expected to reduce radio revenues in 1940 by half. The recent AFM situation had been resolved through the added expenditure of more than three million dollars. The cost of developing television and FM could not possibly be covered by anticipated revenues. Buoyed by the first general boom in a decade in sheet-music and record sales, music men looked for the same in music licensing, and talked a 100 percent increase.

The new ASCAP contract unexpectedly solidified the radio business. Rather than deal with radio's designated agent, ASCAP President Buck called in representatives of the three networks, several major affiliates, and one independent. Radio was divided into four catagories: stations whose gross billings were under $50,000 would pay 3 percent of sales receipts, less agency and other deductions, and a one-dollar-a-month sustaining fee; an immediate group that grossed between $50,000 and

$250,000 annually would pay 4 percent and a sustaining fee reduced by 25 percent; the large network stations would pay the present 5 percent charge; and, for the first time, regional and national chains would be charged 7½ percent. The duration of the agreement was not fixed.

Edward Klauber, executive vice president for CBS, stated that only the NAB would negotiate for his organization. The others said that broadcasters would grind the society into the dust. Paley rejected "ASCAP's attempt to split our industry into hostile camps," and added that Klauber would continue to represent CBS. Throughout the next year, Paley and Klauber were BMI's most ardent champions.

In competing press releases, ASCAP stated that 540 stations would pay between 20 and 50 percent less than in 1939, while the networks countered that they would pay an additional $3.25 million. *Variety* calculated that, using 1939 as a sample year, ASCAP would have collected $7.1 million, instead of $4.3 million, if the new system had applied, $4.125 million from the networks.

The support of the small stations was essential for industrywide support of BMI. Carl Haverlin, an experienced salesman and former broadcaster, was hired to direct station relations. His task of selling BMI to broadcasters was a formidable one. Half of the network stations and nearly two-thirds of the independents had not yet closed ranks. Haverlin raised the number of subscribing stations and brought in more than 85 percent of the industry's time-sales income. His efforts were substantially assisted by the new ASCAP contract. Its terms were for five years, but television rights were excluded, as was any reference to either a per-piece or per-program alternative.

BMI officers were installed in New York. Neville Miller, president of the NAB, became BMI president, while Kaye was elected operating head and attorney. For general manager, Kaye chose Merritt Thompkins, former head of Schirmer Music, and most recently president of the AMP.

BMI was printing one new popular song a day and twenty-five new arrangements of public-domain, non-ASCAP musical numbers a week. With financial backing from CBS and support from NBC, negotiations were conducted to purchase the MGM Big Three music catalogues, for Robbins, Feist, and Miller copyrights provided one-seventh of all ASCAP music used in 1939. By September BMI had 220 full-time employees, who turned out fourteen printed popular songs and thirty-five arrangements of BMI-licensed familiar public-domain music each week. In addition to regular new releases of transcribed selections in the BMI Bonus Library, 400,000 units of printed music were shipped weekly. Before the year ended, BMI executed long-term contracts with Ralph Peer and Edward B. Marks, their copyrights proving to be crucial to BMI's survival during its first year.

BMI's continuing efforts to ally itself with a dissident but well-established ASCAP house culminated in late 1940 with the signing of the E. B. Marks music business. The networks' financing of this move reflected their apprehension about the effect a switch from almost all ASCAP music to the BMI repertory would have on sponsors, advertising agencies, and name bandleaders.

Remembering the averted Warner Brothers strike in late 1935, the networks required a gradual reduction of ASCAP music from sustaining programs and threatened to cut off remote pick-ups of uncompliant bands. Bandleaders complied until they found they could make more money plugging their own music and opening small publishing companies, licensed through BMI.

Kaye's blueprint briefly mentioned compensation to songwriters and composers. Kaye thought that customers would be better off if they "received no compensation for performing rights, provided they received a fairer share of the revenue incidental to music publication and to the sale of mechanical rights."

Contracts with BMI's thirty affiliated publishers did not obligate them to pay their authors and composers for performing rights, though it was expected they would. BMI-published songwriters and composers were paid one penny per performance, either live or recorded, per station, based on a new system which involved monthly examination of 60,000 hours of program logs supplied to BMI by 150 stations across the country. One striking difference from the ASCAP method was payment for recorded performances, on both transcriptions and commercial disks.

On the day after Christmas in 1940, spokesmen for the Justice Department announced that it was entering the music war, and would add NAB, BMI, NBC, and CBS to the defendants in a criminal antitrust action. Thurmond Arnold, the new assistant attorney general in charge of the antitrust division, was furious with ASCAP's "foot-dragging" in signing the 1934 consent decree. Denouncing the "ASCAP welchers and double-crossers," according to *Variety*, Arnold insisted that ASCAP meet every one of the six points raised in the 1934 complaint.

Any last-minute compromise with ASCAP, such as renewing the present form of licensing, would put them in additional jeopardy. Radio was now prepared to take all possible precautions to avoid infringements. The society had set up listening posts in major centers to document illegal use of its music. A total of 660 stations, out of the 796 commercial facilities, were signed to BMI. Among the last-minute stragglers were members of the National Independent Broadcasters, who voted en masse to support BMI.

The last week of December was the first without any ASCAP music on the networks and their affiliates. RCA Victor and Columbia Records, owned by the networks, were printing legends affirming the ASCAP or BMI status of each selection on all new releases, and small new companies were doing the same.

The ASCAP versus radio war was beginning in earnest.

The creation of BMI was the end result of an uneasy relationship between ASCAP and the broadcasters. Those tensions affected every portion of the music industry. Tin Pan Alley music publishers acknowledged that the depressed popular sheet-music business would have to change. Kresge's, the largest sheet-music outlet in America, reported a 25 percent drop in sales. Publishers no longer provided complete free scores to dance bands unless they were featured on a sustaining radio program. The chief source of income for members of the MPPA was from mechanical licensing—the twenty-five cents collected by the association for four minutes of radio play on electrical transcriptions.

The sale of printed dance arrangements, however, tripled. Bandleaders played one-night stands on a straight percentage of receipts, ready to perform free of charge, hoping one engagement might lead to others. Cutting in bandleaders as the authors of new songs, in return for on-the-air plugs, flourished. Popular bandleaders and singing stars regarded collecting royalties as part of the trade.

A controversial resolution aimed at the practice, calling for a fine of half a year's ASCAP distribution, was unanimously approved at the annual meeting in 1941. With their chief source of revenue—sheet music and record royalties—about to be unimportant byproducts, publishers made deals with new artists and bandleaders before the regulation became effective. The first formal charge was made against Roy Turk and Fred Ahlert, who cut in Bing Crosby on their song "When the Blue of the Night Meets the Gold of the Day," in order to get him to use it as his CBS theme song.

Disagreements arose about restoring the division of performing rights to the two-to-one writer-to-publisher ratio abandoned in 1921, and a closed shop for members of the Songwriters. The publishers cut all sheet-music royalties to a penny a copy, leading to a new Songwriters Protective Association that opened its rolls to ASCAP members and elected Billy Rose president. His friend Arthur Garfield Hays, attorney for the Authors' League, framed a suitable new contract putting songwriters on the same footing as authors and dramatists, who could hold copyrights in their own names throughout the two terms of protection. The SPA hoped to clarify who owned the small rights, in favor of songwriters, who could then license each of them: publication, mechanical reproduction, public performance. To demonstrate their concern for the future of ASCAP, all SPA members would grant the last right to the society for life. The MPPA would not make a similar offer.

Possible affiliation with the American Federation of Labor stiffened the publishers' opposition to the SPA. Lobbying in Washington led to an amendment to the Copyright Act calling for registration in the name of the creator. The SPA leaders agreed with those members who believed the SPA might have gone too far. Demands were modified, and ASCAP was asked to arbitrate.

After seven months, a "standard uniform popular songwriters' contract" won the approval of the boards of both ASCAP and the MPPA. It called for individual negotiation of royalties. A minimum 33⅓ share of mechanical and foreign rights was fixed, as well as of "receipts from any other source or right now known or which may hereafter come into existence." Its vague language permitted publishers to withhold any share of radio transcription disk fees collected by the MPPA or from the talking-picture synchronization fees. The contract also introduced regular accounting periods and allowed writers to examine the publishers' books. It did not deal with the issue on which many writers intended to stand firm: Who owns the small rights?

ASCAP's general manager, J. C. Rosenthal, died in 1931, and was replaced by Claude Mills. NBC's six-million-dollar venture into the music business, with the purchase of eight ASCAP music companies, proved to be less profitable than Warners' ten-million-dollar purchase, and both were ridding themselves of all or part of these properties. Mills had been a principal in the network's music operation, but when subsidiaries were sold back to their original owners, he was asked to remain as a consultant to NBC. Over the protests of songwriters who remembered his role in arranging their capitulation to the publishers in 1920, he returned to ASCAP and began to change the method of distribution and license-fee collections.

The society rearranged the publisher payment ratios and added a new double-letter classification to the top three classes. The amount of payment to each class was also changed. Protest came from all but those in Classes AA and A. Jack Robbins was the loudest. The only publisher to have been punished for a violation of the cut-in rules, his income was reduced by 75 percent when he was moved from Class A to Class D.

In 1932, Robbins filed an antitrust suit against ASCAP, which made public the workings of publisher classification and exposed the extent of control of the board of directors by motion-picture interests. By voting together, Warner Brothers, Paramount, the music houses of Irving Berlin and the Santley brothers, which had agreements with Universal Pictures, and Max Dreyfus's half-dozen firms that represented major Broadway production writers dominated the society. Soon after, Robbins was returned to Class A, given retroactive royalties, and the suit was dropped.

When it became obvious that income from radio would fall below the 1931 levels, an increase from the business that killed music through repeated use was clearly necessary. A modest 25 percent raise was rejected by the National Association of Broadcasters. Anticipating a fight to the finish with ASCAP, the NAB planned another tax-free library of music to replace the society's copyrights. The prospect of a radio war loomed.

Discussions with network executives stumbled when recognition was demanded of their vast expenditures for the development of radio through furnishing sustaining programs to the affiliates and for experimental work not reimbursed by their affiliates. Rather than tax the networks "at the source," ASCAP proposed collecting its royalties from the income of all local stations, including their owned-and-operated facilities.

An agreement was made for fees that would rise annually from 3 to 5 percent of all commercial time sales, less specified deductions, over the next three years. The networks' first contracts with ASCAP contained what was to become known as the "twilight zone exclusions," which allowed the chains to escape from any ASCAP fees for commercial programs other than those by their owned-and-operated stations. Unaware of the full ramifications of this provision, the ASCAP board and members approved.

The improved agreement focused on proceeds collected directly by the radio stations. It was generally anticipated that the chains would receive around $40 million in 1933, and the other broadcasters an additional $35 million. However, the networks' $40 million was immediately reduced to $6 million because the affiliates were paid only 15 percent of all time sales from sponsors for network programs. As for the remaining $41 million, on which the society's expectations were based—$6 million paid to stations, plus the $35 million earned by local stations on their own—that figure was reduced to $25,840,000 after approved exemptions, deductions, and discounts.

Within a few weeks it was clear that the music business had outsmarted the broadcasters, and some network affiliates complained of exploitation by CBS and NBC. NBC affiliates were charging from $150 to more than $550 for an hour of electrically transcribed programs. When they were connected to the network, however, they received no more than $50 an hour. No option on their time was paid for by NBC. The situation was different at CBS. It offered sustaining programming without charge to all affiliates in return for a confirmed option on any part of an affiliate's time on the air. All income from CBS network broadcasts was shared with the participating stations after expenses, which nonetheless left the major portion of income to the network.

By 1940, according to a Mutual Broadcasting System's *White Paper*, twilight-zone income, "that portion of network receipts which is not paid over to affiliate stations or credited to the network's own stations, but is retained by the networks," represented about $34 million out of an estimated total of slightly more than $60.8 million of network time sales, and of slightly less than $129 million of the entire industry's total net time sales.

To counteract ASCAP's poor public image Claude Mills recommended that the press might treat the society better if they were offered more advantageous contracts, which allowed the NAB to point out ASCAP's discriminatory practices. Nevertheless, nearly 250 stations had signed with ASCAP by Christmas of 1932.

While *Variety* publicized the society's error in accepting the networks' exclusion of their own time-sale income, an ASCAP committee studied reclassification. Made up only of Class AA and A publishers, it recommended that radio income be divided among writers and publishers only on the basis of performance. The proposal was immediately adopted by the very same men who had made it and stood to gain the most from it.

In 1932, the thirteen leading music publishers merged their distribution and bookkeeping departments into a single cooperative entity, the Music Dealers Service. They hoped to control distribution, fix prices, and freeze the middleman jobber out of handling printed popular, standard, and production music by servicing the retailer directly. Each member firm invested $1,000 in the MDS, but operating expenses came from the proceeds from a penny added to wholesale prices. The wholesale price of popular, two-page songs was fixed at fifteen cents, providing a uniform 40 percent margin of profit to retailers.

Sheet-music sales did not increase, but Maurice Richmond's former partner Max Mayer, in Richmond, Mayer Music Supply, filed a $1.125-million antitrust action against the MDS, charging that it had combined and conspired unlawfully to control the sheet-music business, and had sought to eliminate him as a competitor. John Paine, Mills's successor as chairman of the MPPA board, was named a defendant, described as the organizer, representative, and agent for the MDS.

Immediately after *Mayer v. MDS et al.* went to trial, in March 1934, attorneys for the Warner firms and Irving Berlin's company offered Mayer a cash payment and the promise to dissolve the MDS if he dropped the case. Other firms joined them, and soon only three publishers remained to fight the action. The trial concluded with MDS receiving only a tainted bill of health. Though he lost the suit, Mayer was victorious in hastening the dissolution of the MDS.

At the 1933 annual meeting newcomers to ASCAP criticized how the publishers' control of the board gave preference to the no-longer-productive old-timers. The constant stream of complaints led to a popularly elected writer review board. Many now believed that the SPA could do a better job of representing songwriters than the society.

Hard times were getting harder. Many syndicate stores on which the publishers depended shut down. Record royalties were at the lowest point ever, but synchronization fees turned out to be higher, due to the $825,000 settlement of the "bootleg seat" tax against ERPI, which had omitted to account for their number as screen theaters proliferated before the Depression.

Mills took on more of the ASCAP office operations and attended to the problems of the younger writers. A special fund of $12,500 was set aside each quarter and distributed among the writers of the ten most-played songs for that period, which first were listed on a weekly basis by *Variety* on September 5, 1933.

Mills's poor reputation was unimproved when the ASCAP contract he predicted would bring in two million dollars showed only a $300,000 gain, and indicated that the networks had paid on only 18 percent of time sales. Tempers were defused when Mills announced pending negotiations with AT&T to use its ERPI subsidiary to collect from all radio stations for a 25 percent commission, less than the thirty-five cents out of every dollar currently spent. However, AT&T removed itself from consideration on the grounds of conflict of interests. The music business was reminded of Claude Mills's easy access to the business world, where rumors flew that a Justice Department probe of ASCAP was being instigated by Newton D. Baker, the new NAB counsel.

Mills indicated that ASCAP would dissolve if necessary, leaving the networks and broadcasters to sort out the inevitable chaos. This led NBC officials to assure Mills that they would not support the NAB's tax-free bureau and might be amenable to maintaining a 5 percent rate throughout the decade.

Tin Pan Alley was concerned by the interest shown by the SPA and Hollywood-connected music houses in revising ASCAP's Articles of Association. Writers wanted assurance that they would share equally with publishers in all music rights. Determined to enlarge their share of distribution, the Warner music group indicated that it would not renew with ASCAP in 1935, feeling that it might be cheaper to license their music separately.

During the winter of 1933–34, Sigmund Romberg wrote a confidential memorandum to the SPA, which proposed ensuring ASCAP's survival by removing the restrictions imposed on the writer by the Copyright Act of 1909. As to who owned the small rights, he urged support of an amendment to the Copyright Act introduced by William Sirovich, which provided that "the author or composer may, to the extent of his ownership, license all or any part of the rights of such author or composer." When the memorandum was offered to the ASCAP board, publishers opposed any significant changes. Instead, separate classification review boards were established, whose findings would be final.

The publisher of many of the best-known SPA members, Max Dreyfus was a financial backer of the lobby working to kill the Sirovich bill. After its defeat, the SPA considered dissolving ASCAP and taking over its licensing function. More than 500 authors and composers and most small music houses were ready to renew their ASCAP affiliation for five years, but the large publishing firms held out, fearing the vague promises in the agreement to effect changes in ASCAP's distribution process and the compromise in the small-rights impasse being urged on them by their own most successful authors and composers.

Meantime, the eighteen-year war conducted by the MPPA over "pay for play" continued. John Paine immediately levied a fine of $1,000 on the first occasion a publisher was found paying gratuities. A $200 fine was exacted for each succeeding offense. A new point system was introduced at ASCAP, removing the AAA classification and the AA. A performance on NBC or CBS was credited with one point; every use in a major motion picture, with one quarter.

When the *Mayer v. MDS* lawsuit was settled, the plaintiff, Max Mayer, gave evidence to the Federal Trade Commission purporting to prove the interlocking interests of ASCAP, the MPPA, and the MDS. The split between the networks and their affiliates and the majority of the business now extended to fixed opposing positions on music licensing. The chains feared dissolution of ASCAP, though each felt the society was in technical violation of the law, but most members of the NAB wanted it dissolved. In response to countless complaints from broadcasters, the Justice Department's antitrust division began investigating ASCAP in 1933.

On August 31, 1934, a formal complaint was filed in New York District Court charging ASCAP, and MPPA, and the MDA with having interlocking directorates and agreements in a conspiracy to monopolize the music business. An injunction was sought to terminate agreements between the defendants and also with record companies and broadcasters. The MPPA board resigned immediately, but was replaced by other officers of the dominant member firms who remained on the ASCAP board.

The speedy trial was attributed to the approaching NAB convention and a possible conflicting action by the 15,000 motion-picture theater owners just before the expiration of their ASCAP contracts. ASCAP now offered them a new seat-tax agreement, which was expected to bring in as much as an additional four million dollars, but in fact resulted in only a 50 percent increase in fees and much less in actual income. This was followed by a forty-two-page reply to charges. Just before New Year's Day, 1935, the society was confident of the successful outcome of the trial, though faced with mounting problems on the Warner front.

Harry Warner had turned over the task of increasing Warner's ASCAP income to Herman Starr, Warner Brothers treasurer, who had instructed Warner representatives on the ASCAP board that they should assume that Warner would leave the society at the end of 1935 unless its share of publisher distributions was greatly increased.

Warners attempted to obtain at least part of the renewal rights of songs written from 1910 to 1913, so they could claim their performing rights. To increase the use of Warner music, fees for radio use were substantially reduced. This use fell under Warner's legally undefined grand rights, though the music business claimed that the performance of three or more selections from a stage work, together with unifying dialog and narration, constituted a grand right not covered by ASCAP. The same principle was applied in licensing electrical transcriptions.

A majority of the ASCAP board fought to keep the Warner firms. Starr would not budge. Fearing ASCAP's dissolution, most publishers signed the five-year renewals. The film publishers did not, even though all television rights were excluded, because they insisted that grand, and not small, rights were involved. Starr's resolve was strengthened by a sound defeat at the annual meeting of an amendment to the bylaws sponsored by Hollywood interests instituting elections by popular vote.

Mills and Buck went to Los Angeles and met with Harry Warner, who threatened to leave ASCAP unless distributions were based completely on performances. To obtain the million dollars a year he wanted, Warner required $2.5 annually. The movie magnate did not mention that he was also talking to Paley about selling CBS all of Warner's music houses. Warner lawyers were seeking a separate consent decree for the company in the antitrust suit against ASCAP, in order to begin separate music licensing. Negotiations were instituted with the two major networks to license the Warner catalog. Starr emphasized that there would be no last-minute reconciliation, and, having joined the ASCAP board, demanded a full accounting of all transactions since 1925.

At the semiannual membership meeting, Burkan warned that 70 percent of the Hollywood writers signed away all rights. While producer-songwriter contracts stipulated that the performance right was subject to an agreement with ASCAP when the present affiliation agreement expired, producers could regain ownership of the right. The issue legally had not been tested, and the Hollywood contracts were subject to the publishers' membership in ASCAP, an advantage on which Starr based his right to withdraw.

Mills simultaneously conferred with the NAB and the networks about a new contract. The decision to renew with ASCAP, at a steady five percent for five more years, stunned most broadcasters. The sudden recess of the government suit indicated the case's untenable foundation. With only Warner as an alternative, and succumbing to pressures from the IRNA, many broadcasters signed, reluctantly. Diehards took advantage of the ninety-day provisional contract offer by Warner, and began to exclude ASCAP music, in order to force Mills to come up with an acceptable per-piece arrangement.

The entire Warner group resigned in early December 1935, leaving other publishers in a quandary. ASCAP attempted to appease them with the installation of a new method of payment: a 50 percent distribution based on performances, and the balance based on availability-seniority. The holdout firms were informed that without them the society would have to relinquish power to the songwriters and the SPA. Burkan was ready to defend the networks in case of infringement suits, now proceeding on the theory that Warner's music was licensed by ASCAP through the songwriters.

Pay for air play was taking a toll in 1936 of publishers and songwriters. Not the MPPA, the FTC, nor "song pluggers" were able to halt its influence. It cost at least $1,000 to start a song on its way to the top of the *Variety* list.

Starr fired all his song pluggers and mailed requests to his 167 licensed customers, asking them to play songs from new Warner Brothers musicals. He then announced a reduction to two cents only for mechanical rights. Other publishers panicked, believing the practice of charging for both performing and mechanical rights in connection with transcriptions might be declared illegal. In a suit defending against alleged infringement of Warner copyrights, CBS questioned the total ownership of all privileges of copyright by music publishers, including that of public performance. Warner's return to ASCAP was made a major priority.

When the vast majority of independent stations had not renewed ASCAP contracts, the society floated several rumors: ASCAP was going out of business and turning its affairs over to the SPA; it was providing the money for some large music houses to buy out Warner Brothers; Claude Mills was moving over to NBC to take charge of music-rights acquisition. Things were not going well for Starr. He was unable fully to collect license fees from NAB stations but did manage to eke out reasonable profits from sheet-music sales and recording royalties paid on old songs. Warner's $170,000 share of the record $935,000 taken in by ASCAP for the first quarter of 1936 was divided among loyal publishers, but most Warner writers turned their checks over to the ASCAP relief fund. ASCAP opened its files for the first time to employees of the NAB, who compiled the first title index to ASCAP music, preparatory to building another tax-free library. Then the certainty of Warner's return and the death of Nathan Buran ended ASCAP's conciliatory policy.

Mills dangled revision of the ASCAP network contract as bait for Starr, interrupting the Warner music head's sudden renewal of negotiations with NBC and CBS for a ten-year contract recognizing payment at the source. This panic-inspired move by the networks was stimulated by their fear of shouldering all legal costs and a possible two-million-dollar judgment against them for copyright violations.

With backing from Gene Buck and at the suggestion of Max Dreyfus, Mills secretly went over Starr's head and sent Sigmund Romberg and Jerome Kern to Harry Warner to promote a network contract with a 10 percent share of all commercial income at the source, predicated on a reunited ASCAP. Publisher distribution had jumped 25 percent in a single quarter, and writers were clamoring for change. Many in classes AA and A could not hear their old songs on the networks, and their new work suffered from slipshod exploitation. Starr reacted furiously and developed the deep personal hatred of Mills that eventually led to the latter's ouster from ASCAP's inner circles. There was also great pressure on Warner from his producers, who told him either to get back into ASCAP or give up making musicals.

On August 4, Warner's music returned to the networks as its companies rejoined ASCAP. Infringement suits against the networks and other stations were withdrawn. Before making the move, Warner attorneys secured approval by the Justice Department, to preclude its being used as grounds for a resumption of the recessed federal suit.

ASCAP's new point system, with its heavy concentration on radio plugs, proved to be an economic boon to writers and publishers in the top brackets and offset the erosion of sheet-music sales, which had declined by 70 percent since 1927. Music publishers were reluctant to publicize any relevant information music dealers might use as a basis for placing sheet-music orders. If song ratings had to be made, the publishers preferred *Variety*'s lists. Though not amenable to an MPPA demand that it drop the Most Played on Radio list, the publication instituted a 15 Best Sheet Music Sellers list, based on information supplied by publishers.

Throughout 1937, ineffective attempts were made to change what was publicly branded Tin Pan Alley's "unhealthy condition." Most problems sprang from the performance-right distribution system introduced in 1936 in order to keep the society united after Warner's departure. In removing the authority to determine how much each publisher would receive from the self-perpetuating publisher half of the ASCAP board, the society had, in effect, turned it over to the most free-spending music firms and the radio bandleaders. Radio's position was solidified as the only vehicle that could make or break a song. To guarantee that distributions were made on a mathematical basis, the determination of the performance and seniority factors on which 70 percent of them were based now rested in the hands of ASCAP bookkeepers, who rated all broadcast performances equally.

The repeal of Prohibition and a slowly rising economy spurred the dance-band business. The chief beneficiaries of ASCAP's changed system were those small and recently purchased publishing houses that established close personal relations with, or were owned by, bandleaders. The "give and take" extended to late-afternoon sustaining programs featuring vocalists, many of whom had private understandings with music firms, another element in the pay-for-play practices *Variety* began to call "payola" in 1938. The networks learned they had little control over the remote shows. Now that others were treading on once-private turf, and with the promise of strengthened anti-bribery provisions in a proposed new MPPA code of private practices, two of the three dominant ASCAP publishers—Warner's and Max Dreyfus's firms—finally joined the organization, leaving only MGM outside. *Variety* introduced a new Breakdown of Network Plugs chart, indicating how many were on commercial broadcasts and how many were vocal interpretations. In 1938, this became a complete tabulation, Network Plugs, 8 p.m. to 1 a.m., and offered advance information, on which forthcoming ASCAP royalties could be anticipated. To exclude cheating orchestra leaders, requirements for admission were tightened by ASCAP, and works offered for registration in the society's catalog were scrutinized to avoid cut-ins.

At a meeting of publishers in the spring of 1938, Gene Buck again was directed to improve the distribution guidelines. For one thing, 84 percent of the crucial availability points went to publishers represented on the board, as did 70 percent of the seniority factor. However, any changes collapsed under pressure from the film-owned houses.

With little hope for a successful outcome of the government suit, the independents-dominated NAB was unprepared to be forced by the networks and the IRNA into a compromise of the music-licensing problem with ASCAP. At its prodding, anti-ASCAP legislative activity escalated. Freed of managerial duties, and now chairman without vote of a special all-publisher administrative committee, Claude Mills attempted to persuade Middle American skeptics of ASCAP's probity. He was replaced by his successor at the MPPA, John Paine, who instituted an auditing department empowered under terms of the standard ASCAP contracts to inspect the books of delinquent customers. Under Paine's direction the examination of radio broadcasts was increased toward an eventual goal of 25,000 live programs. Despite this, 64 percent of publisher distribution continued to go to the 13 Hollywood-affiliated houses.

ASCAP was determined to avoid or postpone dissention in its own ranks. Until a new ten-year contract with radio, at higher rates, was in effect, little more than lip service was paid to the reforms advocated by a voluble minority of writers and publishers. Having been led to believe that the government might shelve the suit, ASCAP gained approval for the renewal of membership affiliation agreements. The grant of television rights was again excluded out of deference to the film-music publishers. A new SPA standard writers' agreement was drawn, calling for no less than a minimum equal distribution of mechanical, synchronization, and foreign receipts. Songwriters were given more voice in merchandising their music: bulk deals required their consent as did the licensing of television performances, the use of a song title, and certain synchronization uses. Publishers were required to issue statements on a quarterly basis.

Only Claude Mills mentioned BMI at the annual ASCAP dinner in 1940. Verbal brickbats were missing from this gathering. Gene Buck hailed the recent rapprochement between the SPA and the MPPA as a symbol of the united front. Tin Pan Alley enjoyed its best year in a decade. Sheet-music sales had gone over the 15-million-copy mark, an increase of 25 percent in a single year. Annual record sales neared the 70-million-unit mark. With all the professional songwriters and music publishers already in ASCAP's ranks, it was difficult to believe that BMI could compete.

Neville Miller, the reorganized-NAB's president, met with John Paine in 1939 to inform him that he wanted two things: a per-piece license and at-the-source collection from NBC and CBS that would provide a reduction in music fees. There was no response for months. Although it did partially meet Miller's request for collection at the source, the aborted March 1940 conference between the two parties indicated the in-house auditing and collection service introduced by Paine had taken a toll, because of fancied or real "imperious" demands on many stations by field representatives. Convinced that ASCAP would not again compromise on the at-the-source collection, the networks, too, were ready for a fight.

The domineering presence of Herman Starr on the ASCAP board and his appointment as chairman of the Radio Negotiating Committee underscored that group's unyielding behavior. Starr made no secret of his "embittering" recollections of 1936, when he sought to get Columbia and NBC to accept a Warner music license. The revenge on Mills's call on Harry Warner came after a meeting of the Radio Negotiating Committee where Mills was shown the new at-the-source network-licensing provision, which had none of the deductions customarily allowed to radio stations. Mills indicated that he believed NBC and CBS would reject the proposal and subsequently was removed from any further participation in the radio negotiations. Soon after, his resignation was requested and granted.

Under the direction of attorney Julian Abeles, MGM intended either to realize a five-million-dollar net profit from the $75,000 it had lent Jack Robbins years before for a majority share in Robbins Music or to obtain from ASCAP in increased royalties the $500,000 it cost each year to operate the Big Three. The one important publisher to withhold the renewal of its ASCAP affiliation, MGM constantly threatened the society's dissolution. When BMI opened its doors, Abeles began negotiating on behalf of the Big Three with Sydney Kaye for their sale for $3.35 million, which William Paley agreed to advance on behalf of CBS, but he backed away when MGM would not guarantee indemnification for all copyright infringement.

The E. B. Marks deal with BMI, also arranged through Abeles, ended MGM's hope of being the first film company to have the advantage of exploitation by radio's music pool. Marks, too, made his decision to leave ASCAP on the anticipation of "BMI's power to make hits with the cooperation of broadcasters." Marks Music collected around $85,000 from the society annually. Under the BMI contract, which guaranteed $225,000 annually, there was no need for promotion, and the firm's income would increase severalfold. BMI agreed to set up a special royalty accounting system to take care of Marks's writers, most of whom had given him their music before being accepted for membership in ASCAP. Because BMI was eager to provide the sort of music some important advertisers wanted, it gave up any indemnification against copyright infringement in the case of Marks, which had been partially responsible for the collapse of the MGM negotiations. NBC and CBS agreed to assume responsibility for the entire $1.7 million involved in the Marks transaction, sharing it on an aliquot basis with their owned-and-operated stations.

Now Abeles began to dicker with the ASCAP board for its best counteroffer to his demand for a guaranteed $500,000 each year to Robbins, Feist & Miller. The board had already rejected an additional $15,000 from the industry's songwriters and film-musical screenwriters to Marks annually and now looked to other Hollywood studios to pressure MGM. However, the society had to return television rights in the Robbins, Feist & Miller catalogues to Loew's.

The intra-industry rift, on which the ASCAP Radio Negotiating Committee counted to break a solidly united front, did not happen. A general feeling of ill will toward ASCAP pervaded all levels of broadcasting in 1939, while the fieldwork done to sell BMI was responsible for industry support of the new organization. BMI's tenacity in recruiting reluctant stock buyer-licensees was crucial to the solidarity against ASCAP when the contract expired at the stroke of midnight, December 31, 1940.

Of more serious effect on the established music business was the government's renewed activity, beginning in 1939 under Thurmond Arnold, which eventually legitimatized ASCAP, with internal reforms, and ensured its existence by curbing the monopolistic and discriminatory practices that threatened its survival. Radio-station owners spurred the drafting of anti-ASCAP legislation. The networks realized that ASCAP was intransigent in the matter of payment at the source and agitated for a renewal of government interest. In May 1939, at Arnold's direction, a move to dismiss *United States v. ASCAP* was dropped.

Arnold, now head of the antitrust division, moved toward a settlement by consent decree. In late 1940, subpoenas were sought to permit access to hitherto-unexamined ASCAP files and records and to acquire lists of all the society's members and customers, to whom was sent a questionnaire regarding ASCAP practices and methods of operation.

When a consent decree was suggested, Starr, in his capacity as head of the Radio Negotiating Committee, and with assent by Buck, sent his attorney to Washington to negotiate the terms. *Variety* reported that the compromise offered by Arnold embodied "an agreement to license performances on a per-piece basis, with abolition of the blanket fee, a different basis for splitting the ASCAP take so that new members might receive fatter checks, limiting the organization to the function of police work in order to detect infringement, insuring the right of individual bargaining by writers and composers, lowering of the membership eligibility bars, and to a more democratic form of control."

By bringing these quasi-voluntary reforms to ASCAP, the government felt that the broadcasters could not refuse to deal with the society. If, however, they continued to use only BMI and non-ASCAP music, it would prove that radio was engaged in a conspiracy to restrain competition. Because ASCAP had violated the law, Arnold said, he did not intend to allow broadcasters to do the same thing.

Speaking for the Hollywood-owned publishers, Starr shrugged off the prospect of admitting a criminal action on a nolo contendere plea by signing the consent decree on Arnold's terms. His lawyer, Milton Diamond, completed "the best deal possible for ASCAP" with Arnold and his staff, who were proceeding on the belief that he was empowered to act for ASCAP. Final papers were to be signed on December 4. An open break between the anti-Hollywood faction and Starr and the major publishing houses was stimulated when Gene Buck indicated to Washington that ASCAP did not intend to sign the consent decree.

Washington announced that an action would be filed in Wisconsin, pressing eight criminal charges against ASCAP, BMI, CBS, NBC, and other parties. Each one expressed its innocence while applauding the government's charges against opponents.

Clearly, there would be no temporary truce in the ASCAP versus radio music war.

6 The Growth of the Independent Record Labels and the Ascendance of the Disk Jockey and A & R Man

THE struggle between broadcasters and ASCAP had little immediate effect on the recording business. However, its economic consequences would change not only the character of all popular music but also the very structure of the music industry.

With the exception of RCA Victor, most record firms used acetate blanks for masters, painted over an aluminum base. Strikes and government priorities led Columbia and Decca temporarily to curtail recording. Victor, however, kept up with orders and constructed a secondary plant in Cincinnati. To shore up sagging Red Seal sales, in 1941 Victor began an ambitious and expensive advertising campaign. Tied in with promotion of a new "Magic Brain" turntable, a two-for-a-dollar sale began of selected disks. The Magic Brain, equipped with two separate tone arms and capable of playing both sides of up to 10 or 12 records, was advertised as bringing the performance of extended works nearer concert-hall reality.

The year 1941 was the industry's best since 1921, producing 130 million records. Certain that the declaration of war would curtail their most important supplies, Columbia and Decca stepped up production. In April 1942, a month after the manufacture of all radio receivers and phonographs for personal use was terminated by the War Production Board, the official production of records was fixed at the 1940 level of about 50 million units, and prices were frozen at the December 1941 levels.

Columbia, Victor, and Decca again confronted the power of Petrillo and the American Federation of Musicians. In 1942, Petrillo announced that the present contract with manufacturers would terminate on August 1, giving only a month's period of grace. No union member would enter a recording studio until his demands were met. Record men had expected the threat and stockpiled against it, awaiting White House intervention in the interest of the war effort. Still, unprecedented sales and desperate, last-minute recording sessions before government restrictions and production ceilings were imposed, enabled Victor to finish with 56 million units sold; Columbia, 39 million; and Decca, 35 million, for a second 130 million total annual sale.

The times clearly were not propitious for a new independent company, as the founders of the Los Angeles-based Capitol Records soon learned. Formed in 1941 by Glenn Wallichs and songwriter Johnny Mercer, Mercer induced Buddy De Sylva to invest $25,000 and serve as president; Mercer was in full charge of recording and song selection; and Wallichs handled all other production and business details. Petrillo's month of grace allowed Mercer to build up a supply of masters, which lasted throughout the next fifteen months. He and Wallichs mailed copies of their releases to disk jockeys at the 500 stations that depended on Petrillo's reviled "canned music" for survival.

Capitol was the first record company regularly to service disk jockeys with free releases. Victor and Columbia, waiting in 1940 for a Supreme Court ruling, were prepared to charge for radio use of their products. Unexpectedly, the court let stand a lower-court decision that said that when a record was purchased at its list price, all property rights belonged to the buyer.

In the first year of their strike, union musicians lost four million dollars, while the record makers prospered. Half-million- and million-copy sellers were common. Victor presses were busy, turning out records as well as special orders for the office of War Information—Records for Our Fighting Men and V-Disks, produced by the Army Special Services.

The first monthly batch of V-Disks was shipped out in 1943. Each set contained twenty double-faced platters and provided four hours of recorded entertainment. Initially, one-third of each shipment was made up of newly recorded music; the balance was taken from air checks, existing masters made available by all companies except Decca, which refused to cooperate, and motion-picture soundtracks. The pressings were first made by Victor, but later also by Columbia, Muzak, and World, the transcription company Kapp had acquired in early 1943.

Kapp purchased the facility not only to secure additional shellac but also in anticipation of a settlement of the musicians' strike. Arranged by Milton Diamond, chief counsel to Petrillo and the AFM, it broke the major manufacturers' hitherto united front and gave Decca an almost year-long advantage, because Victor and Columbia believed that the government would curb Petrillo. Herman Starr, owner of a quarter-interest in Decca, was instrumental, with Diamond, in forging a contract that permitted Kapp to release any music previously or thereafter recorded by World, provided the musicians were paid full union scale.

The Decca resolution with the AFM break and, with it, agreements by some new small, independent labels came in the fall of 1943. Victor began to run out of materials and seriously considered meeting with Petrillo. Their decision was also influenced by Decca's five-to-one advantage in jukeboxes, as well as competition from Capitol Records. The government also appeared ready to make shellac supplies available to the new competition, as the major manufacturers depended upon the fluctuating black-market price of shellac.

New devices to improve sound recording were developed for military use during the war, among them wire and tape recording and reproducing machines, and 16-rpm recording machine whose reusable five-dollar wire spool was capable of recording nearly seven hours of sound. Dr. U. L. DiGhilini's recording and playback machine, demonstrated in 1945, embossed music on blank masters. Running at 16 rpm, the slow-playing records offered slightly more than two hours on a twelve-inch disk, and over five on a sixteen-inch disk.

However, the real future for recorded sound was revealed by the head of the Russians' Radio Berlin in 1945 when he showed a group of radio executives a captured German Magnetophon, whose fourteen-inch reels reproduced music on magnetic tape with extraordinary fidelity of sound far superior to any commercial American recording.

In 1945, the business looked forward to a sales boom that would see a total annual output of 600 million records. Two new popular-music firms opened: Majestic Records and the MGM-Lion Company. The former was owned by the Majestic Radio business, whose proprietors intended to take advantage of a built-in distribution system—the receiver and parts dealers who did business with Majestic. Studios, a pressing plant, and a backlog of masters were purchased from Eli Oberstein, who had resurrected his bankrupt Hit and Classic labels during the war.

Jack Robbins established an MGM label during the war, using Oberstein's studios and pressing plant. Loew's, the MGM parent company, signed Frank Walker to be operating head of the new company. At RCA Victor during the 1930s, he ran its customer-recording division, helped introduce the thirty-five-cent Bluebird label, and supervised the company's popular music activities once Oberstein left.

Most in the business believed that the released flood of pent-up postwar demand would improve conditions, but nobody was ready for the doubling of sales that took place, to $89 million and 350 million records, all but 50 million released by Columbia, Decca, Victor, and Capitol. Thirty-two pressing plants were now operating, providing that additional 50 million records sold by the independents in 1946. A portion of these were the products of small firms specializing in jazz, hillbilly, and race music, which had prospered during the war. They charged about a dollar for their records and won airplay by making new releases available to those small 250-watt stations that could not afford electrically transcribed programs. Many of these small firms had become music publishers, recording only their own copyrights in order to save the copyright's mandatory two cents for each side sold. Once these small firms learned that BMI was offering to pay as much as $250 per side of the airplay rights, many used that money to build a catalogue.

Many of these new firms lacked an efficient distribution system. With 210 record makers now licensed by the AFM, six leading eastern independents formed their own distribution chain, offering a full return privilege as an inducement. Accustomed to a 5 percent return policy after six months, many local suppliers placed large orders, and then returned them once the majors began to crowd out this unwelcome competition.

Another problem was the copying of an up-and-coming release by an artist on one of the Big Four labels. The New York-based National label anticipated a nationwide best-seller with "Open the Door, Richard" by Dusty Fletcher when fourteen versions appeared. The song became number one on *Billboard*'s Honor Roll of Hits in early 1947, principally because of the Decca version, made by Louis Jordan, which crossed over to the long-all-white hit charts.

Aware of growing challenges to their control of distribution, the majors solidified their position. Columbia's first annual dealer convention concentrated on the company's popular product and went one step further than Capitol in the station-service field, opening an in-house publicity and promotion department. It bought time on radio stations for a transcribed show featuring disk jockey Martin Block, who introduced new releases supplied by the local distributors or retailers.

Faced with vigorous competition from the majors, and with pressings of improved quality and expanding distribution, many independents specialized in fields of music that most majors neglected: the hillbilly and western music and race records which reflected the changing life of urban black communities and radically changed the popular culture and vernacular music of the world. Still, a four-month sales slump in 1946 hit even the most successful independents. Shellac prices rose and radio manufacturers' production of radio-phonograph combinations was upset by the failure of the Office of Price Administration to reduce ceiling prices on tubes and other components. Many retail stores did not stock the secondary labels.

The cancellation of wage and price controls in late 1946 had mixed effects on the record business. The majors' popular disks went up to seventy-five cents, classical music to one dollar. Some of the small, underfinanced independents fell into bankruptcy or were absorbed by more successful competitors. Columbia and Victor expanded their album catalogues, while Decca, now fully owned by Jack Kapp, matched the extraordinary success of the pioneering original-cast album of *Oklahoma!* with the first album by a popular artist: *The Al Jolson Story*.

Jack Kapp had retrieved full ownership of Decca by buying back the remaining 25 percent owned by Warner Brothers, making his company the only true independent of all the prewar major firms. Following many months of economic and production setbacks, MGM Records concluded an agreement with EMI for worldwide distribution of recordings by Hollywood personalities and soundtrack music. In the fallow period since its initial success, Capitol emerged from mid-year economic uncertainty with several hit recordings.

The independents' success was due to the attention of the country's disk jockeys, of whom 1,000 were believed to be central to making a hit. Together with a few up-to-date independents, Capitol, Columbia, and Decca had promotion men visit deejays to secure their friendship and air play. *Variety* now included a weekly chart measuring the popularity of songs and recorded talent as reflected by the "most-requested" records. The platter spinners' increasing influence led *Variety*'s Ben Bodec to write, "A goodly percentage of them have moved into upper-income brackets. They have to a widening degree community standing. They have become an integral part of local broadcasting. They carry economic weight with the advertisers. The only way their ranks could be seriously decimated in the next years would be for Congress to amend the copyright law so as to curb their use of records."

Petrillo turned his attention to the welfare of his constituents, which, he argued, should be funded by contributions from the recording business. He prematurely announced that another ban on recording by AFM members would take effect on January 1, 1948. The Taft-Hartley Act, which proscribed such an action, was passed over President Harry Truman's veto. Victor, Columbia, and Decca were confident that the act would be upheld by the Supreme Court and so adopted a more leisurely pace of prestrike recording than they had in 1942.

However, once Columbia and Victor were convinced that Petrillo was in earnest, they invested two million dollars in a frenzy of record cutting because a provision in their contracts with the AFM stipulated that, in the event of a strike, all existing artist contracts with the AFM would be invalidated. There was also a clause in the Taft-Hartley Act that made the payment of royalties to the Performance Trust Fund illegal after June 1, 1947, when the union leader began his struggle against the record companies. Only Decca was better prepared; it had an existing store of unreleased masters and the World Transcription Library, to which Petrillo gave it unlimited access under the 1943 settlement.

The year 1947 proved to exceed 1921, the industry's previous best effort. It was a $214.4-million year at retail; 3.4 million record players of all kinds were produced. The Big Four issued 300 million popular and classical records, and possibly another 75 million were released by the rest of the industry. The business could only improve, it thought, once Petrillo was beaten.

A magnificent leap for recording techniques and equipment occurred in 1948 when the ABC Radio Network went "all-tape" for nighttime programming, thereby underscoring the technological changes since the discovery of Magnetophon. Despite government confiscation, enough parts were smuggled back to the United States to allow electronics laboratories to produce and patent an improved Magnetophon, and develop compatible magnetized plastic tape. Bing Crosby in 1946 switched from doing live shows to ones recorded on tape and then rerecorded on acetate transcriptions. The tape he used was 3M, a product of the Minnesota Mining & Manufacturing Company, and the recorder was an Ampex.

Despite the intriguing possibilities inherent in tape, other matters demanded the industry's attention. Sales were falling. The demand for recorded music dulled the imagination of the heads of most major labels. The older executives at Victor, and to some extent at Columbia, became increasingly conservative. Judgments merely affirmed those of their competitors, and the result was a homogeneous output.

Eli Oberstein was replaced with decisions by a committee of product and merchandising department heads at Victor. The firm recognized, however, that disk jockeys on about 1,200 stations influenced 85 percent of all record sales. Last of the original Big Three to make the move, Victor now sent out promotional Vinylite pressings of all new releases except the Red Seal packages. Free disks were delivered by regional distributors and dealers, who were charged at wholesale for promotional materials.

When Edward Wallerstein moved to the post of chief operating officer of Columbia Records in 1947, he fervently supported the development of a recording head capable of tripling the eighty-five grooves to an inch of standard records and appropriate lightweight, high-fidelity pickup, amplifier, and 33-speed players. In anticipation of Columbia's long-playing records, Wallerstein ordered his staff to record all classical works on sixteen-inch transcriptions and, once Columbia added tape machines, on them as well.

At the annual dealers' meeting, Wallerstein introduced the new technology, standing between a fifteen-inch stack of 101 LP albums and eight feet of the same music on 78s. It was indeed the "revolutionary new product" the press was promised in publicity releases, but some present pointed out that the purchase of special equipment on which to play the microgroove disks might be an economic drawback.

Columbia cancelled all promotion and production of 78-speed popular disks. The first long-playing releases were announced. After four months on the market, Columbia reported that several hundred thousand plug-in LP players, built by Philco, and 1.25 million LPs had been sold. Only Mercury took advantage of Columbia's offer to press its classical microgroove records on a licensing basis.

The American record industry's international-licensing situation was upset when the government sought to end an alleged conspiracy and disk cartel scheme led by EMI and both the British and American Decca companies. They were charged with dividing the world's markets in order to control the manufacture, sale, and licensing of phonograph recordings. The Justice Department charged that, beginning in 1934, the three had allocated all of the Americas to Kapp and American Decca; Australia, New Zealand, and the Far East to EMI; the rest of the world to Lewis and British Decca.

However, as EMI grew stronger it began to dominate. The government asked for cancellation of all existing contracts, an injunction against restraint of trade, and a cessation of all price fixing. This left the matter of international representation of American firms activated since 1934 wide open. Columbia was negotiating with Philips of the Netherlands, and Capitol waited only for the government to confirm an agreement with Telefunken that would supercede the Mercury-Telefunken arrangement. When it was approved, Capitol, for the first time, had outlets throughout Europe, with the exception of Russia, and soon after in Africa, through EMI. Decca, now headed by Milton Rackmill, president since Kapp's death, in March 1949 concluded a reciprocal deal with Deutsche Grammophon, the last European major without an American connection.

Victor's backlog of masters was nearing depletion when the label and Columbia entered into talks with the AFM. The Justice and Labor departments argued over responsibility for approval of the peace plan, whose acceptability under the Taft-Hartley Act might be questionable. Most other manufacturers went back to their recording studios. After eleven and a half months, the Labor Department approved a settlement.

The battle of speeds came to a head when Sarnoff revealed that a new 45-speed disk would go on sale in 1950. Records sales were falling by 25 percent when the 45-speed system was introduced. The new seven-inch disks of unbreakable Vinylite could be played on only two record changers: a plug-in machine priced at $24.95 and a self-contained three-tube model that sold for $39.95. Columbia immediately reduced the price of its player attachment to $9.95, and marketed a Microconverter which enabled a 78-speed turntable to play 33-speed LPs.

Victory in this competition rested on several factors: the number of new players that could be sold; the degree to which records at the new speeds would be accepted by disk jockeys and jukebox operators; and how many of the other major manufacturers would adopt one of the new speeds. The manufacturers of phonograph players and radio-phonograph combinations, except for RCA, made a Solomon-like decision and used both.

The increasing studio use of tape reels and recorders presaged an era of more faithful reproduction and new sounds on commercial records. Columbia was the first of the majors to benefit, redubbing all old material and recording new sessions on improved plastic tape. Columbia spent two million dollars and Victor five million to promote and merchandise their new systems before a three-speed formulation was accepted. Victor's Extended Play classical 45 disks were not generally accepted, for they required a break in the movements of an extended work. To resolve the dilemma Victor demonstrated a new line of three-speed machines. Victor was willing to go along with a classical repertory on 33s, but it held out for its own favorite creation as the medium for all nonclassical music.

Record speeds continued to drop and inventories piled up. The fact that only one out of every twenty-five popular releases was successful indicated the dearth of qualified executives at the major labels. Of the nearly 300 new songs offered each week, fewer than a dozen got any serious attention. It cost a major publisher between $25,000 and $30,000 for promotion and exploitation of a song that eventually got near the top of *Variety*'s Most Played on the Networks list. Uncertain of their ability to pick hit songs, publishers simultaneously offered six to twelve manuscripts to the record companies with promise of the exclusive rights for recording.

Columbia and Victor now enjoyed better sales with new, high-quality, unbreakable disks. The surfaces of 33s and 45s improved because of a new thermal engraving process that used a heated stylus and provided a quieter record. Their combined sales accounted for nearly one-third of 1949's $172-million retail gross, and represented 50 percent of the sale of conventional 78s, LPs, and 45s. Discount selling, the acceptance of the 33 as standard by several hundred new record companies that specialized in classical music, and a growing audience reflected a broadening market.

The father of discount record selling, Sam Goody of New York, sold new LPs at a 30 percent discount. A $3.98 ten-inch record, which Goody sold for $2.80, actually cost him $1.85. Ballyhoo, an inventory of 300,000 LPs, and an overhead strictly maintained at 15 percent made Goody the talk of the business but led to litigation, instituted by Columbia.

Several attempts were made to stop Goody. He was refused a special 10 percent return privilege and denied the extra credit granted to a few favored major outlets. Columbia suffered a serious defeat in a hearing before a New York Supreme Court referee, who ruled that the state's Fair Trade Act could not be applied to record price discounting. The ruling immediately effected discounting in large metropolitan centers, where Goody's operation and selling policies were copied. A boycott by Columbia and Victor of these operations followed, but was countered by Goody with transshipping. It was customary to allocate new record releases on the basis of past sales. Many distributors and dealers were left with unsold overstock that could not be returned until six months had elapsed. Goody had dealers transship LPs and 45s to him at a 40 to 50 percent discount. He could still enjoy a 4 percent profit. Such sales practices made discounters indispensable to consumers and built a devoted following for classical music and the outpouring of original-cast albums.

Songs from current musical productions had long been recorded, but until Jack Kapp released *Oklahoma!* in 1943 there had been no complete recording of a musical made by an original cast. It became the general practice to advance a musical's producer several thousands for first-rejection rights, in return for which, and a royalty up to 10 percent on all sales, the original cast was made available for recording sessions.

At the same time, the market was witnessing the world-wide popularity of hillbilly music, now become known as country and western. It accounted for a third of 1950's popular-music sales. The biggest popular hit in twenty years, "Tennessee Waltz," sung by Patti Page, had already sold two million copies, and an additional 2.2 million sung by others. Artists such as Ernest Tubb and Hank Williams generally averaged more than three-quarters of a million disks. Sales included a myriad of fans in urban centers. In 1951 nearly half of all pop single Decca records sold were rooted in country music. Columbia announced that country-and-western records represented 40 percent of all 78-rpm sales.

Black America's "race music"—officially rhythm and blues, or R & B, once *Billboard* adopted the phrase in 1949—was enjoying a growth paralleling that of country and western without its commercial returns. By 1951, labels specializing in rhythm and blues—Atlantic, Savoy, King, Imperial, Apollo, Aladdin—experienced an occasional half-million or better seller. Each had its own publishing company, formed to avoid paying song royalties or to take advantage of contracts offered by BMI. Money was laid out only for promotion, or sometimes to make deals with distributors and realiters. Only the best-selling artists received a royalty.

R & B distributors put the most important local deejay on their payrolls, provided boxes of free records, or paid a royalty on each disk sold within the listening area. Expenses of this type were passed back to the manufacturer, who gave a higher discount, a cash subsidy, or one free record for every three ordered.

Aware of the economic significance of R & B music, many of the major companies opened a "race," "sepia," or "ebony" department and adopted the marketing pattern of the independents by harnessing the power of disk jockeys. Subsidiary labels were formed to handle the rhythm and blues lines—OKeh by Columbia, Coral and, later, Brunswick by Decca—each with independent distribution and promotion. In spite of them, the independents continued to prosper. New black talent and their agents preferred independent labels with compatible owners, artists-and-repertory men and musical directors who were familiar with R & B music.

The majors changed some of their business practices, particularly their returns policy. Because the new distributors and retailers feared being left with unsold inventories by unfamiliar artists, they insisted on a 100 percent return privilege. Panicked by declining sales in 1951, some majors shipped out as many as 100,000 copies of a new release by an unknown performer to major retailers, on a guaranteed 100 percent return basis. With the cost to the manufacturer of pressing, packaging, and shipping at around twenty cents, the loss could be considerable when a record was unsuccessful.

Impressive royalty statements of successful newcomers annoyed many executives, who demanded a revised standard recording contract, which called for a 2½ percent royalty on retail price, with a 10 percent return or breakage privilege. The artist was also asked to pay all costs for a recording session, taken out of royalties. Their income was predicated upon concerts and personal appearances. Local disk jockeys became more important than ever to the recording artist, playing their new releases and promoting live performances.

The new role of disk jockeys provoked cries about payola, for promotion men pressured them by dispensing gifts or buying time on their stations. Through the 1950s *Variety* took up the matter in a series of editorials. The solution lay, *Variety* concluded, in a revision of the Copyright Act that would end free use of recorded music by broadcasters and coin operators.

Free records to disk jockeys, payola, and Sam Goody's discounting led to the formation of the record manufacturers' most important postwar trade association and lobbying group: the Recording Industry Association of America, established in 1951. It was officially designed to deal with legislation, the allocation of materials, preparation of industry statistics, and such matters as the government's request for a voluntary price cutback. Fortunately for the manufacturers, it came just after they had boosted prices in fear of a Vinylite shortage due to the military action in Korea. More recently, a 10 percent excise tax had been added on all records. About 800 record companies now were registered with the AFM, but fewer than forty-five did an annual business in excess of $20,000. More than half were about to go out of business.

In the early 1950s, EMI faced serious competition. Victor was making new arrangements for international distribution when its contract with EMI expired in 1957. The well-financed Philips, Europe's largest electronics manufacturer, bought out French Polydor, and, ready to distribute America's Columbia, moved into England. American independents finally got the opportunity for distribution in Europe. A consent decree, signed in New York in December 1952, dissolved the last vestige of EMI's cartel agreement in the United States and permitted Decca to make its masters available to any foreign distributor if EMI did not accept them within ninety days after release. Within a year, EMI was doing business in the United States through the resuscitated Angel label.

CBS classical records made a dramatic showing during the early 1950s—a sale of $37.8 million, representing about 20 percent of total industry volume, which rose to almost 35 percent in 1952, an all-time high. Independent labels that specialized in generally unrecorded repertory, such as the Haydn Society, Remington, Vox, and Westminster, went to Europe to record hitherto unavailable music. In 1949, a classically oriented company could buy a professional Ampex tape machine for $1,000 or $2,000 and with an additional $8,000 put enough recordings on the market to realize a small profit.

Freed from the antitrust EMI-cartel suit, Decca agressively entered television production. Its new president, Milton Rackmill, believed that a marriage between theatrical film production and television was inevitable, and he moved toward the purchase of a substantial portion of Universal Pictures. Universal stepped up production of three television series to be sold as a package of twenty-six programs. The profits lay, Rackmill believed, not in selling the films but in residual rights, which would continue as long as a series remained in use.

The first official breakdown of industry figures, issued as one of the RIAA's initial public statements, reported a $200-million national gross sale of 186 million disks in 1951, purchased by the owners of nearly 22 million record players of all types. The traditional pre-1939 largest market for singles, the nation's 550,000 jukeboxes, now accounted for only about one-tenth of total production. Production of 45s surpassed 33-rpm records, almost doubling their sale.

The undisputed king of artists-and-repertory men in 1952, Columbia's Mitch Miller, introduced musical instruments and vocal effects for which tape seemed to have been created. The favorable profit situation for which Miller was chiefly responsible helped to underwrite the expansion of the CBS television network and the development of a CBS color receiver, also expanding the smoldering feud between Paley and Sarnoff.

The influence of the disk jockey, however, continued. Not even the RIAA was in a position to compile an accurate list of their numbers. The figure 2,000 had been bandied about, yet smart record-company promotion men concentrated on about 100 "real song breakers," men with national reputations as effective song boosters who regularly received some form of payola. Bill Randle of Cincinnati, regarded throughout the industry as one of the most powerful and honorable record spinners, told *Variety* in 1952 that no more than 100 disk jockeys really controlled the popular-music business. Without their concentrated action, no song could become a major song and no artist could remain a major artist, nor could a new name be made.

The world of rhythm and blues was subject to the disk jockey's sway, too. It had remained in economic isolation until the independents made it sufficiently attractive to the major manufacturers. In spite of them, in 1953 R & B was still outside the white-dominated music business, just as black radio.

Since 1942, the income of the average black family had tripled, while that of a white family had doubled. At least four New York radio stations, mainly featuring R & B, were targeted at this market, and 260 more throughout the country. Listeners responded most favorably to the products of the Aladdin, Apollo, Atlantic, Herald, King, Savoy, and Specialty labels. Hit disks, by Ruth Brown, the Clovers, The Five Royales, Amos Milburn, and Little Esther rarely sold fewer than 150,000 copies, and the Clovers had already sold two million disks of a single hit.

In an eight-page brochure, *The 50 Year Story of RCA Victor Records,* put out in 1953 to celebrate the company's anniversary and the diamond jubilee of Edison's talking machine, the company freely admitted that popular records underwrote all losses from the Red Seal catalogue. Victor's total sales since 1902 were approaching two billion units, including 325 million Red Seals, one billion since 1946. Mannie Sachs believed a 15 percent increase from the year's $200 million loomed for 1954 due to the entrance of chain stores, supermarkets, and major department stores into the retail record business, particularly with children's records and LPs; the steady growth of the teen-age market; and the high-fidelity boom.

The improved sound that came from the usurpation of FM frequencies for telecasting plus the introduction of binaural and stereophonic sound into moving-picture theaters were influential in creating interest in high fidelity. Only a superior machine could reveal the full glory of registered trade names the manufacturers used for their improved, noise-free unbreakable vinyl plastic records.

At the same time, one million wire and tape-recording machines had been sold since 1946, many of them now antiquated. Ampex offered a low-price machine, and its eastern distributor, Audio-Visual, announced a new library of 160 prerecorded tapes. Citing the possibility of overnight obsolescence of a major portion of their Red Seal and Master Works catalogues, Victor and Columbia delayed their entrance into the taped-music field.

There was little response from either Columbia or Victor when a $150-million civil antitrust suit was announced on November 9. Speaking for songwriters, the thirty-three plaintiffs alleged that CBS and RCA, with their recording companies, and a host of other defendants, including the major radio and television networks, were responsible for the present condition of the music business. The defendants were engaged, the complaint stated, in a conspiracy to create a monopoly in the production, exploitation, and use of music, favoring that available from the broadcaster-owned music organization, BMI.

Under government controls, there was little change in cost of living until the restraints were lifted in the late 1945. Heavy taxes on luxury items and entertainment, the government's war bond campaigns, and gas rationing kept Americans at home, and were factors in the return of sheet-music sales to proportions of the golden days. Despite a paper shortage, sales doubled. The Music Publishers Protective Association was chiefly responsible for the growth in this old source of music publishing and licensed song-lyric reprinting in the two legitimate song-sheet magazines, *The Hit Parader* and *Song Hits*. Because these publications would reprint only MPPA-licensed lyrics, Broadcast Music Inc. was obliged to fund competing magazines.

The music-rack business was now the principal means of selling sheet music. In early 1940, the MPPA entered into a formal agreement with the Hearst Corporation's International Circulation Co., which guaranteed the placement of metal display racks in about 500 national locations. Two copies of each new song that found a place among the top twenty sellers were automatically put into each rack. In four years, 13,000 racks had been installed.

The printed-music business's newfound prosperity during the war years was surpassed only by the economic gains enjoyed by ASCAP. Distributions to members nearly tripled during the war years. Still, problems remained. In 1941, ASCAP music virtually vanished from the airwaves. Sheet-music sales plummetted for all ASCAP-affiliated music houses from a total 16 million copies, at the rate of about 300,000 a week, to 120,000 a month. Realizing that the networks would hold out for months, the large firms waited out the dispute.

Assistant attorney general Thurmond Arnold wanted to hasten a settlement of the fight between ASCAP, BMI, and the broadcasters. Feeling the case against them lacked any merit, William Paley and David Sarnoff unsuccessfully met in early 1941 with members of the boards of BMI, the NAB, and the IRNA. Paley and his chief assistant, Edward Klauber, adamantly opposed the 7½ percent network fee which Arnold and ASCAP supported. Sarnoff and the NBC contingent believed a favorable decree would damage the existing radio-industry structure or BMI's continued operation. With counsel for CBS, Sydney and Geoffrey Goldmark and the NAB president, Neville Miller, they negotiated a consent decree signed by the NAB board on behalf of BMI. Among other provisions, it gave music users a choice of blanket, per-program, or per-use access to BMI's repertory; required payment at the source from the networks; and was conditional on the acceptance and signing of a similar order by ASCAP. A year of grace was permitted in order to revise network-affiliate contracts to reflect the pay-at-the-source requirement.

Reactions to the decree varied among ASCAP writers and publishers. Hollywood interests favored the blanket-license concept as a per-performance provision could affect screen-theater licensing. Small publishers thought the per-piece concept aided them in competition for access to radio. Many of the top-classed production writers and concert-music composers resented that it treated all ASCAP music on an equal basis, regardless of its artistic value or merit. All, however, agreed on the permanence of BMI's existence as a source of competitive music licensing.

Annoyed when Louis Schwartz, ASCAP's counsel, accompanied by associate Herman Finklestein, returned to Washington to resume discussion of the settlement, members of the antitrust division prepared to charge the Society with violations of the Sherman Act and other offenses. The new charges altered thinking in ASCAP's New York offices. Herman Starr's inflexible anti-network attitude was rejected. Claude Mills's resignation was shelved, and the veteran negotiator rejoined the executive management committee.

Arnold had been blunt. In plain language, he said:

> ASCAP is charged with exploiting composers by preventing them from selling their music except on terms dictated by a self-perpetuating board of directors. That board has had power arbitrarily to determine on what basis various members of ASCAP shall share in the royalties from ASCAP selections. In addition to discriminating against composers, ASCAP has begun using its monopoly powers to charge the users of music for songs they do not play. This is done by compelling the user to pay a percentage of his gross receipts on programs where other music is used or where no music is used. By this method, anyone who does not belong to ASCAP is excluded from the market. These practices we consider not only illegal, but unjustifiable on any ground of fair dealing. Our proceeding is aimed only to compell ASCAP to stop these abuses. It does not desire to prevent ASCAP from protecting the copyright privileges of its members.

A new, outside counsel, Charles Poletti, firmly held the need to revise the language that would effectively strip the society of its discretionary power to pool and distribute income and insisted on network payment at the source. Surrender on the first would be a veritable act of suicide and reduce ASCAP merely to a policing organization; the second was needed to increase revenues, and to correct an old abuse. An acceptable consent decree was drawn up, in which Poletti and ASCAP prevailed on both points.

Over the NAB's objections, the consent decree was approved by Judge Henry Goddard, of the federal court in New York, on March 4, 1941. This dismissed a six-year-old civil action, though it left unresolved the Justice Department's criminal case. Henceforth, any new ASCAP contracts required the approval of the New York court, to which complaints and applications for enforcement or modification of the decree must be brought.

Variety on February 26, 1941, printed the complete decree. It barred ASCAP from sole, or exclusive, right to the works of members. In effect, ASCAP had returned to the 1941–20 condition, when most firms licensed directly. However, direct licensing could be done only with the society's permission, and all fees were paid to it for distribution. All users of music were given a choice among blanket, per-program, and per-piece licenses; the self-perpetuating board was ended; one composition was now sufficient for a candidate's admission; and royalty allocations were based on the "number, nature, character and prestige of a member's works, the seniority of a member's works, and popularity and vogue of such works."

Judge F. Ryan Duffy settled the government's criminal action by accepting pleas of nolo contendere from the society and a total fine of $35,250. With the decision that no moral turpitude existed on the defendants' part, Duffy found each of the ASCAP representatives guilty, but declared the charges to involve a misdemeanor, not a felony.

The payment-at-the-source clause in the consent decree allowed CBS and NBC to delay before a final settlement. This twilight-zone issue had already been resolved in the case of BMI, which was also obliged to license networks at the source. BMI's new contract, which became effective on April 1, 1941, called for a flat fee from both chains of 1½ percent of income, less certain deductions.

The respite that followed, before ASCAP offered acceptable terms, was a particular boon to the two major networks. Imminent Justice Department and FCC actions were predicated on the following issues: the monopoly represented by newspaper ownership of broadcasting facilities; multiple-station ownership by individuals and corporations; the vertical integration implicit in the ownership by both CBS and NBC of recording and transcription companies, and of artists' bureaus, as well as their involvement in BMI; and, most significant, the results of the three-year-old FCC monopoly inquiry into network ownership of stations and their contracts with affiliates.

While unproductive network meetings with ASCAP continued, the Mutual Broadcasting System began negotiations to use ASCAP music immediately. Mutual's executives were especially pleased with the plight of their competitors when the FCC turned a formal monopoly report over to the Justice Department. The institution of new regulations ended RCA's dual ownership of the Red and Blue networks with the divestiture and sale of the latter. The network's artists' bureau, which gave them dual roles as employer and agent, was declared "a serious conflict of [fiduciary] interest" and sold to a partnership of several leading independent booking agents and managers, the National Concerts and Artists Corporation. CBS's sale of its talent agency to the Chicago-based Music Corporation of America began MCA's domination of the entertainment, record, and television business. Formed in 1924, MCA was already a leading agent and booker, but the CBS sale added creative writing, directing, and producing radio talent. The Columbia Concerts Corporation, the leading representative of concert and classical-music artists, was completely separated from CBS and purchased by Arthur Judson, the man responsible for the formation of the chain in 1927, and, a decade later, for CBS's acquisition of the American Record Company.

The Mutual Broadcasting System negotiations with ASCAP broke the industry's hitherto united front. Claude Mills's public statement that the new terms represented the best ASCAP would offer further exacerbated intra-industry tensions, as it called for a 3 to 3½ percent fee, less deductions, for all network and single-station commercial programs. In spite of strenuous lobbying by the NAB, CBS, and NBC, the majority of

MBS affiliates approved the agreement, and ASCAP music went over a network.

During the annual NAB convention, Sydney Kaye fired up industry support for BMI by announcing rate reductions of one-third. Local-station fees were cut to ½ to 1 percent. Though CBS and NBC remained at the previously announced 1½ percent rate, Kaye did not discuss the actual basis of accounting in the twilight zone that was permitted to the two chains, which involved the difference between what they received from advertisers and what they paid out to affiliates. BMI's rate for per-program licensing ran from 3 percent of time costs for stations with an income below $50,000 to 5 percent for all others. Kaye also reported that BMI had spent $1.8 million in its first year, and that the new contracts would bring in nearly $2 million for its second.

Nearly one-quarter of BMI's projected income, after expenses, for 1941–42 was marked for the E. B. Marks catalogue, which had the highest cost-per-performance rate for any of its music, and was causing a severe strain on cash outlay. A substantial portion of the Marks catalogue was not under BMI license. Their exact status awaited a declatory ruling by the courts. It was Marks's contention that these songs were not owned by their writers—many of whom were now among ASCAP's most prominent authors—because they had turned over full proprietary rights to the publisher under the pre-1930s contract. Most major publishers feared any court clarification of this practice. If the writers won, they could take control of ASCAP, throw out the publishers, and divide all collections among themselves. If the court upheld Marks's position, songwriters would realize that they shared ASCAP income only on the sufferance of the publishers, withdraw from the society when their contracts expired, and support a strengthened Songwriters Protective Association.

In 1941, Marks and BMI brought suit against ASCAP in the New York Supreme Court. They hoped to resolve the true ownership of small rights, particularly that of public performance for profit. The plaintiffs maintained that Marks owned all rights in the three songs selected for the case. They had been written and copyrighted in Marks's name before their writers joined ASCAP, and therefore they were linked specifically to the society.

The ASCAP answer avoided any assertion of ownership or conclusion of law, leaving that to the court. Had the society done otherwise, it would be used against them in any future lawsuit, brought by songwriters, over the issue of small rights. The SPA lost in its desire that the defense should be based on the argument that the writers vested rights in ASCAP, or that they did so with publishers as joint tenants in common.

Shortly after the first papers were filed in *Marks v. ASCAP*, the society hoped for legal resolution of the music war. ASCAP retained David Podell as special counsel to handle a $20 million antitrust and conspiracy lawsuit against CBS, NBC, BMI, and the NAB, and possibly network sponsors. Within a fortnight, counteroffers to the Mutual agreement were in the hands of a new ASCAP Radio Committee. The proposals offered between 2 (CBS) and 2½ percent (NBC) of net receipts after agency commissions, time-sales fees, line charges, and various other expenses. An immediate rejection by the Radio Committee was quickly tempered by instructions from Hollywood interests to speed up settlement.

With the major houses losing at least $40,000 a month, and business at a complete standstill because of ASCAP's absence from the networks, total losses for the music divisions of Hollywood studios passed the two-million-dollar mark. Six months after peace came, control of the song-plugging mechanism was retrieved from BMI firms and the networks. Hollywood re-examined its ASCAP connection in light of the consent decree. Taking advantage of its new nonexclusive provision, many of the large studios were getting performance rights from their staff songwriters, or employing them on writer-for-hire contracts. In order to get radio plugs, the studios prepared to license their music either without charge or through BMI.

As acting ASCAP president, Louis Bernstein took up the cause of the film companies and got an affirmative vote from the Radio Committee. CBS passed the negotiating lead to those at NBC, who were amenable to ASCAP's most recent compromise offer calling for 2.75 percent from the networks, and 2.25 percent from affiliates. In addition, stations were allowed to pay only for the ASCAP music they used, on a per-program basis. The charge was 8 percent of program time costs for those using ASCAP music and 2 percent for ASCAP music used for themes or as background.

Many NBC and CBS affiliates opposed any agreement. They felt that BMI was supplying sufficient music and wished to continue saving on their ASCAP fees. The two chains found themselves facing prospective litigation. ASCAP officials told the broadcasters that without quick approval, the society would disband and force broadcasters to negotiate with separate copyright owners. The antitrust suit against them would be filed immediately, with an added charge of conspiracy to destroy the society.

Quite suddenly, the NAB ended the music war, and NBC, CBS, and a majority of their affiliates signed the contracts. ASCAP music returned to the air almost immediately. Looking for elusive signs of victory, ASCAP pointed to the vindication of its stand on payment at the network source.

Ben Bodec wrote in *Variety* in January 1942 that the networks had actually "out-foxed ASCAP, preserving the twilight zone on music licensing payment, but in a different form." A rebate of each affiliate's share of ASCAP fees was deducted by the networks from their payments for local participation. "In the case of the Blue Network," Bodec explained, "the deduction can amount to as much as 85% of the gross billings, so that the amount turned over to ASCAP by NBC on a particular program could be even less than the fees that NBC has deducted from the Blue affiliate involved . . . if the network rate on, say, PQW, is $100 the Blue would be required to pay ASCAP only on the basis of $15." He continued:

> The rate that NBC would pay that same Blue station is $30. The station is committed to permit the network to deduct 2.75% from the $30 as its share of the fee going to ASCAP. That deduction would amount to 55 cents. The Blue's residue figures in this illustrative instance as $15. The ASCAP network fee of 2.75% applied to the $15 figures 41 cents. Out of this arithmetical operation there may be derived the observation that whereas NBC has deducted 55 cents from the station's money, the network itself is obligated to pay ASCAP only 41 cents on this particular bit of business, leaving the network a profit of 14 cents.
>
> In the case of the NBC-Red and Columbia the results would be somewhat different. With the frequency discount limited to 25%, the most that either network could deduct from the billings is 65%. Assuming that same station PQW were on either of these webs and rated at $100 per hour, the taxable residue would be $35. The ASCAP network fee of 2.75% applied to $35 is 96 cents. Under the circumstances the network's own disbursement for ASCAP music would be 41 cents while the station's figured at 55 cents. Broken down into percentages, the station would be paying ASCAP on the basis of 2.75% while the network would be paying its share on the basis of between 1.8 and 2%.

Unless ASCAP called for changes before January 1, 1949, the agreement would automatically extend after January 1, 1950. In addition, the society capitulated on the arbitration issue, but any fees arrived at through such a procedure could not be reduced.

The conclusion of the music war was a victory of which network officials were proud, but reluctant to discuss in detail the twilight zone and its problems. Instead, they pointed with pride to BMI's accomplishment, promising to support the organization until 1949. BMI was, they said, responsible for a saving of at least $10 million over the next nine years.

ASCAP publishers soon were responsible for virtually all the most-played radio numbers. Clearly, Sydney Kaye and the BMI stockholders would never again have exclusive hold on airplay. Many successful newly BMI-affiliated companies were flirting with the society.

An inept full-service music publisher, BMI's board of directors moved to run it exclusively as a licensing agency and encouraged the formation of independently owned BMI-affiliated companies. Many disliked how BMI's payment system was based on the program logs sent in by broadcast licenses. These were broken down into nine "sample groups," and the number of performances of a sample group was used to compute the total number of performances on all licensed stations.

The original two-cents-per-performance payment to publishers was doubled in time for the second-quarter distribution in 1941. There was disappointment, particularly in view of the large sums BMI stockholders had regularly paid ASCAP in the past. The payment system continued to be refined, and payments increased.

The reforms imposed on ASCAP by the consent decree provided ammunition for continual confrontations. The self-perpetuating board of directors was replaced with many of the same faces, chosen by the new weighted vote called for in the decree, based on income. Due to the costs of the radio war, the society's debits exceeded the $4.2 million in income by $800,000. Growing discontent among the publisher-directors with Gene Buck's lengthy tenure, since 1934, increased, and he was ousted, with the promise of a $25,000 annual pension.

Deems Taylor, composer of symphonic music and a network radio personality, was chosen as the new, unsalaried president. Soon after Claude Mills was retired by the board with the promise of a year's salary, provided he did not engage in any association inimical to ASCAP.

The basic structure of publisher distribution standards was unchanged since 1935: the 50, 30, 20 percent division based on performances, availability, and seniority. Required to do so by the consent decree, the publisher-directors overhauled the royalty classifications. A 5 percent increase for performances was made, availability remained fixed, and seniority was cut to 15 percent. The weighted-vote principle, based on income, was approved, and it resulted in a new, self-perpetuating board, favoring the highest-paid members.

The new distribution plan proposed fourteen classes of writers, ranging from AA to 4. Payments to the upper half were based on an application of 70 percent for "efficiency," meaning the "value" of each catalogue to the society as determined by the bookkeeping department, 15 percent for seniority, and 15 percent for performances. The logging of performances was based on daily network reports of all musical performances, supplemented by the Accurate and Peatman reports.

Despite attempts by the MPPA and the Music Publishers Contact Employees to scuttle the Accurate Reporting Service, it survived. During the music blackout, the Accurate sheet had dropped tabulation of BMI music, a gap that was filled by BMI's subsidization of the *Peatman Report*, which covered all performances. Its operator, Dr. John Gray Peatman, a professor of psychology, believed music publishers should acknowledge a "social psychology of popular taste" rather than the old trial-and-error technique of old-line music men.

In preparation for the new writer distribution system, all members were sent photostatic reprints of their ASCAP copyrights. The new plan called for modernization of the ASCAP bookkeeping department. Full-scale testing of the plan, started in 1945, was the subject of major disputes among various writer factions.

A 1942 FCC survey showed that music filled three-quarters of all air time. The average station operated 112 hours a week, 86 of them devoted to musical programs. Use of music to attract listeners increased. In 1944, $383.9 million was spent to buy time on the air. Network income had increased by 25 percent in a single year. The combined performing-rights income paid to ASCAP and BMI exceeded $7.5 million, more than half of it paid by affiliated stations because of the "Chinese bookkeeping" in the twilight zone practiced by the chains.

A revised writer-payment method was adopted in 1946. For the first time, all writers in the fixed-annual-royalty classes began to receive distributions related to ASCAP income, figuring on a point basis, which gave an opportunity to move up the economic scale. Many felt the move was made only to stop younger songwriters from swinging to BMI. Also Sydney Kaye had protested that his board of directors was ready to study a proposed plan to distribute performance royalties to writers of BMI-licensed songs for the first time.

BMI made successful overtures in late 1940 to the American Composers Alliance and its publisher arm, Arrow Press, and an agreement with the Associated Music Publishers for broadcast rights to the music of leading European serious-music houses. These brought BMI both an effective public-relations cachet and an important repertory of music.

Acting on the behalf of the ACA, Aaron Copland, its president, unsuccessfully attempted to win ASCAP membership for both the organization and himself. The society's agreements with European performing-rights societies did not include licensing rights to serious or symphonic music, leaving those to the American agents for the original publishers.

Copland unsuccessfully wrote Gene Buck to consider the problem. ASCAP believed that licensing this music would lead to serious problems with women's clubs and musical organizations, which would charge that imposition of fees "tended to stifle the development of audiences for serious music in America." As a consequence, Copland and the members of the ACA licensed their music through BMI on a short-term basis.

As BMI actively courted Copland and his associates, a major change was made by ASCAP. ACA was paid to undertake a survey of serious American music, and a special fund of $5,000 was allocated to promote it, so that at least one American work would be performed in every concert. ASCAP announced its intention to license concert halls in 1943, after financial projections indicated potential annual collections of nearly $150,000 a year.

Looking to expand its repertory of serious music, ASCAP met with representatives of European performing-rights societies. The first was the British Performing Rights Society, which dealt in nondramatic public performances of both serious and popular music. The initial agreement between PRS and ASCAP was made in 1929. PRS paid ASCAP half the money ascertained to be due for performances of the music it controlled. The other half was distributed to British co-publishers of

American music, which they again split equally with their partners in the United States.

In 1932, as their agreement was about to expire, PRS negotiated a new understanding that properly compensated performances of British music in the United States. Because ASCAP did not distribute its income solely on the basis of performances, PRS representatives proposed the payment of 6 percent of ASCAP collections of royalties based on actual network use. The second course would bring in about four times the existing payment. ASCAP opted for the second proposal, and PRS regularly received about 4 percent of the society's collections, after expenses, although it did not yet collect fees for its concert-music repertory.

A tentative agreement with the French society SACEM, for its concert music, was similarly negotiated. The new contract was signed in the spring of 1945 for five years on a straight per-performance basis, with no advances of guarantees. ASCAP thus acquired a valuable concert repertory, and it began licensing of concert halls and symphony orchestras in 1945.

By the summer of 1945, ASCAP unfortunately had collected only $20,000 from American concert halls. No method of distributing this income had been devised, but some of the one-fourth of ASCAP membership recognized as "serious composers" unsuccessfully suggested a multiple-payment factor for concert and symphonic music.

Finkelstein and Paine found further ammunition against BMI in the annual rebates given to broadcast licensees. This reduction of fees, made because BMI had operated at less than the maximum called for in its contracts, was first announced by Kaye during the crucial NAB meeting in June 1941. This payback became an annual practice. The BMI board continued to be made up of working broadcasters only, and its president remained the incumbent head of the NAB. Neville Miller, no longer viewed as the small stations' friend, was replaced in 1944 by J. Harold Ryan, of the Storer stations, who was followed in 1945 by Justin Miller, formerly judge of the United States Court of Appeals for the District of Columbia.

ASCAP's income increased, rising to $8.891 million in 1945. The society's defeat at the hands of the NAB and BMI still rankled. Intercession of the *Marks v. ASCAP* suit, seeking clarification of the "joint tenants in common" issue, was regarded as yet another example of duplicitous conduct on the part of Sydney Kaye and the BMI board.

To maintain a stable ASCAP, the SPA joined that action. Developments during Claude Mill's short tenure as general manager of the SPA underscored how major publishers as well as leading SPA officials and councilors had tremendous influence over the ASCAP board.

Shortly after his enforced retirement from ASCAP, Mills was named to the newly created SPA post. Sigmund Romberg, again president of the SPA, explained that with negotiations for a new publishers' contract still several years in the future, Mills would establish reciprocal relations with foreign songwriter groups. Mills intended aggressively to build the SPA so it could retrieve many of the privileges that authors and composers had surrendered in the approaching MPPA negotiations.

The full details of Mill's master plan for the SPA emerged only after he received his last check from ASCAP. Addressing a West Coast SPA meeting in 1943, Mills proposed that songwriters retain control of copyrights and receive two-thirds of all performance and synchronization rights, and that all contracts would require SPA approval. Some MPPA members charged Mills with "trying to upset the ASCAP applecart" and threatened a boycott of all music by SPA members. They were particularly indignant about Mills's suggestion that songwriters get half of the half-million dollars paid for lyric reprint rights by the three best-selling song-lyric magazines.

Romberg hailed Mill's proposals and welcomed a recent Supreme Court decision that found a writer could dispose of his renewal privilege at any time as well as ASCAP's anticipated victory in the Marks case, in which the SPA was also appearing with the defense. To finance the revitalized SPA, Romberg called for the tripling of annual dues, which would raise those of AA writers to $200, only about 1 percent of their performance royalties. But it was from this group that the greatest protest came, fed by publishers. Five SPA councilors, all of whom were also ASCAP directors, opposed Mill's plan, went behind Romberg's back, and negotiated an end to the expected strike. During a special ASCAP board meeting in 1943, the directors unanimously agreed to halt the checkoff system for SPA dues and extend ASCAP membership for an additional fifteen years, ending in 1965. Thereby, no matter what the outcome of the Marks suit was or what reforms the SPA won in a new MPPA contract, all its members, now committed to the society for twenty-two years, would have to share ASCAP royalties on the fifty-fifty basis until then. Within three months, the extension agreement was overwhelmingly approved. In January 1944, Mills resigned as the SPA general manager, and the SPA lapsed into inactivity.

Clarification of the ownership of music on television reappeared as a secondary issue in early 1945 after an appellate court held that *Marks v. ASCAP* should proceed. The presiding judge in that action, Ferdinand Pecora, asked all parties to negotiate. This was rejected by the ASCAP board when BMI offered to license the disputed songs and some 3,000 others in the Marks catalogue on the same basis as it did music written by members of the society.

Edward Marks did not appear as a party in the action, and BMI argued that once a writer assigned his copyright to a publisher, the latter, not the writer, could then assign administration of any portion of the copyright privileges to a third party. ASCAP's position was that, as a member of the society for many years, Edward B. Marks Music Co. never disputed its writers' grant of the administration of public performing rights to ASCAP, and in accepting the equal division of license fees with them would not now assign their administration to BMI on behalf of all parties.

Pecora's verdict, on May 1, 1945, in *Variety*'s words, "threw a bombshell into the music business," giving ASCAP "what's deemed as the most important victory it has ever won in court." It stated that the relationship between a songwriter and a publisher was "basically a joint venture for the commercial exploitation of the performing rights to the songs." The publisher's title to a copyright was "subordinate to the joint venture," the publisher holding title "to that end that the exploitation of the songs might be more advantageously achieved. He is in this respect merely a trustee of ASCAP and its [writer] members." Marks was faulted for having negotiated with BMI without consulting his writers. With the Marks understanding soon up for renewal or the exercise of an option to buy the catalogue for a second million, Pecora deplored BMI's conduct.

Younger, progressive SPA members believed Pecora's surprise ruling proved them to be the sole owners of performing rights. If this might be true, it would seriously affect not only television licensing but also control of film-music catalogues. Many music-business attorneys thought that the decision could be upset if Marks was able to prove that he had "worked the copyrights" by licensing them for recording, publishing, and merchandising sheet music, and other printed music.

Shortly after the Pecora decision, a group of leading Hollywood film-score composers formed their own protective association, the Screen Composers Association, with Max Steiner as president, and began to agitate for distribution of ASCAP income. The staff songwriters employed by the studios had little voice in ASCAP's operation until the consent decree strengthened their position by its requirement that at least 80 percent of the writer vote, based on recent distribution, was necessary to pass changes. But because film music was not counted in establishing classifications, the screen composers were given the lowest ratings.

Studios made one of two deals to get a new film score or set of songs. Major composers retained their copyrights and administered them through their own music companies or publishers to whom they were under exclusive contract. They arranged too for the outright sale of the music, or a flat-fee payment plus a share in sheet-music and record royalties. In both cases, the copyright was vested in the studio's name and then turned over to a subsidiary music house, from which the company received a sheet-music royalty of between two and six cents and half the publisher's record income.

ASCAP publishers were divided over whether music should be licensed to television as a grand right through the Harry Fox Agency or as a small right, available through ASCAP. The society's plea to publishers for a grant of television rights was accepted by most organizations, except the MGM music firms. The proposal was amended to read that authors and composers would retain the right to restrict the use of their music on television. The Hollywood-owned companies objected and in 1946 won.

The significance of this surrender was not lost on the SCA attorney Leonard Zissu, who reiterated that his clients should receive a share of collections from movie-house owners and operators. ASCAP would soon license television for the use of feature films, of which SCA members' music was a vital ingredient. The screen composers were urged to be patient, until after ASCAP had accumulated all television rights and devised a new method of collection.

During the following twelve months, Zissu and the SCA prepared to obtain a greater share of the ASCAP collections from film houses, as well as an improved standard contract between producers and screen composers. Income would be divided solely among those who were responsible for it, not, as with ASCAP, dumped into a single pool. The SCA asked the studios for the retention by the composers of all rights other than synchronization use, except when the film makers' own music firms worked on the music or issued it in printed form. In addition, they would forbid the use of music in more than one feature movie without full compensation each time. Shortly before the SCA board gave Zissu approval to proceed in late 1947 it was undercut by a new power struggle within ASCAP's executive management following the sudden death of John Paine.

After intense pressure from BMI, as well as representatives of the Music Machine Operators Association, the government filed a civil action in June 1947 which charged that ASCAP's exclusive licenses with members of CISAC constituted a monopoly. The society, it was charged, had conspired with similar societies in all the principal nations to prevent other licensing bodies from having access to their music. ASCAP unsuccessfully attempted to persuade CISAC to forbid any member body from doing business with an international society that did not itself belong to the confederation. This rule made it impossible for any CISAC member to deal with BMI.

Even before ASCAP replied to the government's charges, it demanded CISAC resign at the confederation's first full-scale postwar meeting. The resignation was accepted with regret, but CISAC refused to budge from its position. Negotiations between ASCAP and the Justice Department began. ASCAP agreed to sign a decree that would wipe out any exclusive agreements with foreign societies, but only if BMI was required to do the same.

In March 1948, ASCAP filed its formal answer, in which it charged that BMI had been formed to crush the free enterprise that ASCAP represented, and now threatened to use "the same tactics to embrace the entire world . . . to destroy" CISAC and its members, and make the world's composers and authors "subservient to the American broadcasting industry." BMI, ASCAP charged, was seeking to destroy the competitive system. Unless BMI was made a joint defendant, ASCAP unsuccessfully argued that the complaint should be dismissed. The case continued.

In 1948, the government pursued a revised ASCAP consent decree despite continuing fights between the society and the broadcasters, personified by BMI, and in the relationship between ASCAP and its writer and publisher members and ASCAP and its film-theater-owner customers.

BMI's poor showing as a music publisher, when only one BMI-licensed song appeared among the 1945 twenty most-played songs and four among the top thirty-five sheet-music sellers, was due in part to a change made in the traditional measurement of "plugs" secured on New York radio stations and reported by the Accurate Reporting Service and in *Variety*. The once-official Accurate sheet was replaced by Peatman's surveys by ASCAP publishers in order to end "certain music industry evils." Emphasis on radio plugs was for "quality" rather than quantity, because Peatman based his findings on "audience coverage," which was greatest during prime-time network hours, rather than the daytime sustaining shows and after-eleven-o'clock big-band remotes, on both of which payola had been running rampant.

With its share of radio-music use hovering around 10 percent and the cost to broadcasters for music far above that of its competitors, BMI began, according to *Billboard*, to placate its best-paying customers by giving them "everything for free, but music." A staff of five field representatives constantly visited stations which were "bombarded with catalogs, brochures, bulletins, indexes, lists, copyright info, clearance data, and ready-to-use radio scripts."

In order to forestall any abandonment of BMI by broadcasters on the termination of its contracts in 1949, Sydney Kaye played on ASCAP's reluctance to disclose its plans when its agreement expired at the same time. That contract would be renewed automatically for nine years provided an increase was not requested. If one was, however, the option was canceled, and if the two parties could not agree on new terms, arbitration would follow. The society traditionally shunned arbitration, but offered it as an added inducement to end the 1941 impasse, with the proviso that any new rate would not be lower than the present one. Nevertheless, raising the possibility of a second ASCAP boycott, Kaye reminded broadcasters that BMI had made it possible to effect vast savings.

ASCAP's vague intentions sprang from continuing failure to agree on television licenses. More than 80 percent of the membership had assigned representation of these rights to the society, until December 31, 1948, with only MGM's music firms holding out. The major film studios continued to have ambivalent feelings about giving television-station owners the right to use their feature films through an ASCAP contract. ASCAP therefore formed a separate Television Negotiation Committee to meet with the networks and other parties.

Speaking at ASCAP's 1947 membership meeting, John Paine indicated that the $6 million out of a gross income of $9.9 million paid by radio for 1946, the largest sum yet given by broadcasters, was one reason ASCAP would not seek a higher rate after 1949. Income from the screen theaters was only $1.5 million, raising the certainty that the society would soon be asking for more.

With the major burden of the networks' music fees on the backs of their affiliates, the chains were satisfied with the status quo. Further, they wished no more contention with ASCAP, and many upper-echelon executives privately questioned the propriety of continuing the NAB-BMI-broadcasting link, from which NBC and CBS partially extricated themselves by resigning from the NAB.

It was precisely that large group which welcomed Carl Haverlin's appointment as BMI's first full-time, paid president in 1947. Haverlin announced a change of policy: BMI would operate more actively as a publisher, but would not disturb the work of its Tin Pan Alley affiliates. Negotiations were concluded for a takeover of Associated Music Publishers' remaining capital stock, and all publishing and performing rights for a group of fifteen European music houses. Soon after, the British house of Boosey & Hawkes resigned from the AMP, to join the PRS, taking with it the music of many major British composers, but the acquisition represented another saving for stations that used concert music. It was also a valuable gain in the company's international position, particularly in the face of CISAC's alleged boycott.

Meetings between ASCAP's Radio Committee and broadcasters continued. At the annual NAB convention, Theodore Streibert, of Mutual, and head of radio's negotiating committee, assured the meeting that ASCAP did not intend to ask for increased fees from radio. Many present were surprised by Streibert's recommendation that the NAB divorce itself from BMI. Such a move, he pointed out, could only ensure "more kindly business relations" with other music licensors.

NBC, CBS, ABC, and Mutual announced they were ready to extend their present BMI contracts until 1959, to run concurrently with any new ASCAP agreement. Streibert's assurance that the society would not ask for an increase led the networks to feel that BMI should follow suit. Unlike ASCAP, which charged one dollar a year for television rights, BMI had granted them to all stations in its initial contracts, on the same terms as for radio—a percentage of revenue received from the sale of time on the air—a pattern the networks wanted ASCAP to offer.

In November, BMI took advantage of the network support and asked all its broadcasting licensees for a similar extension to 1959. Haverlin pointed out that it was presently impossible for BMI to enter into contracts with songwriters and music publishers for longer than twenty-eight months, while ASCAP was under no such restraint. An overwhelmingly favorable response was indicative of Haverlin's popularity.

The network's move was not guided by charity. ASCAP's current assignments of television rights ran only through 1948. Because of pressure from the AA Broadway musical-production writers, as well as a threat from the Screen Composers Association that it might unite with the Screen Writers Guild, the ASCAP board intended to license television in the future on a year-to-year basis until the medium developed its own music programming patterns. However, production writers and screen composers were reluctant to underestimate their performing rights in this realm.

In addition, the networks feared the results of the rapidly widening confrontation between ASCAP and independent film-theater owners, sparked by the announcement in 1947 that the society would increase taxes on movie-house seats. Within a few weeks, a long-dormant antitrust suit, *Alden-Rochelle v. ASCAP*, initiated in 1942, was resumed by 164 independent theater-owner plaintiffs. It sought triple damages and an injunction restraining the society from charging for music-performance rights.

Persistent antagonism between theater owners and ASCAP increased after the advent of talking pictures, when control over the music heard in movie houses passed from the operators to the studios. In 1937, Claude Mills proposed "clearance at the source," with all ASCAP fees from film houses to be borne by the Hollywood studio suppliers, but was quickly rejected by the MPPA. More than a third of the publishers' participation in ASCAP dividends was being paid to subsidiaries owned or controlled by major motion-picture producers. In 1947, that figure amounted to $1.432 million, at a time when many major studios were expanding their ownership of music-publishing companies, and had no objection to raising the rates ASCAP charged independent theater operators.

Immediately after losing a 1947 motion to dismiss the Alden-Rochelle suit, ASCAP met with the motion-picture companies to ask them to assume the seat tax and to tack those fees onto rental charges, only to be abandoned following John Paine's premature death.

In the lull that followed, Haverlin asked all independent broadcasting licensees to extend the existing BMI contract so that it would expire concurrently with the networks' agreements in 1959, together with ASCAP's still-to-be-formulated contract. Within a month, 800 stations had agreed; by year-end, nearly 1,000.

Following Paine's death, his duties were temporarily divided among Herman Finklestein, house general counsel; Richard Murray, who handled the theater-seat licensing; Herman Greenberg, a Paine assistant; and George Hoffman, who dealt with finances. In an immediate three-way rivalry to succeed Paine, Greenberg made a bold bid for publisher support and the job. He recommended an immediate 300 percent increase in seat taxes, to which Finklestein objected, warning that it would backfire and cause litigation and government intervention. He proved to be correct on both counts. Alden-Rochelle was to begin trial in the spring of 1948.

The music publishing business to which BMI returned, with Haverlin's presidency, was undergoing a dramatic change owing to its influence as a licensing organization. After sixteen years, the record industry had matched and bettered its previous all-time high sales mark with $214.4 million retail ($97 million wholesale) during 1947. There were an estimated 200 record companies registered with the American Federation of Musicians, plus uncounted independents. The logical medium for promoting and exploiting their products was the independent radio stations.

In 1948, FCC Chairman Wayne Coy predicted that there would soon be 3,100 AM stations in operation and 1,100 FM facilities. Seventy million radio sets were already in use, and each family daily tuned in an average of four hours and nineteen minutes. Coy added that the expected competition from the 100 television stations in operation could not possibly reach more than one-eighth of those homes for another two years.

The growth of music publishing and its income from public performance matched that of the recording industry and broadcasting. ASCAP had 300 publisher members, and BMI nearly 1,000 publisher affiliates. The vast majority of the latter were small operations, which published their music on phonograph recordings. Approximately 75 percent of the society's collections from broadcasters was distributed to writer and publisher members; 58 percent of BMI's income to affiliated publishers, foreign societies, and writer-publishers.

The average sheet-music sale of a hit song was about 350,000 copies. Many were "made" by disk jockeys. The proliferation of record spinners and their increased playing of "canned" music caused the AFM to pull out of the recording studios for a second time, all through 1948, during which record sales slumped. Still, disk jockeys' influence was incontrovertible, as surveys indicated that their airplay now motivated 85 percent of all record sales.

Throughout Haverlin's first year as president, BMI enjoyed little prominence on *Variety*'s disk-jockey chart or its Peatman survey, as either a licensing organization or a music publisher. Fully committed to securing industrywide renewal of BMI's contract until 1959, and maintaining that support, he created a new "muscle-flexing" statistical base to emphasize the company's apparently increasing share of the popular-music business. To increase BMI's representation on *Billboard* charts, unaffiliated publishers and record companies were informed that BMI paid for recorded performances of the music it licensed. BMI offered them a cash advance against earnings, credited against a base of four cents for each local airplay.

The proliferation of radio stations since 1941 created a financial problem for BMI as the 1940s ended. Its income from broadcasters had doubled. At the same time, the number of publishers to whom it had guaranteed advanced payments against performance income went from 25 to 125, nearly all of them firms whose music was chiefly available on recordings and initially played on non-network local stations. With the disk-jockey charts being the current key to the true value of a performance, BMI dropped its forty-eight-dollar "payment-by-plug" agreements and favored recorded airplay in major markets. The basic payment for local-station performances had been raised in 1942 to four cents, and to six cents per station for network performances. Although eighty-three new publisher affiliates joined BMI during 1949, most received contracts on a straight performance basis, the "open-door" policy with which BMI gave substantial financial encouragement to promising new music ventures.

For the first time in nine years BMI-licensed music was prominent on all the charts. This did not effect BMI's income, which was fixed by contracts whose terms went back to 1940 and had been extended through 1959. Little of its music was used by the radio networks or on television. Such popular television programs as Milton Berle's *Texaco Star Theatre* and Ed Sullivan's *Toast of the Town* depended on traditional theatrical routines and its ASCAP music. ASCAP's income from all sources was nearly three times that of BMI and mirrored the use of its repertory by the networks and their affiliates.

However, Haverlin's task was facilitated by the public's newly acquired taste for the music in which his company dominated, exemplified in early 1951 with the most successful song in twenty years, "The Tennessee Waltz," published by the Nashville firm Acuff-Rose Publications. The company was owned by Fred Rose and Roy Acuff, who lent his name to the venture, leaving Rose to secure an advance from BMI after ASCAP refused his offer of an inroad into the Nashville scene. By 1947, when his company's annual income from BMI had doubled, Rose had built up a roster of songwriter-performers that included Hank Williams, with whom Fred Rose formed one of the most successful music-publisher-editor-collaborator and songwriter relationships in the history of American popular music.

All of the big-city writers left ASCAP, chiefly because it had mistakenly expected that its publisher affiliates would share performing-rights income. During these years, it regularly announced that a writer-payment plan was under consideration, but not until August 1949 did it move toward that goal. In the next year, twenty writers were signed by BMI as affiliated writers, for terms averaging seven years, with a sliding scale of payment and a $400 maximum annual guarantee, based on the number of performances expected from each. It was hoped that these writers would achieve parity with ASCAP members who had similar success and airplay. However, since a BB ASCAP writer earned $10,000 a year and those in AA double that, BMI did not expect even its most successful writers to achieve such royalties. The society, meanwhile, was apprehensive, plagued by the pending negotiations for radio and television contract renewals, the impasse between the MPPA and the SPA over a new agreement, and resumption of the *Alden-Rochelle v. ASCAP* action.

A historic and potentially disastrous decision in that case found that the society's theater-licensing constituted a monopoly operating in restraint of trade. On July 20, 1948, federal court judge Vincent H. Leibell ordered ASCAP immediately to divest itself of all public-performance rights dealing with motion-picture exhibition, and return them to the original copyright owners. In addition, he restrained the society from obtaining those rights from its writer members and from refusing to permit these members to grant them to film producers directly.

Although the SPA had not intervened in the case, in October 1948 one of its leading members, Milton Ager, represented by the association's general counsel, John Schulman, asked Judge Leibell to reverse himself and permit ASCAP to represent composers and songwriters in licensing screen-theater performances. Failing that, Ager pleaded that such rights revert to the songwriters. Ager's plea frightened publishers. The pending SPA contract already provided that writers were to receive 50 percent of income from all licensing rights, which would include any that might be sold individually in the future. Herman Starr arranged a meeting between leading SPA members and publishers, which ended abruptly when the former asked that they be given a "proprietary interest" in copyrights for the future co-administration of all rights. The request, viewed as part of a secret SPA blueprint to win control of both ASCAP and the music business, was rejected. Max Dreyfus, publisher of the Ager compositions in question, promptly joined the action on behalf of all publishers.

The Ager and Dreyfus motions and an ASCAP brief arguing that the divestiture of theater-performance rights was not in the public interest postponed any resolution. Leibell denied all motions and directed ASCAP to permit licensing of performing rights to motion-picture producers concurrently with licensing of recording rights.

Leibell's decision raised the possibility that composers and songwriters might capture the right to deal directly with broadcasters and film companies, leading to other reforms that the SPA had been seeking, and thus lead to the destruction of ASCAP. Neither wanted this solution. It would create many small groups of copyright owners, from whom performance licenses would have to be obtained individually. Wise heads at ASCAP decided against an appeal of the Leibell decision, and met with the Justice Department to seek a modification of the 1941 consent decree that would offset Alden-Rochelle's effects.

As ASCAP lawyers worked to offset Judge Leibell's decision, several items buried in the trade-paper news proved to be equally as noteworthy. In Memphis, a young Elvis Presley got up enough courage in late 1953 to enter the offices of the Memphis Recording Service to make an acetate record for his mother. Songwriters Max Freedman and James DeKnight persuaded a small rhythm-and-blues recording company to cut a master of their song "Rock Around the Clock." The record received hardly any measurable jukebox or airplay. Todd Storz, looking for a way to control the selection of music on his radio station, KOWH, Omaha, watched customers drop nickels in the jukebox. The frequency with which certain jukebox hits were selected impressed him and inspired the creation of "top 40 programming."

Speaking at the Grand Ole Opry celebration in Nashville, Carl Haverlin spoke of his conviction that "in seeking its own level . . . an amalgamation of music styles has begun, and it will end with American music finding its own unique character."

7 The Merchandising of Television and Accusations of a BMI-Broadcasting Conspiracy

MANY ASCAP members acknowledged the potentially serious developments of the Alden-Rochelle case, including the distressing decision and the SPA's aggressive role. As Abel Green wrote in *Variety* in December 1948, "The curious secrecy about some of the legal negotiations through the years have made every [ASCAP] faction's lawyers wonder what somebody might be getting—or getting away with. Thus, Judge Leibell has done one affirmative thing for ASCAPers. The talk about what 'they' did up at ASCAP has prompted the shirkers into workers."

At the same time, the final decree in ASCAP's negotiations with the major film studios to recoup the $1.25 million in theater-licensing fees gave either ASCAP or individual copyright owners the right to be paid at the source by film companies. Since April 1948, the studios had continued to negotiate with publishers for synchronization licenses, coupling them with permission to perform the works publicly, payment for which would be made when a blanket license was negotiated by the society with the entire film industry.

ASCAP reportedly was asking for half a million dollars annually from each major film studio and individual producer. The consent decree called for a resolution by March 4, 1952, and the film industry was in no rush. In the spring of 1951, the studios suggested that a blanket license for public exhibition include television fed to theaters. Convinced that future income from commercial television would exceed radio, the ASCAP negotiating committee rejected the proposal. ASCAP indicated that, without an agreement, it would be forced to divorce itself of all film-performance rights.

With the coaxial cable nearing completion, it was believed that at least 80 percent of all programs would soon be on film. Eager advertisers could reach families during evening prime time through film. Network affiliates also preferred the medium, because their largest income came from "hitchhiking," the sale of local advertising before and after network broadcasts.

Faced with such competition, the major film companies offered ASCAP a sum representing less than 0.05 percent of motion-picture theaters' annual income of $1 billion in ticket admissions alone. BMI, on the other hand, offered its music without charge to theater exhibitors. In early March 1952, ASCAP and the major studios forged a blanket license settlement, representing about $500,000 annually plus $1.4 million for retroactive performance fees, although only Universal Pictures actually signed the agreement. In August 1952, ASCAP and BMI issued monthly bills to theater exhibitors for the use of recorded music during intermissions.

Such scant collections were of little interest to the songwriter and publisher associations, who presented a third new standard songwriters contract in 1946 to the MPPA and Herman Starr, who represented powerful non-MPPA firms. It called for a full half-share of the standard two-cent royalty on recordings, regardless of any compromise publishers might make with the record manufacturers, and a sliding-scale sheet-music royalty. All proposals were rejected by the publishers, and the existing contract was renewed on a monthly basis, while MPPA representatives met regularly to work out a mutually satisfactory new one.

For much of 1947, Tin Pan Alley witnessed a serious slump in sheet-music sales. The withholding of a recording until the publisher was ready to work on the song was ending. Younger songwriters found a receptive audience in successful A & R men. With a release assured, they could make a beneficial financial arrangement and then publish the song themselves, using an established house to handle sheet-music distribution.

In the summer of 1948, the SPA won a straight minimum three-cent royalty on all sheet sales or a sliding-scale payment, beginning with two and a half cents for the first 100,000 copies, going to a maximum of five cents for all copies in excess of 500,000. In addition, the publisher was required to pay a minimum royalty of 10 percent on all printed music, other than sheet copies, sold domestically. Gradually, other gains were made by the SPA. Within a year after he acquired a new work, the publisher was obliged to issue piano copies, and to secure a recording or issue a dance orchestration, or else return the copyright and pay a minimum penalty of $250. After the initial copyright period of twenty-eight years, all rights returned to the writer. Only ASCAP was specifically designated as custodian of performing rights for SPA members, making it impossible for BMI-affiliated songwriters to join.

By the summer of 1950, 289 publishers accepted the new terms, a few Hollywood-owned firms holding out until ASCAP's status was cleared by the amended consent decree that stemmed from the Alden-Rochelle action. Of equal concern to the film companies were the protracted negotiations with network-affiliated television stations. ASCAP could not license television rights until they were assigned to the society by the publishers, who had first to obtain permission to do so from composers and authors.

While a new television contract was negotiated, the dollar-a-year fee remained in effect, but stations had to clear every piece of music in advance with the society for every single performance. ASCAP retained the right to revoke this license on thirty days' notice anytime a majority of its membership wished, or to impose restrictions on the use of ASCAP music on dramatic shows. In October 1949, the ASCAP board decided that television must begin paying for music. ASCAP's fourth president Fred Ahlert negotiated a five-year blanket license agreement with NBC, CBS, and ABC, retroactive to January 1, 1948. The terms were roughly about 10 percent more than those for radio. A new 25 percent deduction was added to subsidize the cost of the coaxial cable to the West Coast.

Several problems remained, among them extension for an additional three years of the present licensing assignment by writers and publishers, and a per-program contract that would satisfy the vast majority of independent station owners who found the costs too high and questioned whether ASCAP had the right to collect fees from them for the use of motion pictures. The networks agreed to pay for movies on agreement that copyright owners would not be paid "at the source."

The NAB founded the All-Industry Television Committee and intended to take the matter to court. In March 1951, ASCAP mailed all television stations a contract that proposed an 8.5 percent commercial rate and a 2 percent sustaining charge for stations doing an annual business under $150,000 and 9.5 percent and 2.5 percent for those grossing more than $300,000 annually. Because ASCAP based the fees on a one-time advertising-card rate, rather than on actual station income, the offer represented a 100 percent increase over the AM radio rate. Simon Rifkind was retained by the committee to resolve the case under the terms of the amended ASCAP consent decree, which enabled music users to take their complaints to the New York court for resolution. Acting on behalf of all independent television stations, in July 1951 he asked the U.S. district court in New York to set a fair music rate.

Discussions with the All-Industry Radio Committee foundered on the rocks of cooperative network programs, those that emanated from the networks as sustaining broadcasts, but on which affiliated stations sold time to local sponsors. ASCAP insisted that these programs were commercial shows, to be paid for by local stations at the higher network rate. The stations argued they were purely local commercial broadcasts. In May 1953, ASCAP agreed to a 2.25 percent rate for cooperative programs over stations with a blanket license, and 8 percent for those using a per-program formula.

The amended consent decree permitted criticism of the ASCAP distribution system, about which many complaints were received, particularly from the group known as the "Young Turks," who opposed any method that favored those highly rated old-guard writers of old-fashioned songs. They wanted a formula that placed maximum emphasis on their music. Sigmund Timberg, chief of the Justice Department's antitrust judgments enforcement section, took up revision of ASCAP's writer classification and distribution system. The old system called for fourteen classes, ranging from AA to 4, and allotted less than 15 percent of total income for performances. Changes were called for in the current publishers' distribution plan, which allotted 55 percent for performances, 30 for availability, and 15 for seniority.

One of the directives in the amended consent decree required that future ASCAP distribution be made "on a basis which gives primary consideration to the performance of the compositions of members as indicated by objective surveys of performances . . . periodically made by or for ASCAP." How this was to be accomplished was left to the society. Other provisions were more precise. Judge Leibell was upheld in every particular regarding motion-picture music licensing. ASCAP was denied exclusive contracts with international music-licensing societies. Judge Pecora's decision in *Marks v. ASCAP* was validated. Although an ASCAP member could withdraw his catalogue on proper notice, he could not license his music until all existing ASCAP contracts with music users terminated. All groups of similarly situated music users were entitled to uniform rates, with the right to take any dispute to a federal court for final determination. The decree did away with the self-perpetuating ASCAP board by requiring elections every one or two years. Membership eligibility was liberalized so that songwriters who had at least one work "regularly published," or a firm whose musical publications had been distributed for a year, could join the society on a "non-participating or otherwise" basis. In sum, the decree permitted court regulation of those practices and activities that appeared to be in the public interest.

Due to the Leibell decision, ASCAP income for 1949 fell. The board believed that the loss would be more than covered by direct licensing of the film studios, screen theaters that played live and recorded intermission music, and railroad and airline terminals where ASCAP's repertory was used for background music. Ahlert predicted that television's recent growth would mean an additional million dollars for 1950.

The new distribution plan calculated quarterly payments on three factors; the most important was a performance average for the years 1945 to 1949, which would account for 60 percent of the distribution. One-fifth of this, or 12 percent of total writer money, was to be set aside for allocation to the composers of works in the standard- and concert-music fields, whose performances could not be measured by the society's logging operation. Twenty percent would be determined on the basis of current performances during the past year, and the final fifth by a combined seniority and availability factor.

ASCAP's thirty-six-year-old classification system would base 60 percent of the writers' money on five-year performance averaging that calculated writers' ratings. These would be graduated by increments of twenty-five points, from one for Class 4 writers to between 775 and 1,000 for those in AA. Promotion or demotion would be automatic, with safeguards against too rapid a fall.

To augment the new system, the ASCAP board installed a new logging program. Network performances continued to receive most attention, but an effort was made to catch ASCAP music on stations that concentrated on recordings. Independent stations in ten key markets were logged daily for three hours, on a rotating basis. Performances were evaluated on a point basis: one point for a network shot; one-tenth of a point for theme songs and jingles. Concert works of more than five minutes and choral pieces of more than three minutes received six points per performance. Bowing to the Screen Composers Association's protestations, background music used in motion pictures was credited with one-tenth of a point for each performance.

The task of mollifying the writers would fall to ASCAP's new president, seventy-seven-year-old Otto Harbach, while the SPA's chief attorney John Schulman aggressively lobbied for all authors and composers. In 1950, the SPA unsuccessfully attempted to introduce a new secondary standard renewal contract, with increased rates and guarantees, for songs entering the second twenty-eight years of copyright protection. Prior to the institution of the SPA uniform contract in 1932, the major old-line publishers had insisted on the right to renew all copyrights in their own names, with the writers' prior consent. Acceding to SPA terms, most publishers offered a renewal contract that raised the mechanical rate from the 1920s' one-third share to a half, and sheet-music royalties from the former standard two cents to four or five. However the new plan only benefited the highest-ranked writers. Nearly one-third of ASCAP's members had received a smaller check, while the twenty AA writers' income doubled.

The 60-20-20 plan had been submitted to the government and subjected to a three-year test period, but showed no marked difference in the writer distribution. The "Young Turks" insisted on an end to the cushioning of the special top class and demanded the open election of directors. Given a voice in the process for the first time, the young Turks concentrated on a fruitless write-in voting campaign. The election was a clean sweep for the incumbents. Harbach, the only candidate agreeable to all parties, was re-elected president.

The 100 highest-rated ASCAP writers continued to receive 56 percent of the income. Governmental intervention had brought writer distribution closer to the 55-30-15 formula used by the publishers since 1935, with the greatest emphasis on live prime-time commercial radio, and now television, airplay.

The publishers suddenly found themselves confronted in early 1952 with a crisis which threatened to overturn their distribution plan: 55

percent for performances, 30 for availability, and 15 for seniority. Ralph Peer took advantage of the plan's appeal machinery and sent a representative before a newly organized classification arbitration panel on behalf of the major ASCAP holding, Southern Music. The committee found for Peer, raising Southern's availability from 250 to 450 points. It also recommended a broad revision of the availability ratings of all publisher members. This revived a proposal that many of the large firms had long advocated, a straight 100 percent performance payoff, or at least an increase of the performance factor to the 60 percent level of the writer system. In response, ASCAP's executive committee and legal department prepared a new system acceptable to the younger publishers.

The new system, devised to measure the important publisher availability factor, became effective with the October 1952 distribution and disrupted the status quo. It rated the availability of every song more than two years old on the basis of its radio and television performances during the previous five years, or eight quarters. For the first time in ASCAP history the value of each work was reckoned in terms publishers understood. Performance, availability, and seniority were retained. Under the new plan, the powerhouse firms dominated the distribution. Middle-sized firms viewed the new system as an opportunity to grow by securing more airplay. The standard-music houses and new small companies suffered, for they could not increase their availability ratings or their ASCAP earnings until their copyrights went beyond the eight-quarter limitation.

Damaged by distribution that effectively allocated 85 percent of all payments on the basis of broadcast performances, the standard-music publishers had two options. They could fight for an increase in the value of performances of classical, standard, and art music, or they could form BMI-affiliated companies. Many of their best-known younger composers, among them William Schuman and Roger Sessions, moved to BMI, which supported modern concert music. BMI took over all performance licensing on behalf of the American Composers Alliance and paid its members advance guarantees against earnings, generally out of all proportion to actual collections. The ACA library of manuscript scores and parts and its Composers Facsimile Editions were housed with BMI. Tapes and recordings of ACA members' works were sent to radio stations and the conductors of symphony orchestras. Recordings and performances of ACA and other BMI-licensed music were subsidized.

Composers of "serious" American music expected higher income. However, ASCAP's collections from symphony orchestras, concert-hall managers, and local impresarios demonstrated there was little profit in licensing them. The society collected from the country's symphony orchestras for blanket access to its music on a descending scale, while the two major concert-artist management organizations collected a fee of 1 percent on behalf of ASCAP from performers and local entrepreneurs. BMI did not yet license either symphony orchestras or concert halls.

With such small returns, ASCAP subsidized its serious composers by dipping into the availability fund to compensate for the "cultural importance" of their music. When 85 percent of distribution was measured by performance and the value of works between thirty-six and forty minutes in length was fixed at sixteen points, the society's serious composers found their checks smaller than ever. In October 1952, the value of such "unique and prestigious" works was raised to forty-four points. When the annual guarantees offered by BMI to several ASCAP members surpassed the society's fixed budget, few resisted BMI's blandishments.

The second modified ASCAP writer-payment plan was accepted by the Justice Department in June 1952. The proposal was the handiwork of a committee headed by Stanley Adams, with suggestions from several Society critics, one of them Hans Lengsfelder, a Viennese composer who had joined the society in 1942 and became a leading spokesman for the "Young Turks."

This committee changed the 60-20-20 three-fund allocation to a four-fund formula. The 60 percent performance factor was split into equal parts, a sustaining fund and an availability fund. The remaining 40 percent was divided equally into current performance and accumulated earnings funds. Writers had the option of either a ten- or a five-year-basis for the sustained performance rating, and availability was cushioned against a drop in income, remaining fixed for five years, which in effect placed major emphasis on old music.

With at least half the membership now expected to receive higher payments, ASCAP members overwhelmingly approved the scheme. The new distribution method actually increased the take of top writers in the AA group, fueling the "Young Turks'" anger.

In October 1952, Harbach predicted that the society's income for 1952 would reach an all-time high of about $15 million. In spite of his wish to retire, he accepted a third one-year term. Harbach was considered by the membership as the best man to present ASCAP's case in Washington, where hearings were being held to change the law so that the operators of jukeboxes would pay for the music they used.

Music publishers had shown little interest in having the coin-operated phonograph business made liable for payment when it used recorded copyrighted music, unless admission was charged. They felt the machines were a chief means for promoting their music and should not be touched. In response to the jukebox's popularity, the popular-music business campaigned to change the 1909 copyright law in several particulars, chiefly compulsory licensing and the two-cent record royalty. In 1940, ASCAP actively supported a pending bill expected to yield a fee of five dollars a month to all coin-machine operators without charge and urged their cooperation in fighting the society's "practical monopoly of popular music."

In the 1951–52 Congress, Representative Hugh Scott introduced a bill removing the jukebox exemption, only to have its place on the agenda taken away by the Kefauver-Bryson bill, which exempted the owner of a single jukebox, but required all others to obtain a license and pay a weekly penny royalty on every copyrighted work inserted into their machines.

Testifying against the bill, Mitch Miller stated that the coin-machine operators were chiefly responsible for creating hits. His bold assertion that "some of the most successful" songwriters bypassed the music companies and went directly to the record companies to place their material negated much of the MPPA's testimony regarding the important role of the music publisher. Such contradictory testimony allowed the committee to reach no conclusion or pass any legislation.

In 1953, Stanley Adams, an attorney before he became a working songwriter in the early 1930s, replaced Otto Harbach as ASCAP president. His participation in the 1941 antitrust lawsuit and organization of the Songwriters of America put him in the forefront of the reformers. Fred Ahlert, an ASCAP vice president, who participated in previous copyright hearings, was put in charge of the presentation of the Senate Subcommittee on Copyrights.

In July, Sydney Kaye of BMI threw what *Variety* dubbed a "Kaye-bomb" into the proceedings by joining ASCAP in the copyright fight. This marked the first time BMI took an active stand in support of any ASCAP position, and it broke its long cooperative association with the MOA. No new bill emerged from the deliberations.

The sudden reversal of BMI's position on jukebox exemption was due to the equally dramatic reversal of BMI and ASCAP in their position on the nation's coin-music machines. In a poll of all subscribers to *Cashbox,* a trade paper originally intended for the coin operators, but becoming a successful competitor of both *Billboard* and *Variety,* BMI music was put in an 81.8 to 18.2 percent position over the ASCAP repertory in the pop, country-and-western, and rhythm-and-blues categories currently most popular on America's jukeboxes.

Such an astonishing share of the market was first forecast in October 1951, when BMI-licensed music took their first three places on *Billboard*'s "Honor Roll of Hits," as well as the sixth and ninth positions, its best showing since 1941. Four of the five were "legitimate pop songs," firm evidence, according to *Billboard,* that BMI publishers "could hustle and promote in the same league with top ASCAP pubbers."

Haverlin noted to stockholders that the company had paid $2.6 million in performing rights. BMI had also made one of its annual rebates to all broadcast licensees. Haverlin attributed the strong relation with broadcaster customers to the nationwide program clinics begun in the mid-1940s as an educational service to station managers, which addressed developments in programming and strategies to cope with the looming specter of television. ASCAP perceived the clinics as conspiratorial gatherings plotting to push BMI music into popularity and ASCAP music into oblivion.

Throughout 1951, the society's problems proliferated. Hollywood appeared determined to let the blanket licensing of motion pictures hang until ASCAP accepted a token payment. Authors and composers took separate sides over the issue of equitable distribution, and the SPA was trying to take over the role of spokesman for songwriters. In the face of an ongoing government investigation, the ruling block of motion-picture-owned publishers feared that the studios might be forced to divest themselves of their music interests. The All-Industry Television Committee had refused to accept the most recent ASCAP per-program proposal, and, as provided in the consent decree, asked the district court in New York to set a fair fee.

In the late spring of 1951 Stanley Adams and other former plaintiffs in the 1941 action began to do something about BMI. Identifying themselves as the Songwriters of America, a steering committee solicited contributions to a $250,000 fund from the top 100 ASCAP writers. The money would fight not only BMI but also the "unfair treatment by publishers, record company domination of the song business, payola, artist favoritism, moving picture company power in the music industry, and the closing of avenues for the display and performance of a song." With most of its anticipated funding in hand, in late 1951 the group announced that action would await the outcome of a complaint against BMI filed with the Justice Department during the summer.

The petition was made in connection with an appearance by ASCAP in August 1951 before the New York district court, responding to a demand by the All-Industry Television Committee for a fair rate from the society. ASCAP asked Judge Henry W. Goddard to approve the terms it had offered to the committee. ASCAP had stopped accepting interim payments from the litigants, a move agreed upon in a pact made early in 1950, which called for eventual payments retroactive to January 1, 1949. In the meantime, a lump sum, based on each station's income, was being put into escrow.

ASCAP simultaneously filed a 100-page document, known as the Harbach Affidavit, asking for new language in the amended order of 1950. It argued that broadcasters would "skim off" the society's "gems" and devote the remainder of their programming to the BMI repertory. If the plea was granted, the society could refuse a per-program license to any radio or television station that had a blanket BMI license.

The Harbach Affidavit was passed to the Justice Department, which eventually requested further information. Disturbed by inaction, in March 1952 ASCAP filed a formal request for an investigation of the charges incorporated in the affidavit, essentially that BMI operated as a combination in restraint of trade through its relations with the broadcasting industry. Paul Ackerman of *Billboard,* who had long covered the BMI-ASCAP situation, wrote in April that while ASCAP

> has endeavored to set itself up solidly in TV, and while it has been fighting to re-establish itself in films, it has been constantly losing ground on another front—promotion . . . it has steadfastly refused to promote itself to music users. The result has been that BMI . . . has run the latter a very fast race. BMI, of course, is a wholly-owned corporation, the structure of which permits money to be freely expended for promotion. ASCAP has always taken the position that such funds as are collected must be distributed to the membership.

Rising economic stakes and the changing balance of power in the electronic entertainment business strained relations between broadcasters and ASCAP. The economic value of television to the society had mushroomed. In early 1946, television had not yet become a big business, and ASCAP was content with its dollar-a-year arrangements with stations. However, the constantly rising valuation placed on the Blue Network, after the Supreme Court upheld an FCC order to RCA to divest itself of its second network, dramatically illustrated potential profits.

To comply with the court's ruling, NBC formed a separate corporation, the Blue Network, Inc., and then sold it to the Life Saver King, Edward Noble, in 1942. During the protracted negotiations the Blue Network showed a million-dollar profit, which went to Noble. He changed the network's name to the American Broadcasting Company. The number of new radio stations mushroomed after the government lifted its wartime freeze on new construction, ABC growing from 195 affiliates in 1945 to 282 five years later.

In 1950, with a television network of thirteen stations, ABC ran a poor fourth to its rivals, NBC, CBS, and Dumont, and became a target for motion-picture companies. In late 1948, 20th Century-Fox unsuccessfully offered $15 million for the ABC radio and television networks. Other negotiations broke down, including one from the United Paramount Theaters. UPT officials, principally Leonard H. Goldenson, did not want to let ABC get away. Goldenson was convinced that the film industry's future lay with television. Paramount had been the earliest studio to invest in the technology. It lent money to Paley in 1928 to swing his purchase of CBS, had a 50 percent interest in the Dumont Laboratories, and owned television stations in Chicago and Los Angeles. The consent decree had ended block booking and severed the film companies from ownership of vast theater-chain monopolies.

Long before January 1, 1950, deadline for disencumbering itself of all theater holdings, Paramount Pictures split into two distinct corporate entities—Paramount Pictures Corporation and United Paramount Theaters—which Goldenson continued to head. Paramount films were sold to its former outlets through a theater-by-theater arrangement. Any question as to UPT's right to acquire television facilities was cleared with the FCC immediately after the Supreme Court ruling, and Goldenson prepared to buy ABC. He raised $25 million, formed American Broadcasting-Paramount Theaters in 1953, and served as president.

The television business had grown to a $41.6 million profit in 1951. This profit was going to the local stations and not to the networks. In 1951 and 1952, NBC and CBS showed their first profit from video. A single independent station showed profits the same year matching those of the entire NBC and CBS television operations. Although the four networks' profits were relatively small, they and their owned-and-operated stations accounted for 55 percent of total time sales. ASCAP's collections from television in 1952 came chiefly from the networks, fees from the independent stations still being put into escrow.

Income from television more than made up ASCAP's loss of $1.3 million in theater-seat collections, part of which was also recouped by the payment of about $600,000 a year by the studios for at-the-source licensing. Hollywood began to make a lucrative contribution to television. In 1952 Paramount, Columbia, and United Artists began making half-hour filmed series. *Variety* reported that the production of films for video had become a $100-million enterprise. Some 22 percent of all network shows were now on film.

Fearful that movie exhibitors would immediately boycott all products of the first major studio to cooperate with television, the top eight studios sat on the negatives for 4,057 full-length movies and 6,000 one- and two-reel shorts made between 1935 and 1945. They represented a potential quarter-billion-dollar income from television.

ASCAP's failure properly to compensate Hollywood's background-music writers and their Screen Composers Association were crucial to BMI's gaining an early advantage in the control of film music. ASCAP's revised payment system credited the performance of music written expressly as background with .001 of a point. With its flexible payment structure, and free of Justice Department supervision, BMI offered attractive guarantees to film composers and producers for television rights. Several important cue and bridge libraries opened BMI houses, and many leading video production companies formed "file-cabinet" music firms, into which all rights to background music were placed, secured from their composers on a work-for-hire basis.

While license fees from television climbed, the fortunes of network radio were waning. The days of fortune and glory for the NBC, CBS, and ABC chains slowly slipped away to local stations, whose major offering was recorded music.

Meanwhile, new AM and FM stations were springing up. A ruling by the FCC in June 1945 moved frequency broadcasting into a higher-frequency band, for which neither transmitters nor receivers were available. FM stations already on the air obtained temporary permission to operate on the lower band, in order to serve the "interest, convenience, and necessity" of owners of equipment already in use.

The initial breakthrough of FM-receiver manufacture in 1949 failed to meet the demand. Commercial FM broadcasting appeared to stabilize at around 700 stations, few of them expected to survive the next five years, when around 30 million television receivers would lure audiences away. Major advertisers were now buying time on television.

Every evening segment on the four television networks was sold before the 1951–52 season opened. Network radio was the only place in broadcasting left for sponsors who could not find a place on the picture tube. Vigorous competition to sell them time enforced drastic cuts in talent and production fees and major discounts in time rates, which hastened the medium's demise.

Television producers, loath to use songs without proven audience appeal, turned to great standards of the past. The old music being featured on television became the bread and butter of Tin Pan Alley. A television plug had become worth ten times a radio shot. The days of no deals, no returns of standard songs disappeared, and music publishers' salesmen were given a free hand in arranging special discounts.

Rumors of sheet-music price fixing sparked the first government investigation into printed-music pricing in twenty years. Government attention shifted to the connection between the film industry and the music business. The books of several studio-connected music publishing firms were examined by federal investigators. Concurrently the antitrust action that had been filed in March 1950 by E. H. "Buddy" Morris against Warner, Loew's, 20th Century-Fox, Paramount, Universal, and fourteen publishing companies was finally being settled. His complaint pointed out that by such conspiracy the defendants "succeeded in establishing control of at least 60% of all compositions" used in motion pictures.

Morris's out-of-court victory caused publishers now to bid for publication rights during a sixty-day period, provided the writer and producer had not already come to an understanding. Before studios pre-emptorily had assigned all rights to one of its affiliates.

With the first copyright term of many compositions nearing expiration and the SPA fighting to overturn the traditional renewal process, in June 1951 Warner dickered for the sale of its music-holding operation, MPHC, in order to raise cash to purchase company stock in the open market and maintain slipping per-share profits. Between three and a half and four million dollars was being asked, but Warner insisted on retaining first-refusal synchronization rights to the entire catalogue.

With the final divorce of its screen theaters imminent, Loew's simplified the structure of its subsidiary music holdings by buying out all but one of the four remaining original partners in Robbins Music, paying half a million dollars for their combined 17 percent share. Five years earlier, Loew's had paid Jack Robbins $673,000 so that he could pay all federal and New York State taxes and still retain half a million dollars for his 26 percent interest in the Robbins, Feist & Miller catalogue.

The doubling of home record players between 1945 and 1959 elevated the A & R men to the seats of power at whose right hands were the disk jockeys. Once the deejays successfully pushed a record, the publishers resorted to the second line of plugging, the live radio and television performers. In the first half of 1952, the six major manufacturers released 788 recordings of copyrighted music, only 66 of which appeared on *Billboard*'s best-seller charts. Mitch Miller, who chose all music for the current industry leader, Columbia, was the most successful, with 12 percent of all hits.

Publishers scrambled to give A & R men exclusive first-release rights to all promising material. Sophisticated young writers went directly to their favorite A & R men, and when their material was taken, it often was on condition that it go to a publisher of the A & R man's choice, chosen either because of personal friendship or an under-the-desk arrangement.

Veteran music men chafed at their relegation to the role of money men. Because their contracts with the MPCE union precluded them from employing record-promotion men and they needed songpluggers to work the television performers, they paid for promotion under the table or through the recording artist. After Martin Block and other disk jockeys had become so important that they needed personal treatment, independent promotion men emerged, who made certain that disks sent by artist or publisher would not be lost in the shuffle.

The guaranteed-advance contract that BMI continued to offer publishers had been regarded by ASCAP and Starr as a minor problem so long as the performances of BMI music were negligible. All primary responsibility for most new contracts with popular-music publishers had gradually shifted to a young house counsel, Robert J. Burton, who joined BMI in January 1941 and was later named director of publisher relations for BMI and assistant secretary to the board of directors.

BMI had developed no set criteria for handling advances of guarantees to new affiliates. Decisions were made by Burton on the basis of an applicant's past experience or his access to promising non-ASCAP songwriters. Most contracts were for five years, and all carried a sixty-day cancellation clause. One of Carl Haverlin's initial acts as BMI president was to put Burton in charge of repertory accumulation, as vice president for publisher relations. Burton immediately emphasized productivity. Soon after *Variety* shifted to a full-page disk-jockey popularity chart, Burton sought out music men with access to the key record-spinners. Their request programs more truly reflected changing public response to new music, as well as the growing importance of the BMI catalogue.

In 1952, Burton was put in charge of BMI's new combined publisher and writer activities, and the former director of publisher relations, Robert B. Sour, was elected a vice president. Composer of the American lyrics to "Body and Soul," he joined BMI in February 1940 as lyric editor and staff lyricist. As director of writer relations in the late 1940s, Sour supported a writer payment plan and was put in charge of the program when it was adopted in 1949.

Burton took on a much more active role in the operation of BMI Canada, formed in conjunction with the Canadian Association of Broadcasters during the 1941 music war. He was made its general manager in 1947. Under his direction, BMI Canada had a roster of 120 affiliated composers and 27 publishers, and a catalogue of 5,000 works.

BMI's writer-affiliation program was similarly productive, with 115 composers and lyricists under contract in the winter of 1952. During the past year, BMI contract writers had amassed 2.5 million performances, but their share of royalties was not yet in proportion to that paid publishers. There were, in addition, many songwriters who were under contract to BMI publishers on either an exclusive or a song-to-song basis. They received their performance fees, if at all, directly from the publishers, who racked up the majority of BMI performances, and were responsible for a reported four-to-one performance advantage over ASCAP on independent stations.

An increased portion of the music they played came from the independent record companies, whose copyright-owning subsidiaries quickly affiliated with BMI, for the four-cent-a-station royalty. Often a BMI publisher's advance or guarantee was used to wheel and deal with the smaller labels. They paid for recording sessions, accepted reduced mechanical royalties, and picked up the tab for exploitation by disk-jockey or local-distributor payola, all with the expectation that the more local performances BMI logged, the surer one could be that his annual guarantee would be renewed.

When, for the first time since 1941, BMI songs captured the first five places on "Your Hit Parade," in March 1952, many young ASCAP publishers and songwriters re-examined their association with ASCAP. As a result, it was soon difficult to name a younger ASCAP publisher board member who had no sort of affiliation with BMI.

Under pressure from dissidents, ASCAP, in 1950, began expanded spot-check logging of independent stations. In early 1953, John Peatman, of Audience Research Institute, took charge of the society's logging operation. He added 170 local stations to the year-round logging of the networks made by the society from a complete recapitulation of all music played from sign-on to sign-off. Performance credits of one point for each popular song were multiplied by the number of stations carrying each performance. Seventy local stations, scattered around the country, were logged regularly on a spot-check basis. The remaining hundred stations, from ten regional areas, were logged on a rotating basis. Every performance on the seventy fixed stations received a full performance credit, and a song logged on one of the rotating sample received seven and a half credits. Television was logged on a similar principle—the four networks on a census basis, and three selected local stations each day, with a performance receiving three points.

The attraction of BMI's logging system, based entirely on broadcast performances, was brought home in March 1953 with the consummation of a contract between the society's seventeenth-highest-rated music house—Santly, Joy—and BMI. The new contract represented the belief that forty performances on large stations represented more value than a network shot over forty affiliates.

The Santly, Joy deal underscored BMI's logging operation, which except for rates had not changed since it was first put in place. One-tenth of all stations in the United States, representing a cross-section of every type of broadcasting facility, were logged each quarter. The networks were logged, as by ASCAP, on a census basis. Every network performance was credited with a minimum of seventy-five points, regardless of how many stations were hooked up, unless the number was higher, and performance payments were made on the same four- and six-cent basis as on radio. Because BMI logging picked up 150,000 performances on average for a hit song, an aggressive firm could pile up two million performances in a year, producing a quarter-million-dollar payment from BMI.

A new concern was arising that would change the entire course of American music. The large ASCAP music houses finally broke through the A & R men's "iron curtain," and were going directly to radio and TV live plugs. According to the Peatman survey of the year ending June 30, 1953, television was crowding out network radio as a publisher's most effective plugging medium. The survey's top song, "I Believe," an ASCAP song, received 15,738 performance points on radio, but 18,601 on television.

Both small and large music companies made drastic reductions in overhead and operating expenses. The retail price for a new popular song was increased from forty to fifty cents, and from twenty-three to twenty-five cents for jobbers and thirty-five cents for dealers. Most publishers adopted a "no return" policy on new songs until they showed some chart activity. Smaller firms reduced overhead by lowering their sales and assigning selling rights to agents on a 10 to 15 percent basis.

The first accurate estimate of printed music sales in the United States was commissioned in 1952 by the Music Publishers Association, to be submitted to a Congressional committee in connection with a requested reduction of postal rates. Sixteen percent of the $30-million income from printed music sales came from sheet music. The remaining $25.2 million was from standard classical, religious, and educational music. Twenty-one percent of all popular sheet sales were made in retail stores, the balance in syndicate stores or from racks and by mail. The five or six major firms operated on a fairly stable basis, led by the MPHC, with its pillow of $1.2 million in ASCAP royalties, supplemented by an important share of the new millions it, Chappell, the Big Three, and Paramount reaped from educational folios, band instruction books, and choral and solo arrangements of their great standard songs.

Under the noses of established music and record businesses, during 1953, $15 million worth of rhythm-and-blues records were sold. It was the product of about seventy-five manufacturers, among them RCA Victor, Columbia, Decca, and Mercury, who covered the field with almost exact musical duplications of the real thing. Any legal barrier to the practice was removed by the courts in 1951 with a decision in the "Little Bird Told Me" case. The initial recorded version of the Harvey Brooks song was released in 1947 by "sepia" star Paula Watson on the Supreme label, and was covered by the white chanteuse Evelyn Knight on a Decca arrangement, which copied Watson's phrasing exactly and confused even musical experts. Supreme sued, but lost in a verdict that declared musical arrangements were not copyrighted property and therefore not subject to the law's protection.

For the original source, one looked to the products of such R & B independents as Atlantic, Savoy, Duke, and Chess-Checker. Jukebox operators were the first to notice white teenagers' growing appetite for black R & B. Local record retailers caught on, though at first they had difficulty in keeping up with the demand, because most companies only went into that business for an all-black market.

The tremendous potential white audience for live black rhythm-and-blues shows was first made known to white show business by a Cleveland disk jockey, Alan Freed. Freed attracted national attention and almost spent some time in the city jail after a near-riot in connection with an R & B dance in March 1952. Charges against Freed and his partners were dropped, but the national publicity enlarged his radio audience. Freed also promoted the genuine musical article in sold-out dances featuring the newest black stars and attracted audiences that were never less than one-third white.

The growing market was one reason for the 25 percent drop in the output of the six major manufacturers in 1953 to 2,190 popular releases, during a year when 59,371 musical compositions were copyrighted. Ninety popular records appeared on *Billboard*'s best-seller charts in 1953, but only seventy-two were released by RCA Victor, Capitol, Mercury, Columbia, MGM, and Decca, in that order. The balance came from well-financed new independents. Because of smaller output, the majors' success rate fell to one hit out of every thirty releases.

Network television's increasing emphasis on familiar songs from the ASCAP catalogue persuaded the society's ruling hierarchy that there would be no complications in negotiating with the chains for a new contract, holding to the same figure for another four years, to the end of 1957. All members were asked for and assigned their video rights to run concurrently. Most of the rank-and-file membership took little notice of a situation that potentially was crucial to their future financial well-being. The per-program litigants adamantly refused to accept the proffered new license, expecting Judge Goddard to rule in their favor once ASCAP's motion to amend the 1950 ASCAP consent decree was finally disposed of.

To end hostilities between the society and the broadcasters, in the person of BMI, ASCAP suddenly and surprisingly withdrew the motions supported by the Harbach affidavit. However, the formal request for a Justice Department investigation of BMI, filed in March 1952, remained in effect.

The stratagem had little impact on the networks or the independent plaintiff stations. The networks met with ASCAP and argued that the original television licenses had been drawn up when the medium was undeveloped and broadcasters were unaware that the figures demanded could represent a $10-million payment. They suggested a substantial reduction in the new agreement. By the end of the year, there was a possibility that the networks would again lock out ASCAP music and fill in the gap with the BMI repertory, or else would petition Goddard for a new, lower rate, while continuing to use the society's music. The networks would not accept a proposal to base ASCAP fees on net income, instead of gross receipts. This would mean that financial records might become available to the competition, as well as to the affiliates, who unknowingly were bearing most of the networks' ASCAP television load.

ASCAP's new president, Stanley Adams, found the same internal problems when he took office in April 1953. The continuing attraction of BMI's logging practices, its gains in the concert-music field, and its larger share of the trade-press charts made it difficult for Adams to provide a form of pension for veteran writers by basing their payments on the length of their association with ASCAP. At a membership meeting, he predicted that income in excess of $18 million was expected for 1953. Yet ASCAP had relaxed admission standards in order to compete with BMI for promising new talent, and 800 new members had joined that year. Therefore, with 3,200 authors and composers now in the society, there would be no meaningful increase in the quarterly checks.

The autocratic attitude toward the general membership of the ruling bloc of publisher-directors and of those writer-directors and officers who owed their positions to them was exemplified by Herman Starr's remarks at the meeting. Stricken by the untimely death of former president Fred Ahlert, he insisted on the award of a $25,000 pension to Ahlert's widow. The board approved this without any public discussion, contrary to ASCAP's bylaws.

Unable to reform the society from within and frustrated, not only by BMI's seeming control of what was played on the air but also by doubts about their own ability to write what the public wanted, many ASCAP members became eager recruits to the ranks of the Songwriters of America. They would share in the bonanza of $150 million that lay waiting for a decision by a jury who had had enough of BMI's "garbage" and wanted to return to the kind of music they had written.

The television negotiations and the hearing before Goddard moved to the sidelines on November 8, 1953, when a group of thirty-three writers, the leading complainant being Arthur Schwartz, filed a $150-million civil antitrust suit on behalf of 3,000 composers and authors, charging a radio-television-recording company conspiracy, centered upon BMI, NBC, CBS, ABC, and Mutual, RCA Victor and Columbia Records, the NARTB, and a number of other corporations and individuals, including BMI's directors and executives. All were charged with conspiring "to dominate and control the market for the use and exploitation of musical compositions."

The press conference in the Waldorf-Astoria Hotel on November 9, at which *Schwartz v. BMI* was made public, included an explanation of the matter by John Schulman, attorney for the plaintiffs and counsel to SPA. The complainants desired to destroy BMI's giant subsidy scheme, which was financed by the broadcasting industry and pitted 1,300 BMI-affiliated publishers against a mere 600 ASCAP houses, who were not financed by anything but private capital.

Variety later reported a press interview in which Schwartz argued that broadcasters were in a position to

> turn the plugs on and off at will. . . . it's no longer a case where a song can fail or succeed on its merits, because the public is denied the opportunity to judge for itself. . . . when a writer comes to an ASCAP publisher, who also has a BMI affiliate, he'll suggest that "if you collaborate with a BMI writer we'll handle it via our BMI firm," which can only prove indubitably that the BMI affiliation is more positively lucrative [due] to the subsidies from the number of BMI plugs over BMI-affiliated stations and networks.

Starr saw greater worth in ASCAP's repertory than in BMI's "flash-in-the-pan" jukebox hits. "Quality," he said, "is what lasts and what pays off. At the moment BMI is hot . . . but the music business is everything, and dominantly it's ASCAP income. This is the prime source of income and not the byproduct as when Nathan Burkan helped found the Society."

The Schwartz suit came as no surprise to BMI, but did stun its executives and board because the long-pending issue was finally joined in the uncertain arena of trial by jury. Calling the complaint a "rehash of charges ASCAP has been making for years, and has never been able to substantiate," Carl Haverlin pledged a fight to victory. Shortly before Thanksgiving Day, with Robert Burton and others, he made his visit to Nashville. During radio station WSM's second annual Country and Western Disk Jockeys Festival week, Burton presented Citation of

Achievement certificates to the writers and publishers of twenty-four outstanding country-and-western songs of 1953. This presentation was without the big-city hotel glitter of BMI's second annual Popular Music Awards black-tie dinner just before Christmas. Five of the fifteen songs honored there came out of Nashville or were influenced by country music.

The day before the dinner, BMI won its first victory in *Schwartz v. BMI:* the right to examine ten of the thirty-three plaintiffs in pretrial proceedings. The ten were directed by the court to answer specific questions put by BMI counsel, seeking to link ASCAP to them and the action.

Abel Green's 1953 annual review emphasized that while the publishers were not involved in the songwriters' suit, they would welcome restoration of control of the business. A veteran of Tin Pan Alley told Green:

> Everybody but the music publisher, who used to be pretty good at that, nowadays picks songs. And don't tell me that in the final analysis the public really picks 'em. We . . . used to have a pretty good concept of quality and value of songs that we published. . . . Today, we don't dare publish a song until some artist perhaps likes it, or when the whim of an A & R genius decides it should be done. . . . A record should be a by-product of publishing; not the sparkplug of songwriting and publishing.

8 Technological Advances and Economic Growth in the Music Industry

THE figures for 1953 confirmed that 1947's all-time sales high of $204 million had been surpassed. At its best, the industry produced only one hit out of every thirty releases, and its costs could be recouped only by a sale of at least 40,000. The American Federation of Musicians' power had eroded. A third walkout from the recording studios in 1954 was avoided when musicians' wages increased 20 percent, as was the royalty paid to the union's Music Performance Trust Fund.

Television remained the favored medium for "breaking" a new release, leading Decca to wait for Bing Crosby to introduce "Y'All Come" on his TV debut. However, sales of his disk were disappointing, and lead to charges that Milton Rackmill must step down as company president. A courtroom fight and a major stockholder-proxy battle was in progress, the consequences of his purchase of Universal Pictures. During 1954, profits of nearly two million dollars from Universal accrued, sustaining Decca through a fallow period.

That period ended with Bill Haley and the Comets' series of hit records, including "Rock Around the Clock," "Shake, Rattle and Roll," and "Dim, Dim the Lights." Haley provided the first examples of commercial white rock 'n' roll, abetted by Decca A & R-man Milt Gabler. As Gabler told Arnold Shaw, he taught the group the shuffle rhythms of Louis Jordan's Tympani Five, whom he had produced during the 1940s. "You know, dotted eighth notes and sixteenths, and we'd build on it. I'd sing Jordan riffs to the group that would be picked up by the electric guitars and tenor sax."

Optimistic people in the business looked forward to a "fatter platter" year in 1955. However, a breakdown in communication with Chicago-area retail dealers was responsible for the temporary abandonment in the summer of 1954 of the earliest attempt by a major manufacturer to extend the distribution network with mail-order merchandising. The unbreakable LP had brought a number of imaginative persons to the industry, among them the operators of Music Treasures of the World and its subsidiaries, and Concert Hall Records' Musical Masterpiece Society. The inventor of the direct-mail book club, Harry Scherman of the Book-of-the-Month Club, began a test run of its Music Appreciation Record Club. Initially it produced its own recordings, but beginning in early 1955, it purchased them from Angel Records.

The record clubs had a total membership of one million subscribers in 1955 and an annual sale of $20 million. A dozen clubs existed, doing about 15 percent of the entire LP business and responsible for a third of all classical-music record sales.

RCA Victor and Columbia moved to combat the record clubs' threat. RCA Victor added local white-goods appliance retailers to their distribution network, supplementing their radio and phonograph-combination sales with a coupon plan, the SRO (Save on Records) programs. Any one of 400 LPs could be purchased with coupons gotten from the local appliance stores. A $3.98 coupon book entitled the purchaser to three free bonus records each year and the option to buy up to three LPs each month at a dollar discount.

In what *Variety* called "the most controversial move in the disk industry since 'the battle of the speeds,' " in 1956 Columbia offered record retailers a guaranteed share in profits from the members they recruited. Within a few months, 88 percent of all Columbia dealers had agreed to cooperate; the plan initially attracted 409,000 record buyers.

In early 1956, RCA Victor joined with the Book-of-the-Month Company to form the RCA Victory Society of Great Music. Like Columbia, Victor gave a 20 percent commission on all memberships the BOMC secured, thus ensuring the purchase of six classical music selections annually. Some 200,000 members joined within three weeks, prompting Victor to open the Popular Album Club.

Reader's Digest, which eventually became the largest mail-order record packager, tested the market in 1960 in conjunction with RCA Victor. An album of twelve newly recorded classical Reader's Digest-RCA Victor LPs by European artists was offered to *Digest* subscribers for sixteen dollars, with stereo versions available for an additional two dollars. After selling a quarter of a million albums in the United States and 1.5 million worldwide, the publication began its club.

The record-club business boomed. Columbia alone had 1.3 million members, who spent $30.3 million in 1959, and, with Capitol and RCA Victor, accounted for 90 percent of the mail-order record business.

Victor was seeking to create a two-speed market, to make its 45-speed patented technology the industry's standard for all popular singles and the 45 EP and 33 LP standard for albums. It began killing off production of the 78-speed disks, increasing their prices and raising the dealer discount on them and on all 45s. On January 1, 1955, Victor unveiled its Operation TNT. The price of all 45s was set at eighty-nine cents; that of 78s was raised to ninety-eight cents; that for all LPs was reduced by one-third. The move was designed to make recordings a mass consumer item by simplifying the existing pricing system. The other majors reluctantly conformed to Victor's new schedule.

The singles business was not affected by these changes. Decca was riding high with the greatest showing since Kapp's death, as were Mercury and Dot, a newcomer beginning to appear on top-ten charts with regularity. The business had gone, as *Variety* said, "R & B crazy." By 1955, R & B had become a $25 million annual business. With the exception of Capitol and MGM, which returned late to the field, the major firms were heavily into R & B, attempting to compete with now-entrenched independents. Atlantic led the pack, with one-third of the year's top-selling R & B disks. Atlantic, Savoy, Chess, and some others had gone into the packaged-music field and were issuing LPs and EPs by their best-selling black artists. Record-company owners insisted on a full two-cent statutory mechanical fee for their copyrights. Desk-drawer publishing had become a thing of the past.

Several factors contributed to its emergence on the national white-pop-hit charts: the fantastic popularity in 1955 of the motion picture *Blackboard Jungle,* which introduced Haley's "Rock Around the Clock," and the presence on the national scene of R & B's Barnum, Alan Freed, who had moved from Cleveland to New York City.

Rock 'n' roll, as Freed had named the new music, awaited a figure who would glavanize the public. He was found in the unlikeliest of places: the stage of Nashville's Ryman Auditorium, from which the Grand Ole Opry was broadcast. And he was Elvis Presley.

Elvis Presley was familiar with blues, R & B, and country records. He first entered Sam Phillips's Memphis recording studio in late 1953 to make a gift for his mother's birthday. A year later, Presley made his first commercial release: Arthur Crudup's "That's All Right, Mama" and bluegrass musician Bill Monroe's "Blue Moon of Old Kentucky." The formula was ideal and was maintained throughout Presley's stay on Phillips's Sun label, which ended in October 1955.

Presley went to Nashville for the Country and Western Disk Jockey's Festival and found himself a major center of attention. His manager, Colonel Tom Parker, advised Presley to sign with any manufacturer that would meet terms set by Sam Phillips for his contract and his Sun masters. Jean and Julian Aberbach, majority owners of the BMI-affiliated Hill & Range Songs, put together the package that brought Presley to RCA Victor. According to Arnold Shaw, in *The Rocking '50s,* the recording company "ostensibly put up $25,000 for Presley's recording contract while the Aberbach brothers paid $15,000 for publishing rights and the purchase of Hi Lo Music," owner of several songs Presley had recorded for Sun Records. Hill & Range Songs also administered Presley's BMI-affiliated Elvis Presley Music Corp., as well as his ASCAP company,

Gladys Music, through which the majority of songs he recorded were published. The young singer's contract with RCA Victor was for only two years, with a one-year option.

Within a year, Presley had sold over 10 million disks, among them five million-sellers, "Heartbreak Hotel," "Blue Suede Shoes," "Hound Dog," "Love Me Tender," and "Don't Be Cruel." Seven reissued Sun singles were each selling at the rate of 12,000 copies a day, and his first LP sold 300,000 copies. Presley was responsible for two-thirds of Victor's entire singles output.

Income tax demands compelled a restructuring of Presley's original contract. It now guaranteed $1,000 a week for twenty years. The $430,000 in royalties Presley had already earned were spread out, and the original contract was extended to five years, with an option for another five.

Colonel Tom Parker had made Presley a millionaire in just nine months. In addition to the income from records, an estimated $250,000 came from Hollywood deals, starting with $100,000 for the soon-to-be-released *Love Me Tender*, and an agreement with Paramount Pictures involving a guarantee that eventually rose to more than a million dollars per movie. Parker also had contracts in hand guaranteeing at least another $250,000 for personal appearances.

Record sales rose by about $100 million in 1956 over the previous year's $227 million. The number of million-plus single sellers had risen sharply, although only a few were out-and-out rock 'n' roll. EMI had recently acquired Capitol, which enjoyed a 33 percent increase in profits and sold an estimated 5.75 million singles of only three songs in a few months.

The largest sales increase had been in LP records, with only a minor increase in 45s, and a significant decline in 78s. The prevailing factory cost to an independent record company without pressing facilities was around 15 percent for a 45 single, which was then sold to a distributor for about thirty-three and a half cents, out of which came the one-and-a-half to four-cent mechanical royalty. In the event the recording artist was to receive 4½ percent of the eighty-nine-cent retail price that became standard following RCA's Operation TNT, it was paid only after advances and costs had been recouped. The general markup from distributor to retail dealer was twenty cents, but in the case of distributors handling twenty to thirty consistently "hot" companies this would be increased to as much as twenty-eight and a half cents a disk. It was the distributor who realized the largest profit from a hit record.

The film industry continued to find ways to marry movie marketing and record manufacture. Milton Rackmill created a studio-record company combine by purchasing Universal Studios in 1950. The third broadcasting network's plan to join RCA-NBC and CBS in the record business came to a conclusion in June 1955 when ABC-Paramount launched the new company. Sam Clark, its president, set up a distribution chain for Mickey Mouse Records. Mickey's creator, Walt Disney, the first major studio head to go into feature television production, had sold ABC-Paramount television the *Disneyland* and *Zorro* series. Disney got a nationwide medium for the promotion of his amusement park, in which ABC-Paramount had invested $500,000 and guaranteed loans up to $4.5 million for a 34.48 percent share. Clark also incorporated innovative marketing by purchasing smaller labels that retained their own identity, charging them a low fee to cover financing, promotion, and exploitation, with ABC-Paramount participating in the profits.

The choice of repertoire shifted from publishers to company personnel. Trained musicians served as staff musical directors and artists-and-repertory men. A modern A & R chief was in charge of the promotion of his work, and was responsible for the happiness of all artists under contract, as well as auditioning or negotiating with new talent.

The value of music holdings to a film company was underlined by Loew's annual report for 1956, which showed that the company had earned $2.2 million before taxes, $645,000 of it from MGM Records. The label had operated since 1945; net income to date was $3.5 million. MGM Records continued to exploit Loew's film products in the form of soundtrack LPs.

Paramount Pictures won the race to obtain the services of Dot Records' "young genius" president Randy Wood in early 1957. During 1956, Dot had enjoyed a 12 to 14 percent share of the singles market and had fourteen singles on the year's top-seller charts. Wood formed Dot Records in 1951 and concentrated on country records. As orders for R & B recordings proliferated, he started covering rhythm-and-blues hits by records of Pat Boone. Wood's reported insistence that publishers of any song he covered permit the distribution of up to 70,000 royalty-free records for disk-jockey promotion made Boone's cover records major hits. The two million in cash and substantial block of Paramount stock which was paid for Dot Records, as well as a place for Wood on the company's board, was recouped within five years. In the first year, Wood doubled his singles sales and increased those of Dot's LP product by five times.

Paramount's good fortune with Dot prompted the operating heads of major studios without a record-company affiliation to strike an agreement with either of the two currently attractive independent labels, Liberty and Imperial. With a successful stable of young singers, Liberty ranked among the twenty-five leading companies.

Imperial's three-million-dollar gross in the first six months of 1957 represented the best earnings picture since Lou Chudd, its president, formed the business in 1947. In 1950 Chudd signed a young black New Orleans pianist-singer, Fats Domino, and when Chudd eventually disposed of Imperial to United Artists in the early 1960s, Domino had sold 30 million disks, eighteen of them Gold Record award-winners. With the UA sale came a young white teenage "soft" rock-'n'-roller, Rick Nelson, son of Ozzie and Harriet Nelson. After coming up with two Gold records in 1957, Nelson, who proved to be the most successful West

Coast teen idol, received a five-year contract, guaranteeing him $1,000 a week in addition to the 4½ percent royalty.

United Artists had moved into the record business in October 1957, starting from scratch with two subsidiaries, UA Records Corporation and UA Music Corporation, the latter a desk-drawer operation with about fifty copyrights. In its first year United Artists leased masters made outside the company. With an offer of complete financing and a half-share in all profits after costs were recouped, the company attracted tested independent songwriter-producers, on a production-to-production basis, who brought the company into the rock-'n'-roll business.

Other studios followed in their wake. In 1958, 20th Century-Fox chose one of Randy Wood's lieutenants to head its new subsidiary, 20th Century Records, which it intended to develop from the ground up. Columbia Pictures, which had unsuccessfully attempted a merger with Lou Chudd, established its own disk division in June, using the name Colpix. Almost simultaneously, Warner Brothers entered the recording and electronics business. James Conkling, former president of Columbia Records, was brought in as president.

Capitol's success with the *Oklahoma!, Carousel,* and *The King and I* motion-picture sound tracks and Decca's with those from *Around the World in Eighty Days, The Glenn Miller Story,* and *The Benny Goodman Story* sparked the industry's interest in marketing movies and Broadway show music. Spirited bidding for rights enabled the studios and the theatrical producers to enforce higher royalty fees.

All LP production costs were rising. With a profit of fifty-one cents from a $3.98 popular-music LP, and a net of twenty-one cents after all expenses, including 2½ percent of the retail price to the AFM Trust Fund, the standard twenty-four cents to publishers, twenty cents to the artists, and a forty-cent excise tax, record men wondered whether soundtrack LPs were not worth more to the studios than to the manufacturers.

Both the motion-picture and the recording-industry studios in 1958 faced yet another strike by members of the American Federation of Musicians. New President Herman D. Kennin led a West Coast group of dissident AFM members in the sixteenth week of a wildcat strike against the motion-picture and television companies. Among the strikers were the 303 contracted musicians, who earned $2.7 million a year from the film industry and now demanded a $12-million wage increase and other benefits.

After a vote by the strikers in favor of the new Musicians Guild of America, the NLRB recognized the guild as the sole bargaining agent with the motion-picture and television production industry and negotiations began for new contracts. The industries dealt with the dissidents because they performed the music in pre-1948 movies sold to television and in new television productions. Their legal representatives questioned to which union the mandatory 5 percent musicians' royalties from those sales belonged. The thirty-nine-month agreement eliminated payment of contributions to the Music Performance Trust Fund, but not the 5 percent musicians' residuals from TV sales.

The recording business faced another stroke-of-midnight walkout at the end of 1958, but they believed that the AFM would avoid a showdown this time because of competition from the Musicians Guild. Kennin showed his arbitration skills by achieving the largest gain ever for recording musicians in a new five-year contract—an increase in wages from 30 to 47 percent. The Music Performance Trust Fund was discontinued, and a pension fund was instituted instead. Lower recording rates were fixed for established symphony groups, and the basic payment by the manufacturers to the trust fund was increased by 1½ percent.

The proposed sales of post-1948 motion pictures to television precipitated a new conflict between the AFM and the MGA, each threatening to take action unless its members got their share of all 5 percent residual royalties from the transactions. During this impasse, most producers put the money in escrow. The MGA became more active in negotiating contracts with California-based record companies and inked a new agreement, which gave a 30 percent increase over current AFM scale.

Kennin effected a compromise between the AFM and the MGA in 1961. The Musicians Guild was dissolved, and its members were reinstated to the AFM with full rights. A new contract with the record manufacturers, approved in 1964, improved working conditions and fringe benefits. Recording-session musicians would now participate in the proceeds from the sale of records, whose total gross at retail had risen to $698 million in 1963, a large portion of which came from the original-cast LP albums.

The battle for musical theater rights had grown substantially. The box-office success of Rodgers and Hammerstein's *South Pacific*, whose 78-rpm albums and LPs grossed $9.25 million, followed by sales of 1.5 million soundtrack LPs, put many theatrical composers in a position to negotiate directly with the record makers. Goddard Lieberson's faith in Lerner and Loewe's *My Fair Lady* was such that he persuaded the CBS board to invest $360,000, the entire cost of mounting the production, for a 40 percent share. It gave the corporation initial returns in excess of $2 million in the first two years from a record that cost $22,000 to make. CBS also received an option to produce a television version of *My Fair Lady* following the closing of the first Broadway production. During the run, CBS bought Lerner and Loewe's 30 percent share of *My Fair Lady* royalties outright, increasing its interest to 70 percent. CBS then sold movie rights to Warner Brothers for $5.5 million and a 47½ share of earned income over $20 million from distribution.

During the 1958–59 theatrical season, record companies had invested several million dollars in the theater, receiving in return fourteen original-cast albums. However, only four of the musicals were hits. The structure of theatrical financing had changed. The music conglomerates and their A & R heads now provided the necessary backing for a stage musical. After CBS advanced the entire $480,000 to back 1959–60's Lerner and Loewe musical *Camelot,* in return for a 40 percent share of receipts after expenses, Columbia Records still had to bid against RCA Victor for the album rights, and won only after it agreed to pay more than the standard 10 percent of the LP selling price.

However, as the intense competition between labels pushed up producers' royalties, Columbia Records President Goddard Lieberson branded such dealings as "verging on hysteria." The cost of producing a cast album had quadrupled since *My Fair Lady* opened. Still, the possibility of hitting another *My Fair Lady* gold mine lured manufacturers into fighting for cast album rights, but it also helped Columbia accept losses from a string of box-office failures in the 1961–62 season. On the other hand, Columbia did have six of the seven Gold Record cast LPs. The madness Lieberson deplored continued. In the summer of 1963, five major companies invested nearly three million dollars in forthcoming Broadway productions, all of which bombed.

When the federal excise tax on recordings was revoked in 1965, cast-album profits and those of all other LPs rose. The *My Fair Lady* LP eventually sold six million units, a figure no other cast or soundtrack LP surpassed until the filming of *The Sound of Music,* whose soundtrack LP came out in 1966.

During the decade beginning with the 1955–56 season, only 131 musicals were staged, and only one, *Fiddler on the Roof,* was an acknowledged box-office hit in the 1965–66 season. Goddard Lieberson, now chairman of the newly organized CBS-Columbia group, was charged with diversifying the corporation's holdings. He still supervised the production of cast albums, however, though investment in the musical theater was decided by Clive Davis, an attorney who had handled the FTC suit against the Columbia Record Club. As vice president and general manager of CBS Records, Davis was responsible for the loss of a sizable investment in the production of *Superman,* and had supported putting $450,000 into *The Apple Tree,* by *Fiddler* writers Jerry Bock and Sheldon Harnick. It proved to be a failure. Those disasters brought about a $100,000 limit on all theatrical adventures.

Except for a slight dip in 1954, there had been a steady upward progression of retail sales since the beginning of the decade, according to RIAA figures (later revised upward)

1951	$191,000,000	1954	$195,000,000	1957	$400,000,000
1952	$202,000,000	1955	$227,000,000	1958	$438,000,000
1953	$205,000,000	1956	$331,000,000	1959	$514,000,000

Approximately 100 singles were issued weekly in the late 1950s, a glut that overloaded the country's nearly 700 distributors. Increased by 30 percent from 1955–60, they included manufacturer-owned branches belonging to RCA Victor, Columbia, Decca, Capitol, Mercury, and King, and did one-third of their business in New York, New Jersey, Detroit, Cleveland, and Philadelphia. Because of the recent success of independents, who sold three out of every five 45-rpm singles, there had been a surprisingly small mortality rate among distributors.

Overproduction of singles, 1,200 new releases each quarter, also affected retailers. Only between 5 and 10 percent of them sold in measurably significant quantity: between 100,000 and more than one million copies. At the end of 1957, the Big Four increased the price of their 45s from eighty-nine to ninety-eight cents. Now that three singles cost the same as a popular LP, the majors expected the independents to be hurt severely. Instead, production by 600 companies reached a peak in 1958. Twenty-four percent of the 283 sides on the most important *Billboard* charts in 1958 came from the Big Four, the remaining 76 percent belonging to independents. It was one-stop operators, discounters, and rack jobbers who played a major part in bringing the industry nearer to a half-million gross.

It was difficult for the average jukebox owner to keep up with demand. The one-stop store, which had come into existence before the war, had become, in *Billboard*'s words, "the fulcrum of the entire pop singles setup." Both manufacturers and independents offered them giveaways of a free disk for every one ordered, and flexible return policy, and rapid delivery. The one-stop was a convenience not only to jukebox owners but also to local retailers, who, willing to pay the premium nickel for ready access and simplified bookkeeping, got half their singles from them.

RCA Victor's Operation TNT proved to be the making of many cut-rate record stores. The value of transshipped classical-music overstock, for which discounters had an ample supply of ready cash, was made evident when Sam Goody sold four million dollars' worth in 1955. However, some five years later, Sam Goody, Inc., three million dollars in debt, filed for reorganization under the Bankruptcy Act and accepted a plan to pay off forty-eight cents on the dollar over the next ten years. It included a unique provision that tied in future LP prices, increasing the value of Goody's inventory as they rose, and thus his payments on the indebtedness. The business finally went public in 1966, when it had an annual income approaching eight million dollars.

All other measures having failed to curb transshipping, a group of retailers in 1958 formed the National Association of Record Dealers. They hoped to counter the impact of record clubs and rack jobbers. The first rack operation in the United States, offering the top fifteen hits on both 78s and 45s in syndicated, variety, drug, and self-service food stores, was the creation of Elliot Wexler. Each location that installed one of his four-and-a-half-floor metal racks paid five cents above the distributors' price for the records displayed.

Record racks dated back at least to the "Hit of the Week" operation in the early 1930s. Sheet-music racks, licensed by the Music Publishers Protective Association through the Music Dealers Service, provided MPPA members with profits from a guaranteed first order that underwrote all production and promotion costs. However, the rise of rock 'n' roll cut severely into sheet-music sales.

Bell Records followed Wexler in the business, placing about 180,000 racks carrying its own twenty-five-cent seven-inch 45-rpm cover versions of current hits alongside twenty-five-cent book reprints. Columbia, RCA Victor, and Decca started their own quasi-independent 45 labels, free of price control, for sale on racks. Only Columbia cooperated with Wexler, allowing him 15 percent above the standard sub-distributor's discount. Seeking to get the same treatment from the others, Wexler unsuccessfully sued Capitol and lost his Music Merchants business.

Joe and David Handleman's record-rack business subsisted on the usual 10 percent discount allowed to sub-distributors, making up the difference with great volume. As orders increased, they bypassed local distributors and dealt directly with the manufacturers, at a larger discount. They established the policy, which quickly became standard, of fully servicing the racks with top hits and some standard items, in return taking all operating and inventory risk and giving a smaller share of profits than retailers usually received.

The sales potential in chain drugstores' and food supermarkets' displaying record racks was evident. Victor and Columbia now issued their $1.98 Camden and Harmony LPs for rack locations, which began to account for 16 to 18 percent of all LP sales by 1960. The budget LP market, disks selling for $1.49, was controlled by four independent manufacturers: Remington Records, Tops Records of California, Miller International, and Pickwick Sales of New York, which started in 1947 with a children's record line and now dominated the cheap LP market.

Nearly twenty years had passed since a demonstration in April 1940 of practical true stereophonic recording, made by Western Electric-Bell Laboratories technicians at Carnegie Hall in New York. Two hours of stereo music were played over three hidden speakers. It had been recorded on three different tracks on a single continuous reel of film tape and then fed through separate amplifiers to the speakers, which were spread across the stage to provide width and depth to the music.

A decade before, a British engineer at EMI secured the patent for a practical two-channel disk. Five years later, scientists at Bell Laboratories experimented with two-channel stereo, using a similar process. Emory Cook, who made high-fidelity records for the connoisseur market, brought out a binaural record in 1952, with two separate but continuous grooves, one for each channel, played by two pickups attached to a single arm, a technical triumph but a commercial failure. Many studio technicians believed that the technological future rested with packaged prerecorded magnetic tape. Capitol, the first to enter the taped background-music business, transferred its entire transcription library of 3,000 selections to tape for leasing through its Magnetronics subsidiary. To counter that invasion of its territory, Muzak, which had cornered the wired background-music field, transferred its own library of 7,000 titles to magnetic tape. RCA Victor shortly joined the small group of generally independent tape producers whose chief problem was the matter of speed.

No standard had been established, and existing machines ran from 3¾ inches per second to 7½, 15, and 30. Though the market was limited by the number of playback machines, the 3M Company's series of 1955 price cuts of raw magnetic tape increased the possibility of mass production. So did the experimentation by Columbia and RCA Victor with less-expensive tape machines. Working with Bell & Howell, Columbia developed tape machines utilizing the company's latest speaker assembly, the K—for Kilosphere—that provided "the equivalent of 2,000 minute loudspeakers, capable of relaying frequencies up to 20,000 cycles-per-second." A competing RCA plant was opened in 1956 to produce stereo tape machines and tapes. High-quality stereo tape machines that sold for $600 and more soon had serious competition from Victor's latest, which sold for just under $300. Despite their stabilized high prices, RCA Victor's prerecorded tapes and the lower-priced playback machine presented a challenge to the industry, particularly in light of the impending single-groove stereo flat disk.

In 1954, British EMI launched the HMV Stereosonic label, and in 1957, British Decca entered the stereophonic-disk field. This progress nagged at RCA Victor officials, who took the problem of a single-groove stereo record to the Westrex Company, a subsidiary of Western Electric-AT&T. In 1957 they developed a stereo cutting head and single recording stylus.

The Westrex process soon became known as the 45/45 system, because the heads inscribed sound into a single groove at a 45-degree angle to each other. British Decca's stereo recording heads functioned differently, as did those that produced Columbia's initial stereo recordings, which created compatibility between stereo disks and monaural disks, and made use of record players already sold. The speed dilemma was solved by the Recording Industry Association of America's recommendation of industrywide acceptance of the 45/45 stereo. Stereo disks soon accounted for 6 percent of the year's half-billion-dollar gross.

RCA introduced single-reel stereo tapes encased in a plastic cartridge. The speed of the cartridges was reduced from 7½ to 3¾ inches per second, and the width of the tape from one-half to one-quarter inch.

Each tape carried four tracks rather than the earlier two, offering twice the music, and prices were reduced substantially. Columbia provided facilities for joint experimental work on a new cartridge automatic-changer system. The final result, a single-speed reel and playback, was launched in 1962.

The inroad which stereo tape cartridges might have made during the lull in late 1958 and early 1959 was thwarted by Ampex Corporations' fight to make its own 7½-inches-per-second open-reel tape the industry standard. The new battle of speeds between Ampex and RCA confused consumers and helped turn them to the two-channel record and player at a time when their future was in doubt. The smaller-than-expected 10 percent increase in retail record sales during 1958 was attributed to the decline in singles sales and consumer indecision.

Chiefly because it could not produce sufficient cartridge players to meet the demand, RCA Victor added a line of prerecorded 7½-ips tape reels in 1960 and thereby broke the monopoly in raw-tape manufacture enjoyed by 3M and its smaller competitors, Reeves Soundproof and Audio Devices. The addition of another half-million tape-deck players for home use strengthened a market that was taking second place to the stereo-disk business.

The "hype" for stereo music expanded. A few radio stations transmitted stereo music, and Consumers Union, a nonprofit product-research organization, devoted considerable space to audio components. A favorable report on the Shure cartridge sent the small company's earnings skyrocketing, and other manufacturers benefited similarly from Consumers Union's approval. More than 2,500 chain store outlets joined the boom when Newberry, Woolworth, Kress, and J. C. Penney added low-price stereo players, bearing their own names, to their stock.

Double-channel broadcasting over AM and FM outlets and multiplexing—transmitting two signals on a single channel—were factors in the speeded-up programs by major manufacturers to re-record the classical repertoire in stereophonic sound. So too were "gimmick" demonstration records, which provided such illusions as the "bounding brass" that "ping-ponged" melodies from side to side. The most successsful of these were on Enoch Light's Command label. He used expensive thirty-five-millimeter film tape for his master recordings to produce LPs that were the first to provide greater definition and wider range of sound.

Interest in stereo technology mushroomed, sales increasing by 21 percent in 1959, 26 percent in 1960, and 30 percent in 1961. By New Year's Day, 1961, 7 million of the 30 million phonographs in American homes were capable of playing stereo disks. Poor man's stereo hit supermarkets in mid-1960, when Pickwick Records developed a "compatible fidelity" processed disk that retailed for nearly half the price of the major labels.

Single-disk sales dropped 32 percent during 1959. The majority of

America's 4,500 disk jockeys were forced to adhere to a Top 40 program format and played the same hit singles over and over again, so that teenage buyers often tired of the songs before getting into the record stores. Except for superhits, the sales of a successful song tumbled from a million to 200,000 copies. Because of the explosion of releases and the drop in sales, distributors hesitated to promote releases. Eighty-eight labels made all the places on the *Billboard* Hot 100 chart, but only one out of every three was the product of a recently formed firm.

The major record companies now focused on massive LP release programs. Whenever a track from one showed any sign of promise, it was quickly released as a single. Columbia emerged for the first time with the major share of the business, 21 percent, followed by RCA Victor, Capitol, and Decca. Goddard Lieberson ascribed the success to the Columbia Record Club, which had more than one million members. The proliferation of the mass mail-order business, as well as that of record racks, forced the closing of speciality- and department-store record counters and speeded the doom of the small stores. Their sales had increased, but dollar volume was down considerably because of various sales strategies on the part of manufacturers, rack jobbers, and discounters. To push their rival speeds and increase sales of new packaged music, Columbia and RCA Victor offered discounts to the country's 600 distributors and all retail dealers, ranging between 10 and 20 percent above the standard 38 percent, but forcing profits down by a fifth. The changes were the result of a decision to cut the cost of singles and hold the line on manufacturer-established packaged-music prices.

Hoping to restore the singles business, Lieberson proposed the gradual restoration of a one-speed industry through the introduction of new singles. Columbia's Stereo Seven—seven-inch, 33-speed disks—and Victor's Compact 33 singles were introduced almost simultaneously during 1960, followed the next year by Capitol and Mercury. Many trade insiders were certain that it would take between two and five years to switch the teenage market away from 45s, which, coupled with the shortage of an inexpensive automatic 33-speed singles player, discouraged promotion of the single. The picture for 45s brightened dramatically throughout 1961, rising from 182 million units sold that year to 210 million the next. This led to considerable discussion in the RIAA's inner circles. When Congress resumed its debate on revision of the 1909 Copyright Act, the RIAA discovered the mission which would occupy it for the next several decades, ancillary to their war on record pirates.

The RIAA had been involved in a number of projects. It unsuccessfully campaigned in the mid-1950s to remove the 10 percent excise tax on retail record sales, which remained in effect until 1965. Downward adjustments were made reluctantly, for, as the RIAA claimed, the net profits after taxes for record manufacturers had fallen from 6.8 percent to 1.7 in 1964. List prices were cut and proportionate reductions made in dealer-distributor costs. A majority of the industry reduced list prices on LPs by twenty cents and by four cents for 45-rpm singles, passing less than half of the savings along to the consumer.

The RIAA Gold Record award to the manufacturer and the artist for the sale of a million single records was first given in February 1958. To afford recognition to multi-LP collections, the criterion was revised again in 1961, and called for a minimum factory billing of one million dollars, rather than the sale of 500,000 LPs.

The RIAA accelerated its fight on record counterfeiters in 1961, stimulated by the activities of ARMADA—the rival American Record Manufacturers and Distributors Association—among whose associate members were some of the major labels. The RIAA in 1962 helped pass a federal act providing criminal penalties for trafficking in counterfeit disks.

Open war between the RIAA and the music publishers represented by the MPPA over the issue of compulsory licensing broke out violently in 1963. The RIAA warned that to repeal compulsory licensing "blithely ignored" a fifty-year practice. Payments to the Harry Fox Agency of the MPPA had risen from $4.4 million in 1955 to $13.1 million in 1961. Programming of phonograph records supplied by RIAA members now occupied 79.7 percent of airtime on a number of AM stations. The RIAA was in favor only of a provision calling for criminal penalties for counterfeiting phonograph records.

In 1964, Henry Brief, the RIAA's executive secretary, confided that the manufacturers for whom he spoke were investing $68.7 million yearly to create new products. Proposed legislation would increase mechanical-royalty rates paid to writers and publishers by 50 percent and also increase the cost to the manufacturers of a single by two cents and of an LP by twelve, which, he maintained, would increase royalty income to nine times the record industry's entire profit.

An all-star delegation of record-company presidents testified before the House of Representatives' copyright Subcommittee in June 1965 accompanied by Thurmond Arnold, now special Washington counsel for the RIAA, and a Harvard economics professor, John D. Glover. Glover stated that whereas the disk industry's profits after taxes had fallen from $6.1 million in 1960 to $4 million in 1964, royalties paid to copyright owners had risen from $17.4 million to $25.2 million in the same period. He argued that the statutory rate be reduced. Alan Livingstone, Capitol Records president, testified that while the label was the number-one company in terms of retail sales in 1964, it had realized only a 3.3 percent net profit from sales.

The issue of payment by broadcasters for their use of records was raised by Livingstone. It was inequitable, he said, that the record companies, the performers, and the arrangers did not receive a single penny from air performances of their work.

In October, the House Judiciary Committee passed a copyright revision proposal, which, among other provisions, extended copyright protection to the life of the holder plus fifty years, imposed a nineteen-dollar license fee on each jukebox, and raised the mechanical fee to either two and a half cents for each work or half a cent per minute of playing time or fraction thereof.

Prior to his resignation in 1955 from Columbia Records and the RIAA of which he was also president, James Conkling had persistently supported a record-industry awards night. The National Academy of Recording Arts and Sciences was formed to reward artistic creativity in the recording field with a significant symbol and chose Conkling as temporary chairman. He now believed that recognition of achievement should spring from creative people themselves. Sixteen record industry people attended the first membership meeting in 1957.

The first NARAS awards were presented in 1958 in Los Angeles, and the choices were strongly reflective of a bias for West Coast music and against the Top 10. Many artists who had dominated the sales lists were ignored. A metal replica of a talking machine on a wooden base, for which the NARAS became best known to the public, was presented to the winners.

The 1960 Grammy nominations evoked complaint from Goddard Leiberson questioning the manner of selection, which he said could not "provide a true measurement of artistic merit or give any indication of the record industry's accomplishment in the classical, jazz or popular fields." Nominating practices and voting procedures were reassessed and the categories were increased from twenty-eight in 1958 to forty-two in 1966. A formal voting process was taking shape. An initial lengthy list of selections was screened to ensure eligibility, correct any errors, and make certain that nominations appeared in proper categories. These were processed and made public at the annual Grammy Awards presentations, which were first shown on television in 1959.

Members of the Nashville music community organized a local chapter in 1964, the first year that country-and-western artists received Grammys. Rhythm and blues was recognized in only one category, whereas country music appeared in six, a discrimination that was not repeated on the next year's ballot. Further modifications occurred over the years.

Cutthroat competition brought about the formation of a number of record-business merchandising-related associations in 1959, the most effective of them being ARMADA and NARM, the rack jobbers' National Association of Record Merchandisers. Both faced problems created by the larger record firms, which cut prices, offered special discounts to favorite accounts and free records to disk jockeys, jammed the market with tie-ins, and bypassed rack jobbers to do business with "dumpers" who bought up discontinued records and resold them at a large discount to retailers, who returned the shipments to distributors and were credited with the regular discount. Eventually, at ARMADA's insistence, some of the majors foiled the dumpers by scrapping deleted stock at substantial losses or selling it through special mail-order-only programs.

With significant budget-LP catalogues, RCA Victor, Columbia, and Warner Brothers were among the initial twelve manufacturers that joined the NARM in 1958. Supermarkets and self-service and five-and-ten-cent stores were selling one out of every three LPs supplied by rack-jobber NARM members.

Rack merchandisers serviced 18,500 outlets that made half of all rack sales, representing 27.4 percent of 1961's $513-million gross volume. The following year, with their rack customers increased to 24,000 outlets, the NARM's member companies increased their total sales gross by 25 percent. They were becoming the largest single factor in record retailing. The biggest NARM members now owned retail record departments and spread out with installation in post exchanges. A few acquired major distributorships or formed distribution pools, complicating the role NARM held in the national distribution apparatus. The request in early 1963 by some NARM rack-jobber members to be treated as full distributors with the prevailing discount privilege was refused. Instead, CBS Records initiated a price-stabilization policy. Anticipating that a reduction of about twenty cents on LPs and prerecorded tapes would put all retail dealers on an equal footing with discounters and rack jobbers, CBS fixed wholesale prices. Almost every important record company, including RCA Victor, established similar stabilized year-round price policies.

Initially, ARMADA was ignored by the major labels, possibly because it was far more vocal in calling for industry reforms than the RIAA. Many leading independents found a home in ARMADA, founded in 1939. Rack jobbing, the one-stops, disk counterfeiting, the existing price structure, freebies, transshipping, the brewing payola scandal, and other problems occupied 200 independent labels and distributors who serviced retail outlets NARM members did not. Some $20 million of the $100-million single-disk gross went for counterfeit records. ARMADA, not the RIAA, took the lead against record pirates.

The major retailing operations owned by NARM members dealt directly with the large manufacturers, bypassing ARMADA-affiliated distributors. This preferential treatment was clearly in violation of the Wright-Patman Act, and a matter ARMADA did not ignore. Speaking at the 1963 NARMs convention, in July 1963, Earl Kintner, former FTC chairman and now ARMADA's Washington counsel, warned the industry of the threat of government intervention.

The record industry came under serious government scrutiny in 1955, when a New York grand jury served subpoenas on the Big Four manufacturers and a few smaller companies in connection with a Justice Department investigation of the shipment of free records to radio stations, unfair dealings with independent distributors, discriminatory LP price practices and price cutting. Nothing came of the probe, but it was clear that if evidence of antitrust violation was adduced, indictments would be handed down.

In a sudden reaction to the House subcommittee hearings on deceitful broadcasting practices, as well as to the disk-jockey-payola scandal scheduled for investigation in early 1960, the FTC probed record-distributor relationships with disk jockeys. The FTC hurriedly rushed out the first nine of the eventual hundred or more complaints against record manufacturers and their distributors. Most charged violations of Section Five of the FTC act by paying disk jockeys to favor "certain" recordings, thus giving the defendants "the capacity to suppress competition and to divert trade unfairly from competition through deceptive advertising." As though stricken by a sudden revelation of truth, RCA Victor viewed the complaint as "taken to assure the highest standards for the record industry . . . in the best interests of the people, the artists, record distributors and retailers, and the entire industry."

The consent order framed by RCA and the FTC became a model for all succeeding settlements. There was no admission of guilt, only the promise never to do again what had not been admitted. By the time a new chairman was appointed to succeed Kintner, eighty-two complaints and ninety consent orders, involving every major label, had been issued, and, in the words of a commission press release, "successfully attacked this evil at its source."

The record-club business, which continued to account for more than 15 percent of industry volume, had been the subject of FTC investigation in 1962. Invoking the now-dreaded Section Five against CBS and the Columbia Record Club, it alleged monopolistic practices, illegal suppression of competition, and deceptive advertising. The complaint focused upon Columbia's exclusive agreements with a number of companies.

More than a year after the hearings were concluded, an FTC examiner cleared CBS of all charges, having found that the Columbia Record Club did not discriminate against retail dealers. That clubs had not damaged the industry, he added, was manifest in its rapid expansion to 1964, when 200 of an estimated 2,750 companies regularly produced records. The recommendations were rejected by the FTC general counsel, who reopened the case and led to a final FTC consent order in 1971, ending Columbia's exclusive access to the products of nonaffiliated record companies.

ARMADA's determination to address industry inequities preceded the CBS complaint. Appearing at a hearing conducted by the House Subcommittee on Small Business, Amos Heilicher, president of ARMADA, outlined the dual distribution and vertical integration abuses besetting both independent distributors and retail dealers, for which, ARMADA contended, the major firms were responsible. He told the subcommittee that a handful of companies dominated the record business, "from manufacture down through the retail level." The absence of well-defined distribution, he added, "had given rise to a variety of free-wheeling predatory tactics that must be seen to be believed."

The final FTC Trade Practice Rules for the record business went into effect in 1964. The document addressed the rampant dual-distribution and vertical-integration situation. Retailers were granted more protection, and one-stops and rack jobbers were barred from receiving functional discounts. There was no real change in the vertically integrated structure of the largest companies. Without exception, record companies were restrained from deceptive and discriminatory price-differential practices. The new rules appeared to have little effect on economic concentration in the business. MGM Records acquired additional major distributorships on both coasts, and, freed by the FTC examiner's preliminary ruling in the Columbia Record Club complaint, RCA Victor bought back direct control of the Victor Record Clubs from *Reader's Digest*.

The American record business in the mid-1960s was a far cry from the six-million-dollar affair dominated by a single manufacturer belonging to a vast holding company whose international ramifications were subject to government scrutiny. The simple distribution system—manufacturer-retailer-consumer—had been extended into a vicious circle whose members were at one another's throats. The business had become a vast interconnected international complex.

The earliest harbinger of this change had come in 1955, when the Electric & Musical Industries combine acquired ownership of Capitol Records, the first international negotiation to result in control of an American record firm. EMI had picked up the distribution of MGM Records in the late 1940s, specifically for the film company's important soundtrack record library, and King Records of Cincinnati, for some country-and-western hits.

Recordings of American origin accounted for about one-third of Europe's purchases, and the combined total of 75 million disks sold in Great Britain and West Germany in 1954 represented a third of the American market. The high price of records in Europe, antiquated production machinery, and bigger publisher royalties mitigated against increased sales.

Mannie Sachs, who was in charge of all RCA Victor records, and George Marek, who headed the company's A & R department, introduced the 45-speed disk and player in England and completed a reciprocal distribution agreement with Sir Edward Lewis, of British Decca, effective in 1957, when Victor's fifty-year liaison with EMI-HMV terminated. The company maintained its London label in America, but gave RCA access to its concert-music artists. Similar distribution arrangements were made simultaneously with Telefunken, to inaugurate Teldec Records in West Germany, and Musikvertrieb of Switzerland, thus providing the RCA Victor logo worldwide visibility for the first time.

Rights to Soviet music tapes for North America had been assigned in 1950 to the American popular-music publisher Lou Levy. In 1956, EMI got European and Asian rights to all music recorded in Russia, for distribution by its foreign affiliates. American Decca renewed a long-term pact with Deutsche Grammophon for both its classical-music repertory and popular music from its Polydor division.

The competition engendered by RCA International's new major presence in the European market and the rapid spread of record clubs in Britain and later in Germany reduced the cost of European disks. The clubs handled only records withdrawn from domestic and American catalogues. EMI secretly purchased a three-quarter interest in the World Record Club. When the ownership was made public, EMI explained to the dealers that its sales to them had doubled despite the success of its clubs. RCA speeded up the termination of its contract with EMI in the late spring of 1957, in order to activate the partnership with British Decca. The RCA-EMI divorce enabled Angel Records to increase its distribution of classical music in the American market and Capitol to augment its classical catalogue when most HMV concert-music masters were made available to it. A corporate reorganization followed at EMI, and the existing Angel stock was sold to Capitol. Under the leadership of founder-president Dario Soria, Angel had created a catalogue of 500 titles and sold four million LPs.

The expansion of the European record business intensified with the activation of the European Common Market in 1958. The original members agreed to lower and finally eliminate tariff walls between themselves. The "one world" ideology implicit in the ECM would eliminate inefficient and limited record production and lead to large and efficient modern plants.

Bias against American music was then manifesting itself. In England, the Songwriters' Guild, a counterpart to the American SPA, sought to reduce drastically American material on the BBC, where it occupied two-thirds of all musical programming. In France a 45 percent ceiling was imposed on the broadcasting of foreign recorded music on the state radio network, and authors and composers lobbied to increase the quota restriction.

American Consolidated Electronics Industries purchased Mercury Records and its pressing plants in late 1961 on behalf of its owner, Philips's Incandescent Lamp-works Holding Company, representing the second foray into the American record business by a major European combine. The Mercury purchase was a reaction to Columbia's sudden cancellation of their reciprocal understanding, which left Philips without any American connection. The success of RCA International was another factor in Philips's purchase, as was Great Britain's impending admission into the European Common Market.

The formation of a Dutch affiliate, in January 1963, to press and distribute CBS records in the Benelux countries came just before Columbia Records International formally opened its Paris headquarters and introduced Disques CBS in France. The $125,000 paid for a half-interest in an active German firm was expected to provide CRI with a 5 percent share of that market.

BBC programmers still favored foreign music. Some 59 percent of all popular music played by the BBC was American. However, beginning in 1962, the tide turned as twenty-four British records appeared on the best-selling lists in fourteen countries. British record producers were evolving their own "British sound" out of American recording techniques.

Sir Edward Lewis was not pleased with the treatment his American subsidiary, Capitol, gave EMI-produced records in the United States, his largest export market. One example of Capitol's neglect involved the Beatles, who, despite four successful singles, an EP, and an LP on the English charts, with a combined total sale of more than three million units by November 1963, could not attract Capitol to exercise an option to release the British masters in America.

Early in 1963, an American international clearing-house for record masters, Transglobal, obtained rights to several Beatles disks and leased the masters to Vee Jay Records, of Chicago, who put out two unsuccessful singles. Having received no payment from Vee Jay, Transglobal terminated their agreement and returned all rights to EMI. Disregarding the cancellation, Vee Jay released an LP, *Introducing the Beatles*, again producing no chart activity. Responding in 1964 to what he described as "pressure . . . too great for us to hold back any more," Alan Livingstone announced the immediate release of a Beatles single, "I Want to Hold Your Hand," which sold a million copies in three weeks.

Called the "Beatles" after Buddy Holly's Crickets, they were represented by a Liverpool department-store record manager, Brian Epstein. George Martin, in charge of A & R at Parlophone, offered them a contract for four sides, and a one-cent royalty on each single. The Beatles were "very lucky," as Peter Brown and Steven Gains wrote in *The Love You Make*.

> Their alchemy with George Martin synthesized real gold. Although Martin's role in the production of their records changed over the years, he was always their primary conduit, the intermediary who transposed their inarticulate ideas into music. None of the Beatles could read or write music, although Paul was to teach himself. They had no knowledge or command of any instruments except those they already played, and they knew nothing whatsoever about how records were made or the capabilities of the recording studio. . . . The Beatles' first songs were recorded on four-track recorders in monaural, compared with the sixteen- and thirty-two overdubbings of later years. In any event, Martin was to become the interpretive vessel through which they were presented to the world.

The Beatles were selling records worldwide at a monthly rate of $1.2 million when they arrived in New York City in January 1964 to begin a two-week tour with two sold-out performances at Carnegie Hall. United Artists had distributed the Beatles' first movie, *A Hard Day's Night,* for which the group was paid $50,000 plus 7½ percent of the receipts after production and distribution costs. UA did not expect much from the film, but tremendous sales of its soundtrack were certain.

Beatlemania received its first American exposure over CBS television on the *Ed Sullivan Hour,* drowned out by the screams of teenage girls. Representing 6 percent of the total population, 11 million young girls spent 56.3 percent of the entire $650 million paid for recorded music in 1963. The initial Beatle LP, *Meet the Beatles,* outsold their first single, 3.6 million to 3.4 million, the first time an album had sold more than its single counterpart in both units and dollars. Fifty million American dollars in orders for various Beatles merchandise briefly ran neck and neck with those for their Capitol records, which strained the company's pressing facilities and forced Capitol to farm out orders. The company also instituted an unusual uniform 49 percent discount off list price on all its records to all dealers, one-stops, rack jobbers, and sub-distributors.

The "British Invasion" had begun. Led by the Beatles, whose every new release automatically hit the number-one position and eventually quadrupled Capitol's annual revenue, perhaps fifty British groups were enthusiastically welcomed to these shores. Few, however, matched the Beatles' success.

The Rolling Stones were a far cry from the Beatles' clean-cut image. Their blues-influenced music was part of a calculated, created image of rebellious musicians whom parents would love to hate. They cleared the way for pure rhythm-and-blues-oriented British rock musicians and added another dimension to American popular music.

The black heritage on which the Rolling Stones built their act and their music did not motivate the formation of Motown Records by Berry Gordy, Jr. His goal was to provide a safer and smoother kind of R & B that would have no problem crossing over on the trade-paper charts. With a stake of $700 from his family, and a shove from a young Detroit songwriter, William "Smokey" Robinson, Gordy formed Motown Records in 1959, and put all its copyrights into his fully owned Jobete Music Company. Motown had its first two cross-over Top 10 hits in 1961, five the next year, eleven in 1965, and fourteen in 1966. By 1967, seven of every ten releases that produced Motown's $21-million gross that year were sold to the white buyers he had targeted in the beginning.

A & M Records, the other and equally successful newcomer independent of the 1960s, which like Motown remained in the hands of its founders through the 1970s and 1980s, began with a single hit in 1962. A & M aimed at an untapped audience, the buyers of inoffensive easy-to-listen-to instrumental music. The letters of the firm name stood for trumpet player Herb Alpert and Herb Moss, a record promotion man. Their first single, "The Lonely Bull," sold 700,000 copies in 1962. A & M's gross sales jumped by 12,000 percent, from $600,000 in 1964 to

$7.6 million the next year, chiefly from the sale of Alpert's group, the Tijuana Brass, whose five LPs sold 5.3 million and its 45-rpm singles 2.5 million. Alpert-Moss records offered music that successfully crossed the growing chasm between teenagers and adults. In 1966, when the Beatles sold more LPs than anyone, the Tijuana Brass was second, with 13 million copies of their five LPs.

Rock 'n' roll or rhythm and blues no longer controlled the charts. Folk-rock and Bob Dylan caught on with their message songs, which moved others, too, to deal with controversial subjects. Dylan's records and sheet music sold so well that the oldest, and very conservative, music-publishing combine, Warner Brothers' MPHC, bought his catalogue of seventy-eight songs and worked on them with such effect that eight reached the top of the charts. Protest songs destroyed the recording industry's self-imposed strictures on a song's content and playing time.

Other new entrepreneurs entered the music business. They looked for young persons who were infatuated with rhythm and blues or rock 'n' roll to do a multitude of chores, including production of records.

The teenaged team of Jerry Lieber and Mike Stoller began to write in the early 1950s and gradually took over other aspects of production. Presley's earth-shattering "Hound Dog," a cover record of an earlier Lieber-Stoller production, brought the team to the attention of conservative major firms. In 1959, when they were twenty-five, Lieber and Stoller had ten RIAA Gold Record awards, and were enjoying profits from sale of 27 million singles, many of which they had both written and produced. Like other successful independent producers, they leased masters on a 2 or 3 percent override basis, or 8 percent of retail price, from which their artists received a share.

The next most successful independent production operation in New York was Aldon Music, formed by Al Nevins and Don Kirshner in 1958. Nevins, a veteran of Tin Pan Alley, moved Aldon Music and its stable of writers into the Brill Building, the camp of new music's enemies. A stable of the potentially best songwriters in the New York area, including Carole King and Gerry Goffin, Barry Mann and Cynthia Weill, Neil Sedaka, Jeff Barry and Ellie Greenwich, could be found on the Nevins-Kirshner payroll, producing what became known as "Brill Building Pop." Their average age was under twenty-five, and they provided Aldon Music with more than 300 recorded copyrights and 200 chart hits.

Rather than send slapdash demonstration recordings of his writers' material for consideration, Kirshner invested in high-quality production. Kirshner's ability to produce master tapes that could be put on the market as they were led to the formation of two record companies, for which Columbia Pictures paid about $2.5 million in 1963. The package included the future services of all contract songwriters, the Aldon copyrights, and Kirshner as full-time operating head at $75,000 a year to supervise Colpix Records and other music and film-production subsidiaries owned by the studio. The Screen Gems-Columbia combine, which included Aldon Music, published 173 songs, recorded on thirty-seven labels, in Kirshner's first year as president. Twelve of them sold more than a quarter of a million.

Many successful artists formed their own production teams and leased finished master tapes, leaving the full responsibility for sales, distribution, and promotion to the record company. They included Frank Sinatra who, no longer satisfied with his Capitol contract, formed the Reprise label as a subsidiary of his umbrella corporation, Sinatra Enterprises.

Poor times for the feature-movie business were being mitigated by profits from record and music subsidiaries. In its first four years, Warner Brothers Records continued to lose several million dollars annually. Warners' considered shutting down, but was dissuaded when a hoard of unpaid distributors' accounts was discovered. In 1961, John K. "Mike" Maitland became president and conditions gradually improved. The new vogue for comedy LPs made stars of Bob Newhart, parodist-singer Allan Sherman, and Bill Cosby, who in one period accounted for nearly half of all Warner LP sales. Warner's trio Peter, Paul and Mary used Dylan songs and other message material to precipitate the international popularity of folk-rock and sold 13 million LP units over the next eight years. The addition of Reprise's contract artists to Warner's existing roster of twenty-eight complicated matters, but it did bring in the Latin-beat rock singer Trini Lopez, Sinatra's daughter Nancy, and Dean Martin, who struck gold in the country-ballad field.

The union of Warner with Reprise sprang from a desire to shore up sagging feature distribution figures with a guaranteed box-office attraction. Involved in the merger were four new Warner features starring Sinatra, all of which proved to be theatrical failures, and the sale of Reprise's impressive library of master tapes. The singer got a total of $22 million in a capital-gains arrangement, including twenty-two unissued Warner Brothers Records shares, which gave him a one-third share in the new company and of profits from labels that might be acquired in the future.

When NBC-TV planned to build a vast audience for a new comedy series through cross-promotion of recordings by Beatles-like performers, Don Kirshner was brought in to direct the project. The television series *The Monkees,* starring the first successful "bubble-gum group," was calculatedly created to attract young females between the ages of nine and twelve, who would prove their devotion by buying the Monkees' records and products. The Monkees' fame lasted slightly less than two years. Their first single, "Last Train from Clarksville," published and produced by Kirshner, as were all Monkees' song and records, was issued on a new Colgems label, and distributed by RCA Victor. It hit the number-one spot, with over a million sales, in time for the Monkees' first telecast. It had sold more than three million units by the end of 1966, as advance orders for more than a million copies of a second LP were coming in.

A billion-dollar annual gross loomed. Philips and RCA became locked in a battle of tape systems, whose outcome brought the business to a cross-roads. The stalemate between Ampex and RCA in the late 1950s over tape speed and mode had been resolved; the former's reel-to-reel configuration was now standard for high-fidelity sound and reproduction in recording studios. Victor's predilection for the tape cartridge was rooted in a corporate judgment that record buyers were no longer chained to the living room. The total market for duplicated pre-recorded tapes had not yet reached the potential envisioned for it, but it was thought automobile owners would expand it almost five times, to nearly half a billion dollars in 1970.

Four years before the RCA reel-to-reel 3¾ ips cartridges were placed on sale, George Eash developed his own continuous-play tape cartridge, which offered 600 feet of stereo music at a speed of seven inches per second. Eash told *Billboard,* in March 1966, that "the tape was guided across the playback heads by a capstan and pinchwheel assembly," and was based on the old continuous-loop principle. "The problem was in getting a hub and reel shape plus utilization of a lubricant to allow the tape to slide freely over its adjacent layer." With an eye on its use as an automobile, entrepreneur Earl "Madman" Muntz began improving this Fidelipac in 1958, and met Eash in 1961. He opened his Muntz Stereo-Pak headquarters in 1963.

Orders were taken for the installation of a hang-on four-track cartridge playback, which was heard through a car radio speaker. The Muntz library of taped music grew as the cost of loading a continuous tape fell from thirty-five cents to a nickel. Franchise operations and Stereo-Pak distribution centers expanded. By 1965, Muntz was anticipating a retail gross of $18 to $20 million.

William Lear, inventor of the car radio, developer of the Lear Jet, and a Los Angeles area Muntz distributor, doubled the playing time of the Muntz cartridge by increasing the tracks to eight. Limited production of a patented Lear Stereo 8 playback unit began, and improved cartridges were supplied by RCA Victor. A contract was awarded to the Motorola Radio Company, to which Lear had assigned his first car radio, for the manufacture of all Lear radio-tape units. Most leading car makers opted for Stereo 8.

Lear tape components designed to be attached to existing stereo phonograph systems went on sale in 1966, simultaneously with Philips's compact cartridge system, distributed through Mercury Records. The twin spools, encased in what Philips called a "cassette," used one-eighth-inch-wide-tape, which played at 1⅞ ips. Unlike the Stereo 8 and the Muntz Pak, the Philips players permitted both playback *and* recording, and were available in home units as well as for automobiles.

The original cassette deck had been introduced in Germany in 1962, and the following year in the United States by Norelco, Philips's North American Electric Company. The first Philips Carry-Corder had inferior sound quality as well as tape hiss, which was not fully erased until the introduction of the Dolby Noise Reduction System in 1970. New American companies were organized to supply an anticipated five million four- and eight-track stereo tape units by the end of 1967.

Television's switch to all color, starting in the 1965–66 season, meant larger profits for the film companies, whose vaults were packed with Technicolor movies, awaiting the networks' highest bids.

During a period of internal reorganization and diversification, the Music Corporation of America—by then the largest talent agency in the world—paid $11.25 million in cash for the Universal-International lot, the largest in the world. Universal retained a portion of the main facilities for ten years. Now both a landlord and a producer, MCA retained a larger share of the $150 million spent by the networks for television film production in 1961–62.

Those holdings served as the basic leverage for a 1967 proposed exchange of stock that would merge Decca-Universal, as separate divisions, with MCA. The initial plan was to exchange one share of Decca capital stock for one new MCA preferred voting stock and a third of a share of MCA common stock. Opponents argued that Decca was contributing 47 percent of its earnings to acquire 24 percent of the combined operations, whereas MCA would hold 76 percent in exchange for the balance of combined earnings. Nevertheless, after approval by the SEC, stockholders of both companies approved the merger, during which time MCA stock fell about forty-five points, in anticipation of government intervention.

A month after the MCA-Decca-Universal merger was consummated, the Justice Department charged MCA with violation of the antitrust laws. MCA quickly surrendered its franchise as a talent representative, leaving 1,400 entertainers with no major representation. In the consent decree, MCA was forbidden, for seven years, to acquire any major film or television production company or any recording operation without permission from the Justice Department.

The growth of Revue Productions and Universal-International Pictures continued. Paramount Pictures' $127-million gross in 1965, including profits from Dot Record Company and music-publishing houses, caught the fancy of executives of the Gulf & Western Corporation. Charles Bludhorn, the G & W chairman, was determined to effect a merger with the studio, whose major real estate holdings could serve as collateral for immediate cash. Paramount's large library of feature movies was another asset. G & W already owned 18.5 percent of outstanding Paramount stock. The merger was consummated in 1966, on an exchange-of-stock basis.

The increasing interest in the film industry by big business set off a

40 percent increase in the price of stock of the eleven companies listed on the New York and American stock exchanges. Warner's record-business holdings, almost bankrupt, rose to a place among the world's giant manufacturers in 1966, through merger with Seven Arts Corporation. Among the assets acquired by Seven Arts were the Burbank studios and adjacent real estate, whose worth was incalculable. *Variety* reported in January 1967 that Frank Sinatra was furious that Warner had not given him a chance to buy company stock at the same twenty-dollar price paid by Seven Arts. Shortly after, he sold part of his one-third control of Reprise Records, leaving him a 20 percent share in all profits.

During the Warner/Seven Arts negotiations, a merger was concluded between the Transamerica Corporation, a holding company with insurance, car-financing, and small manufacturing interests, and United Artists, then the most successful American picture distributor, and owner of UA Records, the most successful manufacturer of soundtrack LPs. In the past six years UA had distributed six feature movies that earned more than $25 million each in worldwide rental fees. With Transamerica's $2.5 billion in assets to finance new feature pictures, the film-distribution company could enlarge and diversify the operation of a leisure-time music division.

"The times they are a-changin'," Bob Dylan sang. And so they were.

9 Payola and the Celler Hearings

THE center of music publishing had become New York's television and radio networks where songwriters could demonstrate their material. New songs were taken directly to the A & R men, who determined what would be recorded and promoted. Next in importance were the disk jockeys and radio stations, which had become the primary conduits of exposure to the public.

ASCAP made its quarterly payments for 1953 on the basis of 19 million performances for writers and 16 million for publishers, paying eight cents per airplay to the former and twenty-four cents to the latter. Current performances accounted for 55 percent of the publishers' distribution; availability—performance of songs more than two years old—for 30; and seniority for 15.

At BMI, where all affiliated publishers, but only writers in the company's writer plan, shared in the bulk of distributions, measured by broadcast performances, publishers received six cents for each performance on a network and four cents for those on independent stations. None of the additional factors taken into account by ASCAP were recognized. A song whose performances were mostly those of its recordings earned approximately $6,000 from BMI, whereas one that enjoyed both live network and recorded airplay received more than $15,000.

The popular-music business played the angles as it never had before. Successful young publishers used their BMI royalties to break the gentlemen's code among ASCAP publishers, which observed proprietary interests in copyright renewals that had existed for years. In such a deal, the established price was between seven and ten times the song's ASCAP earnings. But where such transactions had been concluded when the songs were first written "for hire," the publishers possessed all rights. Paying larger sums than usual for even a share in an important copyright, the BMI publishers built up future availability and seniority rights.

The public increasingly was receptive to music of all types. However, sheet-music sales hit rock-bottom. Songs that were number two or three on "Your Hit Parade" sold fewer than 10,000 sheet copies, the number one around 15,000. Most publishers refused to supply the sheet-music racks with an initial order of 75,000 copies, fearful they would be buried by the returns.

Conditions at ASCAP were far from serene. During the first year of his presidency, Stanley Adams played a key role in the 1953–54 negotiations with television broadcasters. Prevailing rates from independent broadcasters were reduced to 10 percent below those prevailing in radio. The networks would enjoy a similar reduction in every year in which their income did not fall below that in 1953. ASCAP's fortieth anniversary was enthusiastically celebrated in early 1954. The top publishers formed a solid block of support and reinstated Adams as ASCAP's chief executive, leaving the rest of the membership little voice. In general, the membership approved the society's operations, though there were some unhappy older members who wanted ASCAP to protect their declining incomes by disregarding all elements, other than seniority, of the four-fund distribution plan instituted by the 1950 consent order.

In order to end the collaboration between BMI and ASCAP members, in April 1955, ASCAP relinquished all rights to such works. Having won approval in the amended consent decree to gain access to music written by ASCAP members on such a nonexclusive basis, in 1952 BMI withdrew its longstanding appeal in *Marks v. ASCAP*. Simultaneously the $200,000 annual guarantee paid to Marks since 1941 was reduced and a new understanding was extended until 1959. The million-dollar option to buy the Marks catalogue was voided. BMI now began to pay all writers, regardless of affiliation.

At the same time, Hans Lengsfelder, proponent of distribution based on a straight 100 percent current-performance basis, gained support from young writers who wished to enlarge the logging procedures. Until the amended decree, only the networks had been logged exhaustively. One hundred regional nonaffiliated stations were added and covered on a rotational sampling basis for two and a half hours a day. This increased the local sample base by over half and airplay credit by a quarter. The points assigned to publishers and writers for performances on a sustaining network show were reduced considerably.

By such actions, ASCAP hoped to attract those smaller publishers whose records dominated airtime on independent stations. But ASCAP could not match BMI's payments in this area; the publisher of a "record" hit that received little or no play on the network shows got $15,000 from BMI. By contrast, 80 percent of ASCAP distribution of income from radio was based on network performances.

Having served the maximum three terms, Adams was replaced by Paul Cunningham, a veteran ASCAP public-relations and lobbying operative in Washington. Cunningham reported the society would distribute $20 million in 1956 and emphasized the self-imposed ceiling on their ratings made by Berlin, Rodgers, Hammerstein, and others. He urged an immediate return to agreements extending television rights through 1961.

With many society members opposing the revised payment system, *Variety* correctly observed in 1956 that an ASCAP storm was brewing. A hastily prepared revision superceding the 1954 Adams plan failed to satisfy the opposition. Lengsfelder and others successfully appealed for a governmental investigation.

Their complaint centered on the weighted-vote issue: a writer got one vote for every twenty dollars he earned. This gave sovereignty to the major publishing houses that served the most prolific writers. Oscar Hammerstein observed in an early 1957 meeting that "there is no reason why a man who owns a thousand copyrights should only have the same voting power as the man who owns one copyright."

The statement was cheered by Arthur Schwartz, composer of such standards as "That's Entertainment" and "Dancing in the Dark." Schwartz, encouraged by anti-BMI materials gathered by the Songwriters of America activists, established the *Schwartz v. BMI* case. Many of these materials referred to BMI president Carl Haverlin's program to maintain the support of BMI stockholders and other broadcasters during a period when BMI's share of performances was a fraction of ASCAP's. The so-called Haverlin "five-year plan," "Your Stake in BMI," offered to the NAB convention in 1948, pointed out that the company had saved the radio industry some $67 million over the contract proposed by ASCAP in late 1940. Haverlin proposed a drive by all stations to achieve an average of 12,000 performances per station in 1953. Schwartz and others believed that there was an inherent preference for BMI music and against ASCAP's, arising from the ownership of BMI stock by broadcasters. The prime thrust of *Schwartz v. BMI* was to prosecute "an action against BMI, the interlocking radio and television broadcasters, and networks, which control it, and all those others who have directly, and indirectly injured writers by placing American music in a strait jacket manipulated through BMI."

BMI's original strategy was to establish a direct link between the Schwartz action, ASCAP, the Songwriters Protective Association, and the Songwriters of America, whose steering committee was made up entirely of past and present SPA officials and councilors. For the next two years, there was little action other than pretrial depositions.

Coming on the heels of the industry's general acceptance of a new standard songwriters contract, lasting until 1957, the close relationship between the SPA and the SOA was solidified by the former's plan to widen its membership and business influence while reducing that of the MPPA. Significant revisions of the existing contract were planned, all of them to the publishers' disadvantage. The SPA also intended to get BMI and ASCAP members the same benefits and petitioned the National Labor Relations Board to become the official bargaining agent for all songwriters.

Meanwhile, BMI acquired repertory and expanded its use. BMI now had more than 9,000 writer members, to whom royalties were paid. Annual guarantees to BMI publishers were used only to keep major firms from going over to ASCAP. Such growth caused Robert Sour, head of publisher relations, to create two new classes of BMI music companies. The first got the standard rate of four and six cents, plus a 25 percent bonus. The second, instituted in August 1955, was paid full logging royalties plus an additional 38 percent, offered in lieu of a guarantee to record companies with publishing firms.

Almost overnight, the focus of BMI's concern was diverted to a single individual—Congressman Emanuel Celler, chairman of the House Judiciary Committee and one of ASCAP's best friends. He ordered new hearings into monopolistic network practices by the Antitrust Subcommittee, which would address the involvement of radio and television networks in music publishing and promotion.

Accompanied by John Schulman, attorney for both the SPA and the SOA, former president of ASCAP, Stanley Adams, set the tone for the proceedings. Adams asserted that the networks "have sought to and do control the faucets through which music flows." The committee's counsel took an openly anti-BMI line of questioning, citing the material provided by Schwartz and the SPA.

Carl Haverlin countered that the networks owned less than 20 percent of BMI, while the majority of BMI's fourteen directors admittedly were from network-affiliated stations. As to the faucet Adams invoked, he offered statistics showing that 75 percent of the songs played on radio and television stations were licensed by ASCAP, 15 percent by BMI. Furthermore, he said the $22 million paid to ASCAP in the past year was three times the amount paid to BMI.

Pointing out that broadcasters could have their ASCAP rates fixed by the court, the society's general counsel, Herman Finkelstein, told the committee that there was no longer any justification for BMI to continue under broadcaster ownership. He concluded that broadcasters would dare to take ASCAP music off the air when the current contracts expired. Celler stated that action by the Justice Department against BMI was inevitable.

Voluntary statements by CBS president Frank Stanton and NBC president Robert Sarnoff that they "would take another look" at their holdings in BMI, provided the lawsuit that been pending for the past three years was dropped, were received enthusiastically by Celler. A spokesman for the SPA, however, felt the divestiture of BMI stock by the entire broadcasting industry would "not give sufficient relief to songwriters."

Celler's antipathy toward BMI and the music business continued. Moreover, his misperception of contemporary popular music, in particular rock 'n' roll, seemed to come from a deeply rooted Puritanism and fear of the black man in the American psyche, a perspective sadly shared by others. Clergyman of several denominations attempted to censor the lyrics of R & B material. The most politically liberal among the SPA membership had little compunction in turning the wave of racial hatred, masked by concern for the nation's youth, against BMI.

In a *Billboard* editorial in April 1955, Ackerman wrote: "Rhythm and blues struck the big time not because of obscene lyrics. It did it through the musical talent of its A & R men and artists. . . . It ill behooves pop publishers, mechanical men, and artists to demean R & B, and it is unfortunate that some will cease only after they latch on to an R & B hit."

The communications industries' expansion into the record business made the ownership of established music firms an equally attractive investment. MGM's Big Three bought the British B. Feldman music interests, for about $225,000. That transaction was followed by a merger of Robbins, Feist & Miller with the British firm Francis, Day & Hunter. The Howard S. Richmond operation expanded in a major way in Europe and Latin America. Richmond believed that placing music with his own self-contained foreign subsidiaries was more remunerative than assigning the American copyrights to a single European publisher, who then subleased it through the world.

It was no longer the quality of the song that counted, but its sound on recordings. The sheet-music business collapsed again. The number-one song of 1958, "Volare," sold fewer than 250,000 copies. The old-line houses' relations with recording artists and A & R men had changed. Most one-hit artists and one-shot record companies ignored the exclusive recording contract. Sixty percent of the records played by the Top 40 stations came from independently produced masters, whose "spontaneous sound," according to Pat Ballard, an ASCAP writer, was "a reckless disavowal of the old rules, and [had] enough dissonance to appeal to the youthful record buyers who feel more at home with sounds that aren't parlor-perfect."

The stalemate in 1956 between ASCAP and every major studio, except Universal, in securing blanket licenses for the use of copyrighted music on film further split the organized songwriters represented by the SPA and the MPPA publishers. Some MPPA members, whose music was prominent in the pre-1948 material being sold to television, bypassed ASCAP and licensed directly on a per-song basis. Maintaining that reruns on TV required a new license, the Harry Fox Agency negotiated with the videofilm distributors and syndicators in order not to disturb their relations with telecasters.

The SPA contract due to expire at the end of 1956 was extended by mutual agreement with 204 publishers, and in considering reforms, the association planned to assume the duties of both the Fox Agency and ASCAP and collect performance money from the motion-picture and television industries. In July 1955, the SPA offered licenses to some Hollywood studios, but quickly learned that it needed the support of the Screen Composers Guild, the only body certified by the National Labor Relations Board to represent background-music composers. The older and more conservative SCA had flourished since its formation in the 1940s. Foremost among those protesting ASCAP's proposed change in the distribution formulas affecting background music, it warned that BMI already was the "biggest threat . . . in the background music field" and that unless proper consideration was given its members, more of them would join BMI.

When Leith Stevens, president of the Composers and Lyricists Guild of America, claimed that his organization alone represented those working within the broadcasting industry, the SPA and the guild asked the NLRB to certify one of them in that capacity. With the exception of those employed by the networks, screen-music composers and arrangers were regarded as hired contractors. Some SPA authors joined the guild in anticipation of a merger of the two or the absorption of the guild by the SPA. When the CLGA insisted on preserving its autonomy, the SPA joined the networks and proposed a dismissal of the guild's petition, and then resumed the fight to administer film-performance money.

The MPPA wished to form a film-performance collection agency to license the public-exhibition rights of motion pictures, taking them away from ASCAP. Most publishers objected that it would give songwriters and composers an equal role in fixing license fees. The reason for the support of the agency plan was the presence of film-company men on the ASCAP board. Many felt ASCAP would improve its legal position if it gave up film performing rights. However, its influence and income would thereby be greatly reduced.

The release of pre-1948 motion pictures continued to enrich the networks as well as the studios, the latter maintaining that music rights in a motion picture were vested in the film producers. The publishers held that a performance right differed from the synchronization or film-recording right for which the Fox Agency was paid, and that an ASCAP license was conditioned on and subject to the clearance of all other rights. Clearly the issue needed federal clarification at the behest of either the former SPA (as of 1958, the American Guild of Authors and Composers) and its militant new president, Burton Lane, or the MPPA. Legal tests were unresolved, leaving in doubt ASCAP's right to collect from television for the use of theatrical motion pictures.

The years 1957 through 1959 were victorious ones for the SPA and its public-relations front, the SOA. Unremitting militancy against BMI was expected to speed up the networks' divestiture of their holdings in BMI as well as of their record companies. When most MPPA members would not relinquish control of their copyrights, the SPA moved to whittle down the publishers' control. A majority of SPA members accepted a new centralized collection agency to deal directly with the publishers in all areas except performance royalties.

The change in May 1958 from Songwriters Protective Association to American Guild of Authors and Composers was explained by Lane as due to the incorporation of serious music composers and their publishers. Because the AGAC insisted on meeting with the publishers before putting its demands on paper, it was not until November that the writers drafted a replacement for the several-times-extended 1946 document. In 1958 the AGAC prepared to make the initial distribution of collections 5 percent, minus a fee, from approximately 1,500 publishers. The AGAC councilors agreed to an indefinite extension of the existing songwriters' contract and prepared to enforce reforms.

To broaden the local radio logging base, ASCAP's consultant, Dr. John Peatman, put all taping of radio stations in the hands of thirteen auditors under his direction. The change was made to assuage some smaller publishers, who had received lower payments. A more difficult problem was the fact that, except for the powerhouse publishers, 80 percent of the ASCAP publishers had some form of affiliation with BMI and were arguing for the renewal of collaboration between ASCAP and BMI writers.

ASCAP began yet another heated encounter with telecasters. It had taken the television industry four and a half years and $100,000 in expenses before the ASCAP contract due to expire at the end of 1957 was accepted. The All-Industry Television Committee was confronted with the probability that the charges made in the Schwartz case might be used by the networks against ASCAP. While Finkelstein and Hammerstein sought to dissociate ASCAP from the lawsuit, its own president, Paul Cunningham, was one of the plaintiffs.

Under terms of the 1950 consent order, broadcasters could ask the court to fix rates, but they feared an increase because it had been admitted that ASCAP licensed 80 percent of the music on TV networks but was paid on the ratio of only two to one over BMI. Late in 1957, ASCAP agreed to renew for four years at the same rates.

The 1957–58 television season began with almost two dozen programs that plundered Tin Pan Alley's vaults. Popular comedians had been replaced by affable singing personalities. Perry Como knocked Jackie Gleason's *Honeymooners* from the Neilson ratings, and Dinah Shore's southern charm mowed down the opposition. Yet before winter arrived, only Como and Shore remained. Clearly, television was not a medium to which people looked for their popular music, although there were exceptions, such as the appearances of Elvis Presley and other rock-'n'-roll stars. Television's dependence on ASCAP was essentially for original background music and the standard songs on which adults doted.

The last major technological revolution in television—from the kinescope copies of a live show, made by filming from the monitor and ready in three hours, to high-fidelity videotape in color, ready to be played back immediately on the same machine—was in the wings. Color tape was first demonstrated in 1955 by Bing Crosby Enterprises, which, like RCA, had been experimenting with magnetic tape for TV since 1951. Tape on television alleviated the need for delayed broadcast. It also made possible easy files of fourteen-inch spindles of two-inch high-quality tape that could be used again and again. Overnight, stock of the Ampex Corporation rose. Soon after, Crosby Enterprises got out of the videotape business, selling all inventory and equipment to 3M.

In 1957, Ampex announced that a compatible Video Tape Recorder would be ready in less than two years. Interchangeability—playing a tape recorded on one machine on a different one—was solved, making it possible to tape a show in New York and ship it anywhere for local broadcast. Shortly after, Ampex exchanged patents with RCA, which controlled the most significant advances in color videotape, and the two electronic companies got control of the most important TV-tape patents and their manufacture.

An all-color future was inevitable. A four-year-old civil antitrust action charging RCA with monopoly of radio and television patents was settled by a consent order. More than 12,000 patents were put in a pool that was available to the entire industry. In 1959 new RCA color and black-and-white VTRs were introduced as was the Ampex Videotape Cruiser, which took the VTR out of the studio and enabled it to do anything a movie camera could.

Most network programming moved nearer the all-tape era. During the 1959–60 season, $150 million worth of taped shows occupied 70 percent of all nighttime programming. For the first time, television's total revenues went over the billion-dollar mark, but profits were down because of increased operating costs.

Radio's soaring expenses provided the same mixed financial picture. Industry profits had declined in 1958 by 31 percent, to $37 million, while revenue rose only 1 percent, to $535 million, approximately half that of television. ABC and CBS managed to retain a loyal daytime following with personality-variety programming and soap operas. NBC went in for a daytime music-and-news format, with live dance bands around midday. Audience share and corporate earnings for the networks stabilized in 1957. Owing to rising operating costs, nighttime network television turned into a mass-entertainment medium, and virtually abandoned the entire national music and national news field.

Local radio-station owners resented paying about 10 percent more than television stations did to ASCAP in 1958, a situation brought about when the All-Industry Television Committee and the networks convinced the society that television's expansion eventually would provide ASCAP with more money and hence justified a reduction from existing "radio plus 10 percent" to "radio minus 10 percent." The Radio Committee wanted a substantial decrease in its ASCAP and BMI rates. Particularly adamant were the stations that employed a Top 40 format, which used more BMI music than ASCAP, and provided BMI in late 1958 with a third of all local radio performances.

Not all broadcasters supported ASCAP's one-year renewal on the old terms which was signed by the networks and many stations. Four choices were open to the industry: contest the fee in the U.S. Southern District Court in New York, which administered the consent decree; get a five-year contract under the existing most-favored-nation clause; sign a one-year renewal; or take out a blanket commercial and a per-program sustaining license. The Radio Committee chose the first alternative, and for the first time prepared to ask a court to fix rates. ASCAP refused to drop its radio rate, arguing that because of the prevailing music and news policy, the society's repertory played a more important role than it had in 1941, when the contract was written.

With talks at an impasse, the committee filed a petition in 1959 asking Federal Judge Sylvester J. Ryan to fix "reasonable rates." The judge ordered the society to renew an interim contract which reduced radio payments by a general 9 percent. The society insisted, however, that the more than 3,000 stations that had already signed would be held to the old rates. A final settlement was made in January 1960 which reorganized ASCAP practices in order to avert new antitrust action.

The solidly entrenched old-line music houses were to lose far more than the $800,000 gained annually by radio broadcasters. In 1957 Hans Lengsfelder and his dissident supporters brought a suit against ASCAP, attacking the weighted vote as unfair to smaller publisher and writer members. Since 1942, he charged, voting control had been vested in less than 5 percent of the writer members and 15 percent of the publishers. Representative James Roosevelt held hearings in 1958, where Oscar Hammerstein, as representative of the society's board of directors, defended his colleagues and affirmed that no one owning one copyright was equal in rights to those with many. Others attacked the weighted voting system, ASCAP's performance surveys and logging procedures, the distribution formulas, the maintainance and availability of minutes and directives, and an appeals process that discouraged appeals.

The subcommittee directed the Justice Department to act "to effectuate the terms and spirit of the consent decree of March 14, 1950." Negotiations between the Antitrust Division and the society began on a consent decree.

President Paul Cunningham hinted to the 1958 membership meeting that changes in the distribution scheme would be at Washington's direction. Paul Ackerman of *Billboard* learned that "ASCAP would like to settle out of court if entrenched interests could maintain control." It was said that the biggest copyright owners "would resign in the event of the drastic change in the weighted vote." Cunningham's offhand remarks raised the possibility that an 80 percent performance factor would force major changes in the operation of large firms.

In 1959, Congressman Roosevelt began the first of a series of public demands on the Justice Department to take corrective action. New terms were discussed with the dissidents, who learned that the government did not agree with them on the matter of a straight-performance pay system. ASCAP and the Justice Department finally reached a tentative agreement on an amended consent order, and final hearings were announced for October 1959.

A 4 percent increase in ASCAP income was confirmed at the spring 1959 meeting by the society's new president, Stanley Adams. As *Variety* had noted, ASCAP needed half a billion dollars in the bank at a nominal 4 percent interest to achieve such an annual income. In 1959, it climbed to $30 million.

Objecting to what they viewed as a greatly watered-down revised consent order, some ASCAP writers prepared for the October hearings before Judge Ryan, ASCAP's legal overseer. Weighted voting was modified so that no writer or publisher would have more than 100 votes. All subsidiaries of major publishers were regarded as individual members, and any increase in the votes of the top ten publishers was limited. Writers had the option of either a 100 percent current-performance base or a "four-fund" mixed base, which would continue the 20-30-30-20 ratings for current performance, five-year performance average, recognized works (formerly "availability"), and seniority. Four-fund writers could switch over to straight performance royalties at the end of any fiscal year, but those paid on that basis had to wait two years before making a switch to the other option. The publishers' 15 percent seniority factor would be phased out over the next five years, leaving a formula of 70 percent for performances and 30 percent for recognized works. Payments for performances of concert or classical music were raised, as were credits for background music, themes, cues, jingles from 100 to 1 to 10 to 1.

Some young members objected that they would not receive a penny for a new song until four distribution quarters passed. Others maintained that large companies still controlled the board. There were reservations, too, about the chance to give smaller publishers a place among the directors by allowing a group that controlled one-twelfth of the total vote the right to elect one member, which might end secret balloting.

Among the reforms proposed was the provision that ASCAP employ an independent outside expert to design a scientific performance measurement system. ASCAP had consistently refused to use the station logs, arguing that the conspiracy to favor their rival's music extended to them. The overhauled logging procedures of Joel Dean, a Columbia University professor and statistician who replaced Peatman, substantially increased the local radio sample and placed it on a more scientifically random basis.

With majority approval, Judge Ryan signed the second amended ASCAP consent decree. Former Supreme Court Justice John E. McGeehan and former Senator Irving M. Ives were appointed to oversee the distribution of ASCAP royalties. It marked the first introduction of outsiders, other than government personnel, to the society's operation.

A house committee concurrently investigated allegedly widespread payola to disk jockeys, the result of lobbying by Burton Lane and the SPA. In 1956, John Schulman, attorney for the SPA and the SOA, approached a representative of NBC to set up settlement talks with a third-party arbitrator. NBC could not escape responsibility for having created BMI, "a cancer on the music industry," Schulman said. Discussions stalled, but they did hint at the networks' fear that BMI's being paid one-third as much as ASCAP for 10 to 15 percent of the performances would become public knowledge.

BMI's seventeenth anniversary coincided with the greatest number of popular hits in its history. The upsurge of success troubled ASCAP, who hoped that its per-performance costs might bankrupt BMI. It was a possibility BMI had recognized and hoped to cure by discouraging any further guarantees against future earnings.

While awaiting the Celler Committee report, SPA members actively sought a new forum for complaints against BMI and the networks in Washington. A brief charging that a combination of all the leading broadcasters and the largest recording companies conspired to gain full control of all music and demanding legislative action was handed to Senator Warren G. Magnuson, chairman of the Commerce Committee, which had jurisdiction over the broadcasting industry. BMI songwriter members of the SPA protested vehemently but uselessly. Though the standard SPA contract contained an exemption for ASCAP performance payments, it did not apply to BMI, which was unable to induce the SPA to include similar provisions.

The Celler report was issued in 1957. It recommended that the Justice Department undertake "complete and extensive investigation of the music field, to determine whether the antitrust laws have been or are being violated." The Celler Committee's predisposition against BMI was epitomized by a summary of the report which stated that the committee urged an investigation "to determine whether the antitrust laws have been or are being violated by BMI." The last two words did not appear in the printed report.

The focus of lobbying in Washington—the $150-million antitrust suit brought by thirty-three ASCAP members—had ground on relentlessly for forty-two months. Tensions subsided when BMI and other defendants were permitted to examine performance cards of 150,000 ASCAP songs from 1934 to the present. However, only a special dispensation of the court could speed up actual trial proceedings.

Senator George Smathers of Florida in 1957 introduced a bill that would prohibit the granting or holding of a license for a radio or television station by any person or corporation engaged directly or indirectly in publishing music, or manufacturing or selling musical recordings. Smathers stated that BMI had subsidized "hundreds of publishing firms [and] today this musical empire consists of 2,000 such firms."

The high cost of meeting the SPA's constant attacks, mounting legal fees, and steadily increasing performances of its music forced BMI to draw back on its expenses. BMI's radio-broadcaster program clinics were suspended. Song-plugging branches of the company's music-publishing operation were closed. Clerical staffs were reduced. Sale of the BMI publishing department were considered.

Meanwhile, executive positions shuffled at BMI. Robert Burton was made vice president in charge of domestic performing-rights administration. Robert Sour took charge of writer relations while Theodora Zavin replaced Sour as assistant vice president in charge of publisher relations.

From the start, BMI had paid distributions on a local- and network-station performance basis. To build up a repertory of music before the 1941 music war started, the company had made many fixed-guarantee contracts. After the 1954 crisis, arrangements for logging plus 25 percent or 38 percent were made. At the bottom of its publisher roster were about a thousand firms that got the standard four- and six-cent contracts, among them houses that had failed to earn earlier guarantees or had their guarantees reduced. With more publishers applying for affiliation, record-company publishing houses increasing, and a larger share of hit songs in all fields of music, BMI's problem in 1958 was its success. The difference between income and outgo was shrinking dangerously.

BMI's guarantee contracts had first come under scrutiny during the Celler hearings, when the renewal, in 1949, of an agreement with Hill & Range Songs, owned by the Aberbach family, was raised by plaintiffs in *Schwartz v. BMI*. Hill & Range had originally been formed in 1945 by four stockholders, brother Julian and father Adolph Aberbach, Milton Blink, and Gerald King, both of Standard Radio Transcription Service. By 1949, the Aberbachs had formed three ASCAP companies, and it was those that were referred to in a clause in the document that permitted them to operate the firms but did not permit them to exploit songs published through the ASCAP firms. BMI inserted the restriction at the demand of its directors, who felt that a protective provision would ensure the Aberbachs' best efforts. The contract also called for a $100,000 annual guarantee and $250 for each copyrighted song recorded by any of a number of major labels. By 1956, Hill & Range was the seventh-highest-paid BMI publisher affiliate. The "anti-ASCAP" clause, which BMI insisted at the Celler hearings had never been invoked, came under particularly heavy questioning and was used by Schwartz.

Jean Aberbach wrote the Celler Committee that though three ASCAP firms were mentioned during the hearings, only one of them was important. He stated, "As a result of the efforts of this group the ASCAP income rose from $850 a quarter in 1949 to $66,000 this year. . . . This dramatic growth . . . under the well-known difficulties of a young publisher to get any money at all out of ASCAP within the short period of five years is, I am sure, unparalleled. . . . I am informed that . . . our group has emerged as the one showing the highest rate of growth during the years in question."

Similar restrictive clauses appeared in BMI contracts with other major companies affiliated with both organizations. They called for quotas of ASCAP songs vis-à-vis BMI music; agreement not to render any services to a non-BMI firm; payment of the salary of a professional manager to devote full activity to the BMI house; money for special exploitation that would not be deemed to be payment for performances.

Senator Smathers assured his broadcaster constituents that the sole target of his bill was the networks and their recording business. He did not intend to destroy BMI but insisted on public hearings. They were announced for March 1958 by Senator John O. Pastore, chairman of the Communications subcommittee. Arthur Schwartz and Oscar Hammerstein assailed the frightening power of the broadcasting business "over the entire music industry and over the listening habits of the American public." Best-selling social analyst Vance Packard asserted that a "gross degradation in the quality of music supplied to the public over the airwaves" was taking place, due to the hillbilly and rock 'n' roll available through BMI, a mistake that returned to haunt Packard and infuriate Schwartz.

Senator Albert Gore of Tennessee read a telegram from his state's governor, Frank Clement, in response to Packard. Clement found Packard's remarks "a gratuitous insult to thousands of our fellow Tennesseans both in and out of the field of country music." It was not the first time country music, Tennessee, and BMI's involvement with both came to the aid of the beleaguered licensing organization. Not only were politicians attracted to country music but the general public was as well. Programs featuring C & W artists proliferated. ABC presented the *Midwestern Hayride* from Chicago and the "Pee Wee King Show"; Tennessee Ernie Ford was on both CBS-TV and AM radio; and *Grand Ole Opry* was an NBC radio and television feature. In that most sophisticated of cities, New York, country-music sales accounted for 10 percent of revenue.

Nashville's struggle to take charge of country music's destiny in 1956 took on a political tone when Governor Clement rose to the defense of a battered BMI. Clement wired Emanuel Celler that the hearings were part of a "scheme of a small group in New York and California to gain complete control of the music business." He further lashed out at the "plot to stifle competition and country music" by the New York monopolists who "for so many years had prevented free enterprise in the American music industry."

At the second round of Pastore hearings, Sydney M. Kaye, now chairman of the BMI board and a member of the law firm representing BMI, spoke against the proposed legislation. He called the episodes cited by Schwartz and others as proof of a BMI conspiracy "so irrelevant, so trivial, so easily susceptible of explanation, that they lack not only individual but cumulative probative force." He concluded by saying, "It is clear to me that the guilt of which the proponents complain is the guilt of competition. . . ."

Variety reported, "There is no gainsaying the polish of the BMI performance. It was designed as a counter-balance against the ASCAP testimony—lawyer against lawyer, publisher against publisher, composer against composer, singer against singer." BMI also presented broadcasters and spokesmen for state associations, including operators of small radio stations, "to avoid any charge they were representing the powerful web interests."

The motion-picture industry sent its representatives to oppose the bill. At the present time, Paul Raiborn, of Paramount Pictures, said, "adjunct sources of income," such as Paramount's Los Angeles station, KTLA, Dot Records, and two ASCAP music subsidiaries, were vital to his company's existence. ABC indicated that the bill would force divestment of all its owned-and-operated radio and television stations and two music-publishing companies.

After witnesses representing the Motion Picture Association and the Columbia and RCA Victor record companies were heard, Pastore announced a recess. Smathers had shown little interest in his bill, and Gore promised a floor fight if it ever got out of committee.

Only Pastore was present when Schulman read a statement and submitted a 110-page printed addition to his previous testimony. When he spoke disparagingly of the current hit "Yakety Yak," Pastore told Schulman that his own daughter and others had bought the record. The type of legislation Schulman and Schwartz were after, Pastore said, was bad legislation, which would throw "every Tom, Dick, and Harry into the soup," while the songwriters really wanted to get CBS and NBC and their stations.

In direct contrast to his action at the Schulman sessions two weeks earlier, Pastore rarely interrupted Judge Samuel Rosenman when he spoke in surrebuttal of the testimony offered on July 15: "Never in my experience have I seen such sweeping charges made against so many people with such flimsy evidence. . . . They have made the most slanderous kind of charges against a whole industry . . . and they have failed to substantiate these charges with the slightest shred of evidence."

The record of the hearings was sent to the comptroller general, the Justice Department, the FCC, FTC, and SEC, from none of which came any request for further information. Apparently the Smathers bill would have heavy sailing.

In March 1958 the *Schwartz v. BMI* defendants were given access to 225,000 ASCAP cards. A motion to dismiss, filed in March, was not expected to be ruled on for some time. The brief pleaded that, because the songwriter plaintiffs neither published nor licensed their music themselves, they were not directly engaged in any activity that could be injured by an alleged conspiracy, and the suit was dismissed.

All pretrial examination was suspended until Judge Edward Weinfeld of the New York federal court dealt with the dismissal motion. Both sides were pleased when he struck down the principal charge—that BMI and the broadcasters were engaged in a conspiracy against ASCAP—and upheld their contention that the society alone was the proper party to sue in the matter. However, Weinfeld affirmed the plaintiffs' rights to bring a suit claiming that the publishers of some 5,200 of their copyrights, all affiliated with both ASCAP and BMI, had been induced by BMI to discriminate against their music. Once again the possibility of an early trial dimmed.

With the shelving of the Smathers bill more certain, the networks divested themselves of BMI stock. The negotiations with ASCAP for a radio license stalled, but BMI renewed all existing contracts with 3,300 radio stations for five years, and with the television industry on the old basis, a sliding scale from 1.2 percent of gross income to 0.75, depending on a station's revenue. For the first time in BMI's history, a network—CBS—asked for a per-program pact, to pay only when BMI music was used, rather than one calling for fees based on a percentage of all income.

CBS honored Frank Stanton's pledge to Celler to "take a hard look" at its relationship with BMI. Many network executives were unhappy with their BMI association and believed that BMI would be deeply involved in the brewing payola scandal. The networks had withdrawn their representatives on the BMI board. Believing that the hearings had shown there was no merit to the Schwartz charges, CBS was the first to act. At its request, BMI bought back 7,012 shares, representing 8.9 percent of all outstanding stock. NBC followed in September, receiving $21,320 for its 5.8 percent. ABC sold back its 4.5 percent interest a few years later. Sydney Kaye told the Pastore committee that if broadcasters did sell all their stock, and BMI became an association of publishers and writers like ASCAP, competition between the two would ultimately cease. If BMI ownership dwindled to a few stockholders, the profit margin would prevail, and any interest in competition with ASCAP or in increasing the catalogue would disappear.

The divorces had little effect on BMI's precarious financial situation. Annual income from licensees was around seven million dollars, out of which two million went to 4,000 writers affiliated under the Robert Sour plan. A bonus-payment guarantee for writers who consistently showed high performance over a three-year period was in effect, adding about 50 percent to the four and two cents paid all BMI writers for network and local performances. The American Composers Alliance—writers of contemporary concert music—had a separate understanding with BMI, and received an annual guarantee out of which the ACA paid for performances of its music. BMI deducted 10 percent for the administration of income from all foreign societies and then paid its writers the balance on a nation-by-nation basis. Film composers benefited particularly under this arrangement. Most music-performance fees came from screen theaters and were distributed by BMI.

The most serious effort to date was taking shape to collect for the performances of recorded music from the operators of jukeboxes. In 1959, half a million jukeboxes consumed 47 million disks, bought at a discount from one-stops, or supplied free by record companies or their distributors or as payola to give preference to certain selections. The more than $442 million generated was usually split equally between the jukebox owners and the owners of the places where the machines were located. The honored division was already changing in many places, to a front-money or minimum-guarantee system: a sixty-forty split, with the operator getting the larger share, or even an eighty-twenty split.

Celler's 1959 compromise to pass the first copyright law requiring coin-machine operators to pay a collective annual royalty of $2.5 million to the performing-rights societies was halted by the uncommitted industry trade association, the Music Operators of America. Legislation to remove the coin-machine exemption mandated in the 1909 Copyright Act had been introduced in every Congress since 1926, and consistently failed to obtain majority approval.

In 1955, a number of senators urged the MOA to compromise with the licensing societies. Capitol Hill's intervention stemmed from the Senate's ratification of the Universal Copyright Convention, which made the United States a participant in a worldwide copyright convention for the first time. Almost every other nation with coin machines had compulsory legislation.

In Lester Velie's article "Racket in the Juke Box" in the November 1955 issues of *Reader's Digest,* the celebrated crime reporter asserted that the mobster "who owned a piece of an artist" could dictate the use of that performer's records on the machines he controlled. The Chicago criminal establishment also offered ASCAP the passage of a bill removing the copyright exemption if they got back a 20 percent share of the take.

In face of adverse publicity, the MOA retained the accounting firm Price, Waterhouse to send a questionnaire to all MOA members which asked for these hitherto uncollected figures: number of machines in operation, operators' total share of collections in 1955, total expenses less salaries, total value of jukeboxes and other equipment, and total record purchases in 1955.

A bill was introduced in 1956 proposing an annual fifty-two-dollar tax on every machine. It served only to tighten the MOA's ranks, according to its president, George Miller. A lobbying campaign was initiated by the newly formed National Tax Council, organized by the MOA to fight a jukebox tax. A performing-rights society sponsored by the MOA would handle the copyrights used by a new recording company funded by the MOA should there be a nationwide boycott of all ASCAP and BMI music.

With the indictment filed by a Chicago grand jury charging the country's largest coin-machine manufacturer, the J. P. Seeburg Company, with violation of the Sherman Act, the first break appeared to come in the jukebox situation. Seeburg, which soon negotiated for a consent order to dispose of the indictment, entered into licensing agreements with ASCAP and BMI for a new jukebox background service.

Seeburg's problem had no effect on the MOA, which bluntly informed the Senate that they would not "contemplate industry suicide" by agreeing to pay for music. The performance societies had "nothing to offer, nothing to compromise, and nothing to sacrifice. There is no benefit or boon they can confer on us."

An updating of the Copyright Act of 1909, part of an omnibus copyright-revision bill, had the full support of the Copyright Office of the Library of Congress. It would be based on several studies commissioned by the Register of Copyrights, Abraham L. Kaminstein. The first, written by Harry G. Henn, dean of Cornell University Law School, called for elimination of the compulsory-licensing provision or the statutory two-cent rate, and clarification of such provisions as they affected new forms of recording. Conflicting views on the first charge found BMI's Sydney Kaye in opposition to the MPPA, the SPA, and ASCAP. He maintained that without compulsory licensing most songwriters would hand over exclusive rights to the first record company that offered to cut their music.

A second study dealt with performance royalties and damages for unlicensed performances. It asked whether the present penalties should be retained, eliminated, altered, or left to a court's discretion; whether multiple infringements should be lumped together or treated separately; whether innocent or secondary interests should be absolved of responsibility.

None of the other studies dealt with the jukebox exemption now in the hands of the Senate Judiciary Committee on Patents and Copyrights, which was considering a bill offered by its chairman, Joseph C. O'Mahoney. Hearings were scheduled for early 1958. The jukebox-exemption hearings passed uneventfully, after which the O'Mahoney bill was passed by the entire Senate Judiciary Committee and sent through the legislative process, only to be abandoned when the Senate adjourned. As to whether the jukeboxes were important in the exploitation of songs, the accompanying report said that "the evidence indicates that in certain instances, the jukebox does popularize music, that on the whole this is a very minor gain to the composer and author in the overall picture. As a matter of fact, disk jockeys, TV and radio programs are the biggest medium for the popularizing of musical compositions."

The jukebox business now faced a probe by the Senate Rackets Committee, headed by Senator John L. McClellan, as well as the FTC.

The Justice Department had brought antitrust proceedings against major coin-machine manufacturers and ordered them to stop limiting distribution territories and the right and purchase of resale.

The music industry saw the McClellan rackets hearings on the jukebox business and its alleged infiltration by organized crime as the final means to prod a discredited MOA into accepting a licensing deal. Testimony to the McClellan committee highlighted numerous violations of the law—accompanied by frequent interjections of recourse to the Fifth Amendment.

The final report said that the majority of operators were "honest, hard-working citizens," who had been the victims of an "astounding number of racketeers, posing as both businessmen and union officials." There was no rush to push through a bill after the hearings concluded, nor did the MOA back down. A patchwork bill replaced O'Mahoney's original proposal, with a provision to double record royalties for disks sold to jukeboxes. Emmanuel Celler introduced in the House for the first time since 1952 a bill to remove the jukebox exemption from the law. "Every time the composers of America have petitioned the Congress to remedy this rank injustice," he said, "the jukebox interests have pleaded poverty and good citizenship and have attacked the composers as being greedy and bad citizens. This farce of the poor but honest jukebox interests is about played out."

Every element of the world of copyright revision was on hand in June when hearings on the Celler bill started. ASCAP's problems in Washington provided already unfriendly committee members with ammunition against the legislation. As Jay Lewis wrote in *Variety,* the Celler bill was "variously depicted as a way of stopping 'legalized piracy,' and a way of destroying small business, helping monopoly, spreading unemployment, and, even, promoting juvenile delinquency."

The Price, Waterhouse survey which the MOA had revived was based on a sample of 1,285 usable replies and represented a total of 75,756 machines, which produced revenues of $33.4 million to the operators from their half-share. Revenues left after expenses for each operator were an aggregate of $5,871. Assuming an average of sixty-four cents for a record, the 1.285 operators spent $5.5 million to keep each machine annually stocked with an average of 114 new records.

ASCAP countered with its own survey of members' income. The annual average per writer was $2,321; 88 percent of the membership received less than $5,000 in 1958; only 244 earned between $5,000 and $10,000; and only 7 percent received more than $10,000.

Failure to agree on an acceptable bill prompted Celler to offer himself as a mediator between the MOA and jukebox manufacturers on one side and the proponents of termination of the jukebox exemption on the other. His compromise called for a five-dollar royalty per jukebox, the fees to be collected by three trustees, who would distribute the money on the basis of performances of ASCAP and BMI music on the machines. One trustee was to be appointed by the attorney general, one by the licensing organizations, and one by the MOA. The music industry supported the compromise. The MOA asked for a delay getting its members' approval.

While the MOA's final action awaited, a New York grand jury investigated the first intimations of quiz-show "rigging" on network television, particularly the *Twenty One* show. It learned that *Twenty One*, *The $64,000 Question* and *The $64,000 Challenge* had been fixed, in addition to a possible dozen more quiz shows.

A serious legislative debate questioned whether the FCC or the FTC already had the authority to prevent fraudulent programming and whether the hearings chaired by Oren Harris, chairman of the Special Subcommittee on Legislative Oversight, were necessary. BMI's newest nemesis, Burton Lane, president of the AGAC, had unsuccessfully written the FCC and the FTC and addressed the end of the Harris hearings about "the commercial bribery that has become a prime factor in determining what music is played on many broadcast programs." He offered the committee essentially a rehash of quotations from news stories, all apparently new to Harris. Lane furthermore asked the commission to require station owners to get rid of their conflicting music-business interests or have their licenses revoked.

However, he ignored the well-publicized multimillion-dollar television payola business. The rapid growth of television had created "videola," when income from public performances allowed producers to ask for money before they selected music. In one case, reported in *Variety*, a producer asked for $5,000 before he gave a publisher all rights to a theme song. When a change in ASCAP's distribution system reduced the value of theme songs and increased by three times that of a network feature-song television performance, the pitch became, according to *Variety*, "If I don't play your tune, it'll be somebody else's. If you want the plug you gotta give one-third of the publisher's share and cut me in as co-writer." The Harris investigation had not touched on videola on the five-day-a-week morning and afternoon strip shows. The fifteen pieces of music usually performed on such shows offered a potential bonanza of $4,250 weekly to the aggressive publisher.

The Justice Department was aware that in 1958 six million dollars was distributed by ASCAP for background music and theme songs. A new figure entered the scene: the broker representing TV producers. "Some of the indie producers like Barry & Enright [producers of *Twenty One*] saw the gravy potential in tv music and set up their own publishing company," Herm Schonfeld wrote in *Variety*. "Barry & Enright's Melody Music Co. reportedly earns about $100,000 annually from ASCAP."

Harris committee investigators also were questioning disk jockeys to determine the extent of corrupt practices on local radio stations. Local independent broadcasting had bloomed. With libraries of electrical transcriptions or 78-speed disks, stations were reluctant to go along with the 45s supplied to major stations without charge, and to the smaller ones at a low nominal fee. A group of New York stations fought the new technology, citing problems in tracking and cuing on 45s. With the promise of regular shipments of free 45s and two-week priority shipments of new releases, the boycott ended.

To appeal to the thriving new radio audience, Top 40 programming was invented, credited to Robert "Todd" Storz, who began the Storz chain of stations in 1949 by buying KOWH, Omaha. He established a twenty-four-hour "Hit Parade" format that emphasized pop hits, with a careful, small infusion of white R & B covers. There was no ad-libbing, introductions were short and to the point, label credit was never given, nor was there any hint of enthusiasm for any particular piece of music. Advertising jingles and musical station breaks prevailed, as did gimmicks to attract telephone response from listeners, and prize contests to keep them tuned in constantly, with enough public-service announcements to satisfy the FCC. Other stations quickly followed Storz's format.

Surveys of record popularity were in great vogue. ABC devised a "Hit Preview of the Week" feature, in which a potential hit was chosen by producers, singers, musical directors, and air personalities on seven live ABC musical-variety programs. Each tune selected was guaranteed twenty plays during the following week. Other stations followed with similar plans.

The dependence on lists had its inevitable backlash. Mitch Miller said that Top 40 was turning radio stations into "automated disk jockeys." An MGM executive complained that the concentration on single record sales, which teenagers dominated, ignored adults, whose purchases of LPs represented 65 percent of the business. Not every radio station was tied to a Top 40 format, but half of all deejays attending the first annual Pop Music Disk Jockey Festival in 1958 indicated that the top whatever record lists were the most important factor in what they played. It was attended by 1,700 persons, who listened to a number of discussions, including a speech by Mitch Miller, "The Great Abdication."

Columbia Records' A & R chief received the only ovation from the disk jockeys present when he said, with one eye on station owners and managers present:

> You carefully built yourself into the monarchs of radio and abdicated your programming to the corner record shop, to the eight- to 14-year-olds, to the pre-shave crowds that make up 12% of the country's population and zero percent of its buying power. It must be more than a coincidence that single record buying went into a decline at the very time the number of stations that program Top 40 climbed to a new high. . . . I'm not asking you to snuff the musical life of these kids and their followers. I am asking you to put new life into radio; I'm asking you to take radio away from the

> lists and give it back to the people. I'm asking you to give up lazy programming—to play music for every age group and taste. The by-product of such a move will be aesthetic, and you can take pride in the public service. But principally you'll be doing it for your pocketbooks by insuring a broader, healthier audience, and guaranteeing advertisers who are seeking that audience a fair share for their money.

Separate from but coincident with the attack against BMI, many advertisers backed away from rock 'n' roll to the adult market, and many moved to the all-album stations. Disturbances at rock-'n'-roll shows contributed to this move, particularly a Boston show promoted by Alan Freed. After the riot, some stations barred all rock 'n' roll. Freed moved from the number-one New York City station, an independent, to ABC's flagship there. But since ABC-TV already had the top-rated national TV disk-jockey show, *American Bandstand*, with Dick Clark, he had to content himself with a video dance program on a local independent station.

Clark, the most successful television deejay in history, did not start *Philadelphia Bandstand*, as it was originally known, with a format of rock-'n'-roll dance records, guest stars lip-synching their disks, and little talk. When the program's creator, Bob Horn, was discharged by WFIL-TV, Philadelphia, Clark temporarily substituted. Soon his all-American-boy personality helped the program surpass all previous ratings, and he stayed on. In 1957, *American Bandstand* went on the network for ninety minutes each day. Within two years, it was broadcast by 101 affiliates to an audience of 20 million. Clark did not "make" a record, for he rarely played a new release nationally unless it was recorded by a truly hot act.

Most radio stations now programmed either Top 40 or Top LPs and more chart-programming formula radio than ever. It also put a premium on the best record spinners, music programmers, and librarians. This problem was addressed at Todd Storz's Second Annual Radio Programming Seminar and Pop Music Disk Jockey Festival. The forums' emphasis was on formula radio, pro and con, but fewer than 150 attended them. The other thousands recuperated from the effects of an endless round of entertainment supplied by the record companies. There was no Mitch Miller to galvanize the assembled and blast station owners.

A second try was made to form an association of disk jockeys and music programmers by Bill Galvin, a programming consultant from San Francisco whose free "Lucky Lager Top 60" tip sheet was mailed nationally. The new Disk Jockey Association made its first order of business a national campaign to enhance the status of members as good citizens and professional entertainers.

In connection with the second Storz convention, *Broadcasting* magazine investigated the extent of payola. The final 1958 report was accompanied by an editorial, which said in part:

> Readers . . . can hardly fail to reach the conclusion that the opportunities for payola are abundant. It is also obvious that some personnel succumb

> to the blandishments of pay-for-play promoters. But there is no evidence that the practice is anywhere near as widespread as the nagging rumor would make it seem. To the contrary, there is much evidence that music policies, recording-selection systems and other controls employed at many major stations constitute a barrier which can be surmounted only with ingenuity if at all.
>
> No general investigation, however, no matter what the findings, can do the job that needs doing on payola. The task can only be finished by the stations themselves. Each management should undertake a thorough investigation of its own. If payola is found, it should be eliminated and safeguards erected against recurrence. Programming responsibility belongs to the stations, so too does commercial responsibility. Management cannot shirk one, and it should not wish to have diluted the profits that go with the other.

Neither this editorial nor the payola report appeared in the AGAC's memorandum to the Harris committee, which was, instead, replete with quotations from *Variety* and *Billboard,* and record-company memorandums. However, it did divert the committee's attention to the music business and, as expected, BMI. Responding to reports that Dick Clark was to be investigated by the committee, BMI notified all stations that it no longer licensed some songs purportedly co-written and published by the producer of Clark's television programs. Soon after, ABC announced that Clark would divest himself of all possible conflicts of interest as his television connections were involved. The FTC began a hurried investigation of corrupt practices by record companies involving disk jockeys.

Payola's national exposure was expected by many in the music business to speed rock 'n' roll's demise. It was already losing ground. In some markets, only one of twelve stations played current hits; the others played albums of sweet music and singles. Records were bought at full retail price, and the hundreds of free items from record firms were sent back. The small independent companies suffered most, since they were no longer able to get their disks auditioned.

Paul Ackerman, of *Billboard,* appeared before the committee to illuminate the music business's long involvement with pay-for-play. "Much of the investigation of the music business has centered around the so-called singles record business, which is a small part of the total record business," he said. "The singles' business is a declining one and in December [1959] represented only 20% of the industry's dollar volume [or over $40 million]. . . . Long play records currently account for approximately 80% of the total dollar volume."

The very abundance of product, he continued, had effectively removed control of hit-making from Tin Pan Alley and put it in the hands of local radio programmers and record manufacturers. A major disk jockey annually received more than 1,000 LPs and 5,000 singles. Competition for exposure had given rise to all types of payola.

The hearings adduced findings from 335 jockeys that they recently had been paid $263,245 as "consultant fees." Testimony already heard by the committee eventually led to the resignations of the FCC chairman, a commissioner, and an assistant to the President, whose pressures on the FTC brought about his downfall.

Dick Clark testified that he had never received payola and added that he felt "convicted and condemned" before he had had an opportunity to tell his story. Documents submitted by his attorneys confirmed that he had made half a million dollars in twenty-seven months from music publishing, talent representation, record manufacturing, pressing, and distributing, before he disposed of all these interests in November 1959 at the recommendation of ABC. Alan Freed testified that he had been paid $40,000 a year by WABC-AM but had given back $30,000 of it to advertise his rock-'n'-roll shows.

The upshot of the hearings was that Congress amended the Communications Act of 1934, and outlawed all pay-for-play, with a year's jail term and a maximum fine of $10,000. While these actions did not solve the problem or address political payola, they temporarily allayed the music industry's anxieties.

The second prong of the Songwriters of America and the AGAC's attack on BMI was revealed at a session of the FCC's inquiry into network and radio programming and practices. Burton Lane again urged the divestiture of BMI stock by its remaining stockholders. Owners of stations were as guilty as their disk jockeys in the matter of payola, he said.

At another session of the FCC inquiry, ASCAP made its first public attack on BMI since the 1951 Harbach affidavit. A ten-month-long suit by 800 radio stations to get a 9 percent decrease in license fees had been settled to their advantage, and the consent order reorganizing ASCAP affairs had been signed by Judge Sylvester J. Ryan, the society's overseer. ASCAP president Stanley Adams referred to the "creation and collective ownership of BMI by the broadcasting industry at large and the incentive thus created for broadcasters to perform BMI music not on its comparative merits, but because BMI is operated for the exclusive benefit of owners of radio and TV stations." He added that the artificial ratings for songs created by BMI manipulation could be substantially reduced if the FCC acted to prevent further payola. Adams was unaware how Top 10 tunes were selected, but he did credit the presence of ASCAP songs among them to "sheer merit, breaking through the BMI barrier."

Sydney Kaye answered the charges by arguing that BMI music ranked lower on charts reflecting broadcaster-disk-jockey choice than on those based on retail record sales. The performance of BMI music had been higher in 1957 than in 1958, one of the two years when payola was alleged to have been at its peak. He then addressed the current videola situation at ASCAP. "Under ASCAP's system, the emphasis on background music is accelerated by the exaggerated advantages given to even a few seconds' use of an established hit accompanying the closing of a door, which will yield as much as if an artist had featured the same song in a full performance on the same program." And in 1957 some 42.5 percent of all ASCAP's payments were based on TV network performances.

ASCAP later filed a supplementary statement to Stanley Adams's "strongly confirming" that "the practice of payola payments to disk jockeys goes much deeper than a few isolated cases." Some 53 percent of 1958's Top 50 records were released by companies already charged with being involved in the payment of payola. Further, ASCAP believed that the commission's "disclosure" rule, which took the form of a routine statement that "This performance is a paid performance," did not go far enough.

The FCC required radio stations either to identify free records or purchase the records used on their broadcasts. This became a burning issue not only for broadcasters but also for the smaller record companies. Under it, only established labels and artists could be heard. Broadcasters protested that few stations could stock new releases. The NAB maintained that such rules would substantially reduce the amount of broadcast music. A compromise bill exempted stations from making the disclosure required by the FCC and spelled out requirements for the use of free records. In the final bill, payola remained a criminal offense, quiz-show fraud was forbidden, application procedures were revised, and deals among station applicants were limited. Station fines were reduced to $1,000 a day, and the proposal to give the FCC authority to suspend licenses was eliminated.

The renewed pressure of Fred Fox and others to the Supreme Court to intervene in *United States v. ASCAP*, together with Kaye's emphasis on ASCAP publisher involvement in videola, attracted new attention to corrupt practices within the society. The FCC withheld the license renewal of a Florida station owned by the producers of *Twenty One* and *Tic Tac Dough*, whose ASCAP house had prospered on the basis of performances of music on their own programs. A new rule curbed "incentives for artificial stimulation of performance payoff" on network programs for "unusual use of theme and background music."

The Supreme Court's decision to hear the Fox case brought ASCAP and the dissidents before that body for the first time, and speeded up other internal reforms. The performance payment for current hit songs was increased. A song whose current performances equaled those of the society's top fifty songs over the past five years was immediately granted "recognized work" status and eligibility for the 30 percent share of the four funds. An improvement in distributions was also made for concert, choral, and symphonic works. Having expected the October 1960 distributions to be larger as the result of the changes, some CWC members threatened to join BMI as had a number of prominent screen composers, all members of the CLGA. On seeing in the October statements the result of their own decision to accept the four-fund reforms, many of ASCAP's 117 highest-paid members questioned their sacrifice.

Many leading publishers also protested the inequities of the first distribution under the new consent decree. Dissident publishers threatened major publisher defections from ASCAP, while others looked to a restructured ASCAP board, now possible because of the change in weighed voting that reduced the maximum number of votes for a single member to 100.

ASCAP and Herman Finkelstein now focused on BMI's practice of giving long-term contracts and substantial cash guarantees to screen composers whose music was used widely on television. The practice was forbidden to the society, it was alleged, because the consent order required ASCAP to distribute royalties on the basis of earned performances. The NAB and its All-Industry Television Committee were preparing to negotiate a new license. Finkelstein believed that ASCAP had in its favor the fact that the industry's own licensing body gave more money for music on television, and the telecasters were not paying enough to ASCAP.

The scope of damages in the $150-million Schwartz case, now nearly nine years old, had been limited in 1960. In ruling against Judge Weinfeld's decision to give BMI access to half a million ASCAP performance cards, overseer Sylvester J. Ryan limited the damages sought by the thirty-five plaintiffs to those actually suffered by them.

ASCAP continued to neglect its West Coast screen composers. AGAC's fight with the CLGA widened the breach between the organizations and their members, a situation the 1960 consent order failed to mend. West Coast composers found BMI offered the best financial deal, which made payments for the performance in all foreign countries of motion pictures for which they had composed the background music. By the start of 1961, several dozen important ASCAP CLGA members had joined BMI.

The dramatic and comedy shows increasingly featured live music. The precedent-setting agreement with the AFM by MCA's Revue Productions to use only live music on all future videofilms speeded the demise of the taped music libraries. So too did the 1961 forty-month contract signed by the AFM and the Motion Picture Producers Association under which "canned music" was banned in all films produced in the United States and Canada. The CLGA also completed negotiations with the motion-picture and TV producers for a new four-year agreement that guaranteed members minimum weekly compensation as well as all existing pension and health plans.

Representative Celler's rekindled ire, fanned higher by the pleas of ASCAP to do something about BMI and television music, was vented on the Justice Department and the FCC in 1961. With revenues from TV the largest single source of ASCAP income, the society was fearful of the pending independent-TV station negotiations.

Celler demanded to know whether the commission was ready to require the divorce from BMI of every broadcasting licensee and asked why the Justice Department's Antitrust Division had been foot-dragging on recommendations made in 1956. Aware of Celler's power, neither agency disagreed with his charges against BMI and broadcasters nor promised any action against either.

The society had been mending some fences. New writer and publisher adviser committees were established. Writer performance credits were increased by stabilizing the total number of credited performances at 25 million a year. Writers and publishers were asked to extend the assignment of their television rights through 1966. A half-million-dollar "prestige melon" was divided among authors and composers whose music had "unique prestige value for which adequate compensation would not be otherwise received" and for music not surveyed by the new monitoring method.

ASCAP's right to license movies of any kind to television was unresolved. Several new suits were pending in the courts. All challenged the film studios' right to sell moving pictures to television without their permission to use the musical compositions synchronized with them.

These actions caught the attention of the legal representatives of the All-Industry Television Committee, chaired by an employee of the Shenandoah Broadcasting Group. In October, a Shenandoah committee asked Judge Ryan to grant independent television stations at-the-source clearance of all rights to music licensed by the society and played in the feature films and syndicated programs they had purchased or were renting. The court was also asked to determine "reasonable fees" and set interim rates.

In the proposed form of clearance, film and taped-program producers would acquire synchronization *and* performance rights to the music they selected. The costs would be included in the final package price. Representatives of the committee said that because music "was 'in the can' after it had been recorded and produced," stations buying prerecorded material were given no choice about the music involved or its price. Counsel for the Shenandoah committee explained that "the producer often arranges to be the 'publisher' of the original music he selects, and thus is able to share in the monies ASCAP collects from television and disburses to its members." If the producer paid the composer up to the same amount he got now through his ASCAP membership, "the total music costs would be substantially lower since the composer receives only 50% of the tv performing right revenues distributed by ASCAP. The other 50% is paid to the publisher, who in most cases is the producer himself."

Pending a decision by Ryan, TV licenses were extended automatically past their December 31, 1961, expiration. When Ryan stated that he could not rule on clearance at the source because of the ASCAP consent decree, the negotiations stalled until an offer described as "Judge

Ryan's idea all the way through" was thrown on the table by ASCAP. It allowed television stations a 17 percent reduction for ten years, and the networks one of about 10 percent against revenues exceeding those in 1961, in return for their complete divorce from ownership of BMI stock or control of the organization. BMI would have to accept the same sort of consent order as that signed by the society in 1950, which moved all rate-fixing negotiations and disputes to Ryan.

In response BMI launched a full-scale information program. *Variety, Billboard, Cash Box,* and, especially, *Broadcasting* spread the message to the industry, most of the television part of which had already decided that the saving of ASCAP fees was more important than support of the present BMI structure. The hottest issue in music licensing since 1941 split the industry, and caused long-lasting enmity among station owners and attorneys. Those independent broadcasters who looked on BMI as an insurance policy against the rebirth of an ASCAP monopoly saw the Shenandoah faction of the All-Industry Television Committee as traitors.

There was a ray of good news when, in a confidential letter to all plaintiffs and supporters of the Schwartz case, John Schulman wrote about the "good possibility of settlement," and with it the necessity for a "compromise on the question of the amount payable by the defendants even to the point where the cost of the suit will not be entirely recouped."

BMI rejoiced when the openly disunited Shenandoah faction announced at the NAB convention that they had succumbed to the united front of radio-station BMI stockholders and rejected the ASCAP-Ryan offer. All talk about television licensing returned to the Ryan court. With the Schwartz case seemingly grinding to an end, BMI returned to its fundamental function—the music business.

The music and record industries found the new rules dealing with them in Section 317 of the amended Communications Act easier to cope with than had been anticipated. Disk jockeys now identified both label and artist of the records they played. Pop singles enjoyed a new surge of popularity signaled by the return in March 1960 of Elvis Presley from military service. Youngsters who grew up with rock 'n' roll were now adult, looking for their own evergreen memories. "Oldie" and "Golden Oldie" programming had found its place alongside the Top 40 hits, broadening the base of the music heard on most stations.

Radio programming fragmented. In large markets, competition between stations was rife and the specialized format continued: talk-music, special features, interviews, news only, classical and semiclassical music, popular music, Big Beat, and, at the lowest end, background music.

The growth of FM stations in 1961 and the increased use of FM multiplex stereocasting encouraged the advent of softer sounds. In New York City, WINS, long the top rocker, switched to pop standards, celebrating the move by playing sixty-six hours of Sinatra records. However, rock-'n'-roll-station ratings remained at the top of the Nielson and Pulse studies. WINS abandoned Sinatra and returned to the fold with a slightly "less raucous sound." Other stations switched to all-talk and then back. The only real formula for success in radio appeared to be constant experimentation.

Judge Ryan refused in 1962 to grant a request for clearance at the source of all filmed programming. Speaking for ASCAP, Finkelstein reaffirmed its contention that the present form of licensing should not be tampered with; that, instead, rates should be restored to their 1954 level by effecting a 20 percent increase. He repeated that the parties' basic problems could be solved by the regulation of some BMI practices, specifically the advances to writers and publishers. Judge Rosenman offered a three-point policy statement on behalf of BMI, in lieu of any divestiture, which Finkelstein rejected. The Shenandoah committee had recourse now only to the Supreme Court, in the form of a petition to reverse the Ryan decision. Request was denied on the same grounds that the society was not required to accept a license which was neither permitted nor prohibited by the various consent orders governing its operation.

The removal of the jukebox exemption and the duration of copyright protection were subjects of considerable interest at ASCAP membership meetings. Stanley Adams's frequent public comments that "public domain is the chief competition of ASCAP" were based on the impending and relentless termination of copyright protection for works written more than fifty-six years earlier.

The 1961 report by the register of copyrights unsuccessfully urged a radical overhaul of the 1909 act. It asked for a twenty-year extension of the maximum copyright term; elimination of compulsory licensing; statutory protection for works as soon as they were registered or made public; enactment of separate legislation in the meantime to remove the jukebox exemption; protection of choreographic works as well as the principle of protection against record piracy; a provision making copyrights divisible, so that parts of a work could be assigned separately; and lifting of the ceiling on damages.

Representative Celler immediately launched a package of interim legislation. One proposal would add five years to the life of an estimated 47,000 works. He attacked the jukebox operators with proposed legislation to remove the jukebox exemption and impose coin-machine license fees. Celler's new friends at ARMADA were especially pleased with his bill to criminalize record counterfeiting.

Surprise opposition to the proposed extension of copyright came from the Justice Department. However, an amended bill reducing extension to three years passed and was signed by President Kennedy. The Celler legislation did not fare as well. Manufacturers and the publishers split over the bill's provisions. The RIAA favored the criminal-punishment clause, which made it a felony, rather than a misdemeanor, to counterfeit disks. The MPPA wanted to allow the recovery of full damages for copyright infringement. The watered-down legislation that emerged, finally, was the country's first legislation against record counterfeiting, but, with the damages clause missing, it promised little relief to copyright owners.

In the face of continuing bad publicity for the jukebox business, surprisingly, the Music Operators of America were able to delay passage of two Celler bills. A second compromise permitted coin operators to negotiate license fees after a mandatory five-year licensing period during which fees ranged between five and twenty-five dollars a machine. The affair dragged on.

The continuing popularity of country music led to the 1958 formation of the Country Music Association. The initial board included Ernest Tubb, representing artists; Cracker Jim Brooker, disk jockeys; Wesley Rose, the chairman, music publishers; Ken Nelson, of Capitol, record companies; Walter Kilpatrick, radio and TV stations; Charlie Lamb, the trade press. Robert Burton, of BMI, was director-at-large.

Thirty-six percent of the country's 3,327 AM stations devoted some of their airtime to country music in 1961, the year the first immortals, Jimmie Rodgers, Hank Williams, and Fred Rose, were elected to the CMA Hall of Fame. Nearly 45 percent of all the pop singles made the following year came from Nashville. The trend magnified when the following year, Ray Charles, the "Genius of R & B," released two enormously successful LPs: *Modern Sounds in Country Music,* Volumes 1 and 2. Country music now accounted for half of all recordings that originated in Nashville. Seventy TV stations employed live country talent, and ninety-seven AM stations were on a full-time country-music schedule. ASCAP opened its first small office in Nashville, where BMI was negotiating for property.

ASCAP's move into Nashville was another step in the plan of younger ASCAP members to develop a more favorable image for the society. The electronic computing machinery installed in the New York headquarters was expected to speed up distributions and reduce complaints while negotiations to effect other changes were being pursued with the Justice Department. The five-year averaging on which the four-fund system was based, covering the period when videola was prevalent among the highest circles and some new publishers, was a major point of discontent. So, too, were the special awards, which the Current Writers Committee believed were better spent to attract the new songwriters who were going to BMI.

On October 1, 1963, changes in the ASCAP distribution system were initiated to improve the lot of younger members. Writers who chose the 100-percent-performance option could now cancel it immediately. The recognized-works base was cut from 30 percent to 20; the difference was applied to the average performance fund, raising that to 40 percent.

Seniority was now based on a ten-year average. Limitations on average-performance and recognized-works ratings were removed to allow an increase in royalties. Credits for theme, bridge, and background music increased.

When an appeals court could not reverse the Ryan ruling against clearance at the source, the Shenandoah committee suggested that independent telecasters pay only for music used on their locally originated telecasts. The producers of feature movies and syndicated programs were expected to deal directly with ASCAP. Music on network programs was already covered by interim ASCAP contracts.

Three years of concurrent negotiations between the TV networks and ASCAP ended in December 1964 when a compromise five-year contract was signed. The 2.05 percent network rate was maintained, and a fee of 1.9 percent was fixed for owned-and-operated stations. The networks would also pay 2 percent on all revenue above that earned in the base year of 1963; their owned-and-operated stations, 1.325 percent.

Television's income grew: 1961 was a $1.3-billion year. The networks earned $675 million, with profits of $87 million. Eighty percent of all VHF stations showed a profit. Aided by 11,000 feature motion pictures purchased from Hollywood, television's gross revenues went above $2 billion. The networks had an income of $1.1 billion and profits of $136 million. A record income was reported for that year by ASCAP: $37.8 million, an estimated $32 million of which came from broadcasters, one-third from radio.

The All-Industry Radio Committee's desire for a reduction in the new contract that would take effect at the beginning of 1964 came at a time when independent radio was almost entirely music and news. The Big Beat and the Big Band Sound predominated nationally, the first from 45-rpm singles and the other from LP albums.

In their last negotiation with the society, radio broadcasters won a reduction of around 9 percent. Now they found it difficult to get representatives of ASCAP to accept an even larger reduction. In March 1964 the committee asked Judge Ryan to fix a $7.7-million ceiling on radio rates, a 22 percent reduction from the 1962 level. Two very different pictures of radio were offered to Ryan. One portrayed that reduction as part of a "Marshall Plan" to infuse money into a "very sick industry" once again on its economic feet. The other offer insisted on a significant additional reduction on the grounds that radio profits were declining and with them the use of the ASCAP repertory. Trade-paper-chart studies showed the decreasing ASCAP presence among the top-rated songs. BMI, which wanted an increase for that reason, agreed to a one-year extension of its contract with radio broadcasters expiring in March 1964. When the ASCAP negotiations concluded, BMI was ready to press for more radio money. Counsel for ASCAP attributed that change to a conspiracy to use more BMI music and reduce that of ASCAP.

At the same time, Judge Ryan was petitioned by CBS-Radio to approve an ASCAP per-program arrangement like that concluded by the network with BMI. The Justice Department filed a contempt suit against the society for its failure to accede to the CBS demand, as required in all consent orders since 1941. In spite of this and other pressures, CBS's demand stalled. No effort was made by Ryan to speed up the matter.

The first formal ASCAP offer to the radio stations increased rates by 22 percent, exactly the reduction sought by the committee. The increase was expected to come from an ascending commercial fee. Network and station per-program contracts remained at the same fees, but were conditioned on a modification along the lines of current television per-program agreements.

Ryan called a new round of meetings to discuss a compromise settlement, for ten years, of all issues, including the Schwartz case, CBS-Radio's per-program request, and bids for reasonable licenses. All negotiations stalled on December 10, 1964. An unexpected civil antitrust suit was filed against BMI by the Justice Department. It was known that the FTC had been investigating both BMI and the Society of European Stage Authors and Composers, the latter because of complaints from southern broadcasters.

The SESAC heretofore had lived a relatively untroubled life. The third licensing body in terms of income, it was the second to be organized, in 1931, by Paul Heinecke, whose family retained full ownership of the organization. In 1959, in the case of *Affiliated Music Enterprises Inc. v. the Society of European Stage Authors and Composers*, Judge Ryan dismissed the lawsuit but said that SESAC's "classic pooling of rights and sharing of revenue" violated the antitrust laws.

Pressure in the early 1960s from southern radio stations first brought members of the Senate and the All Industry Radio Committee into the SESAC situation. All stations were polled about the most persistent complaints against SESAC and a full report on all payments to the SESAC since 1953. The FTC and the Senate Small Business Committee examined SESAC. All investigations ended the following year when SESAC agreed to publish a catalogue of its licensed compositions, so that broadcasters could determine whether they needed a license from the company.

SESAC did not function under a consent decree and represented only publishers, sharing all mechanical, synchronization, and performing-rights revenues with them on a fifty-fifty basis, after the deduction of operating expenses. About 320 publishers belonged to SESAC, with a total repertory of approximately 150,000 works. There was no printed catalogue of all the music SESAC licensed. A list of affiliated publishers was available, with the suggestion that catalogues be obtained directly from them.

SESAC's broadcasting license fees were based on a variety of factors: station location, size of the community it served, hours on the air, and time-card rates. SESAC had reciprocal commitments with twenty-four overseas performing-rights and/or mechanical-rights bodies.

It was apparent that music catalogues and copyrights were highly profitable investments. For years, the value of copyrights was diminished by the threat of confiscatory taxation. The Treasury Department in the 1950s held that all royalty receipts constituted personal holding-company income whenever they represented 80 percent or more of total revenues. When mechanical royalties began to exceed that figure, some publishers were liable to a surtax of 75 percent on the first $2,000 and 85 percent income over that. A bill to relieve the situation was signed in 1960 and strengthened outsiders' interest in publishing.

During 1963 through 1965, millions of dollars were involved in major publishing-house sales. Among the numerous significant acquisitions, the sale of the Leeds and Dutchess music business to MCA and Mills Music to Utilities & Industries Management Corporation of New York City, a $42-million public-utilities holding company, were the most important.

MCA began to build a music-business operation by buying the interests of veteran music man Lou Levy in Leeds Music (ASCAP) and Dutchess Music (BMI), together with subsidiaries in Canada, England, and Australia, and the United States representation of AmRuss's catalogue of modern Russian concert music. Levy's catalog was wide-ranging, including pop material by Sammy Cahn and Saul Chaplin, 1930s and 1940s R & B hits, and some of the Beatles' earliest successes. The Mills catalogue contained 25,000 copyrights by the time of the sale. After $500,000 was put down by Utilities & Industries Corp. for an option to buy, a public offering of 277,712 shares of Mills Music, worth $4.5 million, was fully subscribed at the unit price of $6.50 a share. The balance of $2.25 million was raised through bank loans. A week later, Mills stock was selling at around $17 a share.

The Beatles' recorded copyrights had brought a change in the economic character of the business. Some 1.2 million shares of stock in Northern Songs, the publishing company founded in 1963 and owned by John Lennon, Paul McCartney, Brian Epstein, and Dick James Music, went on sale in February 1965. The company's chief assets in 1965 were fifty-six Beatles copyrights and a contract for six new Lennon and McCartney songs a year for ten years. Actually, the songwriters had assigned their music to Maclen Music, a wholly owned American subsidiary, which assigned all future copyrights to Northern.

The affiliation with BMI of Northern-Maclen Music and other Dick James properties prior to the start of the Beatles' popularity in America helped to bring virtual equality with ASCAP in terms of airplay and prestige, if not in income from music users. The key to this success was the youth market, 20 million teenagers with $10 billion to spend and a special affinity for BMI's music. Attributing BMI's continuing success to the conspiracy and questionable practices charged in the ten-year-old, dormant Schwartz lawsuit, ASCAP shifted its complaints from courts and hearing rooms to the Justice Department, expecting, at the very least, an amended BMI consent order.

Carl Haverlin's sixteen years as president of BMI ended in 1963. At that time the departure of the networks had shifted the balance of power on the BMI board to directors associated with important station groups and successful independent broadcaster stockholders. These men intended to bring a more businesslike administration to the BMI operation, as much control shifted to a newly formed executive committee.

While these men often found Haverlin's predilection for music of all kinds unseemly or whimsical, his actions elevated the public's awareness of the many kinds of music. He had been the only businessman present at a 1951 Ford Foundation meeting to foster performances of modern American concert music. Several concerts subsidized by BMI followed. Haverlin hired Oliver Daniel, a concert-music program director and producer at CBS to head BMI's new Concert Music Department in 1954. Daniel brought many important composers into BMI and helped the formation of Composers Recordings, a commercial record label of which the American Composers Alliance was the majority stockholder. A 1954 survey of broadcasting stations disclosed that 78 percent of them aired some concert music. However, radio stations who bought serious modern and classical music to Americans could not compete with the two giant concert bureaus—Columbia Artist Management and National Concert & Artist Corp.—both spin-offs of the divestiture of the CBS and NBC artist bureaus. The two new bureaus monopolistically allocated to one another the 1,200 American cities, where their Community Concerts and Civic Concerts subsidiaries provided music by the 200 artists each managed. A 1951 government antitrust suit against them ended with a consent decree in 1955, after which the concert management and performance business was ostensibly open to all competitors.

BMI's activities in this connection, plus the proselytizing of radio, played a part in the "massive unstopping of the American ear," according to *Time* in 1957, to hear the widest breadth of live and recorded music. In 1958, more than 4,000 pieces of American music were performed by American orchestras, and a thousand recordings of American music could be purchased. The publication of a score and parts for a modern concert work was feasible, because thirteen public performances paid for it. Among those benefiting from this improvement were the winners of prizes given in another project Haverlin created, BMI's Student Composers Awards, initiated in 1951.

Shifting the load of performing-rights payments to other music users was also evident. The Beatles' 1964 sold-out closed-circuit telecast provided a unique opportunity to enforce a new policy. A 1 percent share of the gross was enforced for access to the BMI repertory, to be divided equally among the writers and publishers of music performed, after a modest collection fee was withheld by BMI. The new policy covered all possible venues. In order to build up a significant catalogue of standard songs from the musical theater and feature motion pictures, BMI inaugurated a new writer-distribution policy, which tripled their payments. A similar system for concert music was already in place, and had led to serious competition with ASCAP for composers of modern music. Screen-music composers benefited particularly as BMI amassed the greatest collection of new movie scores since its formation.

A group of veteran Hollywood composers and independent smaller publishers retained counsel to ascertain why synchronization arrangements, with reduced rates, continued to be made between studio-owned music firms and the major production companies. Most publishers paid a share of the synchronization fees to composers but neglected to share any additional performance royalties. The Justice Department recommended to the society that it base all payments for screen music on its duration, a suggestion counter to the demand of guild members that their music be treated on a par with feature songs.

The failure of the motion-picture and television producers to recognize the bargaining powers of the Composers and Lyricists Guild of America raised the possibility of a strike by all screen composers and copyists in 1964. The NLRB ruled that the power of videofilm producers was "insufficient to establish an employer-employee relationship," and that "all composers in the unit sought are independent contractors." The two-year contract eventually approved by the producers' association in 1965 recognized the guild and set a base of $325 a week for composers under contract and $350 for those employed temporarily; covered all guild members by the industry health and welfare plan; gave composers and lyricists a year's option to buy back any music or songs that had not been used, following a five-year period of grace; affirmed the guild members' right to collect their share of performance royalties through the societies with which they were affiliated; and proved some other major benefits. A few months later, a similar pact was signed by the Society of Independent Producers.

When Judge Ryan failed to act upon the divestiture of BMI, the society campaigned against BMI during membership meetings and the celebration of the society's 50th anniversary. The standard line ran: while ASCAP was made up of composers and publishers, BMI was owned by broadcasters, who subsidized "a vast supply of music, much of it inferior," and attempted to dictate public taste; unlike the society, BMI "would not bind itself to a distribution formula which is published to the world"; BMI did not allow its members to elect the board of directors; BMI continued to "subsidize selected writers, publishers, recording companies, producers, and their employees," and the time had come for the "liberation of BMI writers and publishers from the broadcaster domination that now exists."

The Justice Department's antitrust suit filed against BMI on December 10, 1964, offered the possibility that the 517 broadcasters who owned BMI stock would divest themselves of all holdings. Only RKO General, the largest remaining single stockholder after the networks' departure from BMI, was named as representative and co-defendant, thereby representing the other broadcasters listed in the complaint.

Nearly one year into his presidency of BMI, Robert Burton reminded the press that the company had been formed in 1939 to combat ASCAP's monopolization of licensing. *Variety* pointed out that the statistics accompanying the antitrust charges closely resembled those in the complaint in the Schwartz case. A general denial of all charges was made by BMI and RKO General in March 1965, as was a motion to dismiss the suit.

The All-Industry Television Committee filed documents with Judge Ryan in January 1965 that pleaded that no reasonable rate could be agreed upon by the committee unless it had access to all ASCAP's records dating back to the first television license. The fact that publishers had nothing to do with the background or incidental music that formed nine-tenths of all local-station programming but got half of ASCAP's net income was the central point in the new argument. An 80 percent increase in station payments to ASCAP, against a 33 percent rise in audiences in the 1957–74 period, was offered as additional evidence that the present fees were "becoming increasingly unreasonable and excessive." The society was ordered to turn over various unspecified data.

ASCAP's income continued to soar—$41 million for 1964 and $42.7 million in 1965. The "Big 15" publishers continued to dictate the makeup of the board. Hans Lengsfelder and other dissidents focused on the sampling of local radio-station performances, on which a substantial portion of the society's income was expanded.

Additional sweeping changes in the distribution system were approved in September 1965. Payments were brought nearer to the performance period on which they were based. With elimination of the recognized-works fund, publisher distributions would be based entirely on performances by 1968. The last lingering vestige of videola was attacked. Payments for music on nighttime strip shows were increased. A 20 percent reduction was introduced in the distribution to those at the top.

The entrance of the giant tape holding company, Minnesota Mining & Manufacturing, late in 1965, into the background-music business brought the AGAC again into conflict with large ASCAP publishers. The latter had negotiated with 3M for performing and mechanical rights, bypassing the Fox Agency and AGAC's own collection agency, thus violating, the guild charged, the 1948 Songwriters of America agreement. AGAC's collection agency, founded six years before, over the MPPA's protests, now collected two million dollars a year for 2,200 publishers.

A new 1 7/8-speed, four-track, monaural tape machine, capable of playing twenty-six hours of 700 selections continuously, was being merchandised by 3M. A standard background-music service contract had been negotiated with BMI for its entire repertory, but BMI publishers were expected to deal with the mechanical licensing. Most ASCAP publishers rejected the 3M proposal, but several of the majors granted a license on the basis of a two-cents-per-tune annual payment for performances and a mechanical fee of three cents. AGAC argued that publishers who signed such bulk or block licenses without each writer's approval were violating the 1948 pact.

In January, the AGAC's president, Burton Lane, surprisingly proposed that all new copyrights be taken out in the songwriter's name. Lane said: "Under the new contract, the writer grants the music publisher the right to publish and license his material for records and other entertainment usage. Since the music publisher is given the legal license to market the material, he is protected. He can secure recordings, collect monies, publicize and exploit, without hindrance to his publisher's prerogative." Other proposed conditions would strengthen the rights of writers. An "option agreement" would require a music publisher to obtain a recording within six months, or pay a $250 advance for six months' extension. Contracts would be operative only after a first recording was secured.

Many of the proposed changes were eventually accepted, but the publishers held firm against the transfer of ownership of a song to the author and composer. The 3M licensing arrangement remained in effect, accepted by most publishers, its validity based on a provision in the ASCAP consent order granting publishers a nonexclusive right to license performances.

BMI's second paid president, Robert Burton, suddenly died in March 1965. The largest delegation to attend his funeral came from Nashville, where he had been well known as one of country music's most ardent supporters. In recognition of this, *Billboard* named him "Country Man of the Year" in 1964. The isolation of country music was long gone. The *Grand Ole Opry* radio show was heard in 300 markets, and taped syndicated country-music video shows proliferated. CMA's 1965 survey found 208 radio stations programming country music full-time. The association now had 1,100 members and 111 supporting organizations. The Country Hall of Fame and Museum, which Burton enthusiastically supported, broke ground in the spring of 1966, while BMI was at home in its own building next door.

As BMI searched for a new chief executive, Sydney Kaye instituted the first legal action against promoters of the Beatles' and other live and closed-circuit concerts, who had failed to take out the BMI license requiring 1 percent of gross admissions. ASCAP also stepped up its own drive to collect for live concerts. Credits were increased simultaneously by fifteen times for the performance of works in concert and symphony halls.

BMI's announcement that the 1965–66 season was the best in its history came simultaneously with the election by the BMI board of Robert Sour as president. The revolt of popular-music concert promoters was cut short when the company adopted ASCAP's traditional seating-capacity formula for concert licensing. The payment formula for feature-picture songs was reduced by a third. The $1.08-per-station royalty for a TV network performance fell to 72 cents, and that on a local station from 72 cents to 48 cents. Feature-song performances on non-prime-time television were cut by one-third.

In succeeding weeks, the publishers' share for performances on stations paying BMI less than $1,000 a year was reduced. The company had sustained a pretax loss of $1.7 million for the year ending June 31, 1965, and with performances climbing steadily, and disheartening results from studies of income received for the same songs by ASCAP and BMI publishers, the organization faced a problem of considerable proportions.

Sixty new publishers affiliated monthly with BMI. Only seven publishers affiliated with BMI which were primarily engaged in the popular-music business had five-year guarantee contracts. BMI also had one-year guarantee contracts with twenty-seven other publishers. The amount of a guarantee was computed by averaging a publisher's performances over a three-year period and multiplying by ten cents, so that guarantees exceeded logged earnings.

Thirty-seven publishers of music considered by BMI to be of cultural importance and deserving of support made up a third class of guarantee publishers. These had contracts for between $1,000 and $10,000 and received a total of $403,000 in 1965, or 4 percent of the entire distribution to publishers.

A bonus-payment category of publisher affiliates was created in order to decrease the number of guarantee publishers, a result of the phasing-out of ASCAP's credit for recognized works. Eligibility for bonus payments was determined by performances logged in the prior year. After logging 250,000 performances in the last four quarters, a publisher was paid a bonus of 25 percent on each logged performance for the next year. After the first full year under this arrangement, publishers had to have 300,000 performances for logging plus 50 percent and about 150,000 for logging plus 25 percent.

Of the 9,000 writers affiliated with BMI in early 1965, all but about 400 were paid at standard rates called for in the BMI writer contract. The others had contracts that provided advances or guarantees in direct relation to predictable earnings. BMI had no bonus system for writers comparable with that compensating publishers. There were few straight guarantee agreements. Out of the approximately 400 writer affiliates, 187 had three-year contracts that paid them a guarantee equal to 150 percent of earnings logged during the preceding year. These totaled about one-quarter of all BMI writer distributions in fiscal 1965.

At the end of 1965, an initial 10 percent cut in the value of performances over a three-year spread was instituted, a result of increasing high cost per use of the BMI repertory by radio and television broadcasters. The radio rates in effect since 1941, when BMI's first and only contract was drawn, ranged from 1.2 percent for a station with income over $100,000 to 0.72 percent for those at the bottom, minus specified deductions. Radio's pretax profits for 1965 were $72.8 million. The four radio networks showed an operating loss of one million dollars, but their owned-and-operated stations had revenues of around $36 million.

When rate arbitration languished, BMI sent a three-month cancellation notice to all stations. In 1966, the final agreement raised BMI's blanket license rate by 12.5 percent, the first raise in twenty-six years, during the last twenty of which the average radio station trebled the use of BMI music.

In November, the government ceased attempting to force 517 stockholders to give up their stock in BMI, but did order some changes in the company's practices. It was barred from recording, printing, or distributing music, and prohibited from entering into a contract for more than five years. Affiliated writers and publishers were allowed to issue nonexclusive licenses. The type of contract BMI had made with Hill & Range in 1949, which enjoined a publisher from doing business with another licensing organization, was also prohibited. BMI could not force other parties to record or perform any stated quota of music to which it had performing rights.

ASCAP claimed that the *United States v. BMI* settlement was not in the public interest because it approved the broadcaster's ownership of a music licensing organization. A federal court judge affirmed the consent decree and ruled against ASCAP's plea, as well as one by AGAC. Little had gone well for the society for some time. The Celler bill to repeal the jukebox exemption languished. The Music Operators of America had thrown up a roadblock by proposing an increase in mechanical rates to be paid by operators in lieu of any license from the performing-rights organizations. ASCAP and BMI objected, as did the Recording Industry Association of America, coming out for the first time in opposition to the jukebox industry.

The blue-ribbon Copyright Revision Panel in 1963 assisting the Copyright Office in formulating omnibus revision legislation was a witness to the disparate interests of creators, owners, and users. The RIAA tore into proponents of either the removal of compulsory licensing or a higher statutory rate. The MPPA was for either or both. Music and book publishers fought any reversion to a copyright to the creator of the work after twenty-five years. The MPPA protested that its members lived off the old standards and needed more than a quarter-century hold on their property. The strongest opposition to extension of the life of copyright came from the National Association of Broadcasters and the networks.

The panel's final recommendations retained compulsory licensing and raised the mechanical rate from two to three cents, or one cent per minute of play, whichever was greater. Strengthened civil and criminal penalties against record piracy were established, and the jukebox exemption was removed. The copyright term would be extended to the span of the author's life and fifty years. After a period of thirty-five years, authors and their heirs could, with two years' notice, transfer a copyright. The exemption of payment for "not-for-profit" performance was revoked, and only in-school use, educational radio or television broadcasts, and performances where no admission was charged or the money went to education, churches, or charities would be exempt.

The RIAA's opposition was indicative of the schism between users and creators. Congress was treading most carefully in the face of organized national opposition from educators to copyright revision. In 1966, the House subcommittee wrote a much altered new revision bill. The chief new provisions affecting the music business contained some compromises. The term of copyright remained the author's lifetime plus fifty years. The jukebox operators were faced with a tax of about nineteen dollars on each machine. The statutory mechanical rate was raised by only half a penny, two and a half cents or half a cent for each minute of playing time, whatever amount was larger. The complicated legislative processes required the game to begin again in January 1967.

The Supreme Court's 1965 refusal to hear Metromedia's appeal for a blanket license was one of ASCAP's few victories. The informal agreement made with the TV networks late in 1964 again bogged down in argument. In a 1966 compromise, ASCAP offered a fixed-annual-payment contract to them, to extend over five years. The proposed flat-fee contract applied only to the networks; their owned-and-operated stations continued to pay by the revised 1964 formula. Both NBC's and CBS's principal problem was how to explain the new arrangement to affiliated stations. Their regular contributions of 50 percent and more of the networks' ASCAP fees remained buried in the language of the 1941 contract formula. Under the proposed simplified new method for compensation by the networks, the affiliates eventually would learn the facts.

The TV-station committee considered accepting the new reduced network owned-and-operated-station formula: 1.94 percent on money equal to that earned in 1963, 1.325 on revenues above that. But it then decided to remain on its course toward a "reasonable rate." In the eight years starting in 1957, television-station fees to ASCAP had risen 76 percent. Judge Ryan suggested that the stations and ASCAP quickly work out something "to eliminate auditing and constant friction" before his 1966 mandatory retirement. That occasion came and went, and Ryan remained in charge of all ASCAP matters in the purview of the consent orders.

Changes in the ASCAP distribution allowed the society in 1966 to fight BMI for the new writers of hit songs. The amended consent order now gave ASCAP the right to hand out cash advances to writers, the BMI practice that the society had so long fought. New members could start immediately on the 100 percent current-performance system of payment, and change to four-fund performance averaging after several years, during which these were counted. Performances of BMI-licensed works brought over to ASCAP were counted in the ASCAP surveys, giving dual payment from the societies.

When he officially announced ASCAP's income for 1966, Stanley Adams took advantage of the occasion to brand the new BMI consent decree as "the greatest invitation to payola that exists anywhere." This statement came when many believed that the long cold war between the organizations had thawed.

Some months later, in May 1966, the Music Publishers Protective Association, center of the old-boy network of ASCAP publishers, changed its name to the National Music Publisher Association. Of immediate concern to the NMPA was the overbearing presence of the Harry Fox Agency, which cast a shadow over the MPPA. Though Fox was an employee, he had outright autonomy, won for him by the association's legal counsel. The agency now represented more than 2,300 publishers, whereas the NMPA had a membership of fifty.

Early in 1927, Claude Mills, then chairman of the MPPA board, devised its synchronization licensing contract and collection apparatus, for a commission of 5 percent. Harry Fox, assistant to John Paine, was made responsible for handling the process. When he became head of the agency, it took his name.

The phonograph industry's postwar resurgence raised the Fox Agency's collections to five million dollars a year. In spite of all NMPA initiatives to enforce an understanding of its relationship to the Fox Agency, Harry Fox's autocratic control continued until his death in 1969, when the agency separately incorporated as the Harry Fox Agency, a wholly owned subsidiary of the NMPA, with Al Berman, Fox's former assistant, at its head.

The buying and selling of copyrights and catalogues continued at a stepped-up pace in 1966. E. H. Morris sold back his one-third interest to Bregman, Vocco & Conn for more than he had paid with the expectation of picking up his remaining shares within a few years. From the start, Morris had been a silent founding partner in Bregman, Vocco & Conn. In 1943, he sold out for a few thousand dollars. Music executive Jack Bregman had been a pupil of the late Jack Robbins at the Big Three, in which he remained the only private partner in 1966, with a 4.6 percent interest. Vocco headed the Leo Feist branch in Chicago and worked for the MPHC before becoming a partner in the firm bearing his name. Chester Conn was a successful songwriter-songplugger for Vocco in Chicago and then a Robbins employee under Bregman. The Morris sale was a prelude to 20th Century-Fox's purchase of the firm a few months later, for $4.5 million. An important part of the company's catalogue was the copyrights for the music in twenty-five musicals made by the film company during the war years. Had Morris retained his share of the company, he would have realized a profit of more than a million dollars.

The Aberbach brothers were the winners in spirited bidding for the twenty-seven-year-old ASCAP member Joy Music and some subsidiaries. The Joys held out for two million dollars "on the barrelhead." The Joy Music takeover sweetened an arrangement Hill & Range Songs made with NBC-TV for a jointly owned television-music firm.

The purchase of a half-interest in the late Max Dreyfus's American Jubilee Music and New World Music in Britain by the British Associated Television Corporation was influenced by the growing interrelationships of the worldwide video business and the music industry. Publishers in America recognized the importance of access to promotion of their product, and video film makers were cognizant of the millions paid by television for music. Lew Grade, head of Associated Television, learned of the contemporary value of popular music through the sale of video series to CBS-TV network. After buying rights to the *Secret Agent* series, CBS threw out the original British theme music and substituted an American song, which became an immediate commercial hit. Associated Television was also a large stockholder in the British Pye Record Company, which Grade now perceived as a testing ground for the potential of all original music and songs written for his television series and specials.

The Europeans were getting into the American music business, and not only as rock-'n'-roll performers.

10 Battles Over the Per-Use License and Merchandising of the Superstars

EMANUEL Celler suffered a major legislative setback in 1967 at the hands of the 7,000 jukebox manufacturers who had not attended the conference at the Library of Congress arranged to save at least part of the copyright legislation. The omnibus 1965 copyright revision bill had been reintroduced in 1967 and seemed destined for passage despite its highly controversial provisions: payment for the use of copyrighted material by educational radio and television; a royalty for artists as well as manufacturers for the use of recorded music; payment of royalties for the first time by the young cable television industry; and a tax of about nineteen dollars on every jukebox, representing a potential total of around $10 million in collections.

Celler stated that the coin operators "want to go scot free and pay nothing to the men who give their lives, their sweat and their tears to making these songs. The jukebox would be a junkbox without the song." Pro-jukebox forces sought vainly to return the bill to committee, but found surprising support in the 252 to 126 vote that defeated the proposal. The vote shocked Abraham Kaminstein, Register of Copyrights, as it did ASCAP's Herman Finkelstein, the prime proponent of the nineteen-dollar tax.

The chairman of the House Commerce Committee indicated that he did not intend to remove the cable TV business from his jurisdiction. In response to this and other complaints supporting either the CATV interests or responding to pressures from the National Association of Broadcasters, the industry's lobbyists for restraint of cable television, patchwork amendments were offered, and Celler moved to withdraw the bill and dropped the CATV section. Only compromise could save the jukebox tax.

The Library of Congress conference involved three lawyers representing the jukebox business and the Music Operators of America's chief lobbyist, Nicholas Allen. On the government side were two lawyers and

Kaminstein. No music business or copyright interest had been invited. The coin-machine people conceded they should pay for the use of music, but they felt that nineteen dollars was not reasonable. Kaminstein accepted Allen's promise to make an offer on behalf of the MOA.

It was not Kaminstein but Celler who was blamed for the "Library of Congress surrender" by a majority of the copyright interests, including officials of BMI. It was their conviction that, through Celler, ASCAP was ready to sacrifice most of the nineteen-dollar jukebox license in order to save the "life plus fifty years" copyright provision.

The entire bill was passed by the House and sent to the Senate and the hands of John McClellan, chair of the Subcommittee on Patents, Trademarks, and Copyrights, who was examining a proposition to give artists the same copyright protection for their recorded performance as that enjoyed by composers and publishers.

Alan Livingston, president of Capitol Records, first introduced the possibility of charging broadcasters for the use of recorded music. This ploy successfully brought broadcasters and their national association into active opposition to the bill. The NAB objected to the proposed "double payment" for music.

The Register of Copyrights responded soon after to Livingston's proposal: "There is no doubt in my mind that recorded performances represent 'the writing of the author' in the Constitutional sense, and are as creative and worthy of copyright protection as translations, arrangements, or any class of derivative work. I also believe that the contributions of the record producer to a great many sound recordings also represent true 'authorship' and are just as entitled to protection as motion pictures and photographs." The NAB argued against Livingston's suggestion and pleaded that additional payments to ASCAP and BMI would place "the burden on those least able to afford it—the small stations." In response, the NCRA claimed that approximately 73 percent of all radio airtime consisted of recorded music and produced 81 percent of all radio's income.

McClellan was determined that a new copyright bill not emerge from his committee before the Supreme Court decided whether cable television should pay copyright royalties for the movies it offered. The court was reviewing the case of *Fortnightly v. United Artists Television*, the plaintiff being a small West Virginia cable operation which sought a final ruling that CATV systems did not violate the Copyright Act when they broadcast distant signals. The Supreme Court reversed negative lower court rulings, allowing McClellan to push through a provision calling for a minimum copyright fee from cable.

When the winter of 1967–68 passed without action, Kaminstein urged that a temporary "bare bones" revision bill be introduced. In 1968, he suggested three alternatives: that writers, publishers, and Congress wait a year; that the bill wait for the resolution of *Fortnightly* by the Supreme Court, and then the omnibus revision be pushed through; that

Congress immediately pass the bare-bones revision. McClellan asked his committee to wait for the court. Revision of the 1909 Copyright Act was sidetracked until 1974.

The Supreme Court's reversal in *Fortnightly* allowed CATV to carry copyrighted material without permission or payment. McClellan held off any action affecting the cable business. Instead, five months after the FCC had been granted jurisdiction over cable by the *Fortnightly* ruling, it adopted an interim-procedures rule requiring cable to get permission from copyright owners to use their property. During the next four years, no such consent was asked or granted, and CATV went its own way. Congress meantime passed two-year copyright extensions to protect musical and literary properties that otherwise would have lost their protection and gone into the public domain.

The ongoing antagonism between ASCAP and BMI had been further inflamed by the latter's successes in 1967. BMI Music totted up a 78 percent share of *Billboard*'s Hot 100 chart, 92 percent of its Top Country Hits, 100 percent of the Top R & B Hits, and 48 percent of the Top Easy Listening Singles. The three-year-old lawsuit to achieve "reasonable rates" was concluded. ASCAP accepted a 6.25 percent rate cut from radio stations, whereas its rival received a 12.5 percent increase. ASCAP intended to change its sharply declining share of airtime. Precluded from making payments to its members ahead of distributions, the society instituted a full 100 percent performance payment on a new member's joining, with the option of changing to the four-fund distribution system three years later.

The enticement had been successful. Emphasizing that ASCAP was run by publishers and writers who were motivated to get the largest amount of money from licensees, the society made inroads in Nashville. Wesley Rose of Acuff-Rose Music was elected to the board of directors.

BMI's fourth paid president, Edward M. Cramer, took office in 1968 and was immediately confronted by a growing financial dilemma: BMI's music was being used with more and more frequency, but its income remained fixed while its distributions kept rising. He prepared to negotiate an increase in radio license fees while the All-Industry TV Committee successfully concluded its lawsuit against ASCAP, winning a $53-million saving over the ten-year contract. The old fee of 2.05 percent of station revenue was reduced to 2 percent of average industry income for the years 1964–65 and 1 percent of income over that. The sustaining fee was cut by one-third.

Cramer's insistence that BMI was entitled to higher rates from all radio licenses was quickly characterized as "foolhardy." Further meetings with the Radio Committee broke off when new terms were announced calling for an increase in the maximum rate for all stations with receipts in excess of $100,000. BMI did not ask for any sustaining fee. The committee argued that under the most recent contract BMI's income from radio had already virtually doubled in just two years.

Countering that the increase reflected an increase in radio's revenues, Cramer suggested arbitration. Terms were finally reached in 1968 that raised BMI's fees 10 percent in the first year, to 26 percent over five years for all stations earning in excess of $80,000.

The committee recommended an increase in music costs "not only because BMI is today furnishing more music than any other music-licensing organization, but also because your cost for using BMI music will be at least 25% less than the current sums you will be paying ASCAP during the term of the current ASCAP contract." *Variety* hailed the settlement as "an important victory" for Cramer, because of his insistence "upon a tough stance against the radio broadcasters [which] was upheld by BMI's board of directors, all of whom are broadcasters."

Additional changes in the ASCAP distribution policy were approved by Judge Sylvester Ryan in 1969. Performance credits for ASCAP members regularly associated with across-the-board network programs were reduced, because they could rig the music logs. Other changes included the formalization of cash advances to new publisher members based on trade-paper charts; an increase in credits for concert works of more than four minutes' playing time; and the reduction by 20 percent over two years in the amount flowing down from writers in classes above 1,000 points. The immediate effect was the resignation of another bloc of productive BMI writers and their publishers.

Nearly a year later, the new ASCAP contract with independent television stations was mailed. Both CBS-TV and NBC-TV agreed to make additional cash payment to ASCAP for 1964 through 1968; these waited only for Judge Ryan's approval of a similar flat-fee payment for 1969. The rationale for this new method of payment was the fact that advertisers now bought time in minutes; the fees included both time and program charges. Hence the proposed change from the time-honored percentage of advertising revenue to the fixed rate, which had the additional advantage to networks of knowing in advance what the cost of ASCAP music would be, as well as reduced record-keeping and the elimination of ASCAP auditing.

Contending that, like ASCAP, it had contracts calling for a percentage of income, BMI asked for a correction of the underpayment to which the networks had admitted. With a network-TV rate of about half that of ASCAP, BMI believed the two chains owed around $3.5 million. Interestingly the "Chinese bookkeeping" used by the networks to hide their affiliates' payments of a share of music fees had been abandoned somewhat, and CBS affiliates were notified that their share of the extra payment would be about one-quarter of the total sum.

CBS belligerently argued about the underpayment, and in late October, BMI sent a notice of termination of license to the network. Just before Christmas, CBS dropped a bomb whose repercussions would be felt by the music business well into the next decade. In letters sent to both licensing bodies, CBS asked for a new form of licensing, a per-use arrangement, "under which the amounts to be paid are based on the actual use of copyrighted music."

In ASCAP's case, such a request demanded either a sixty-day period of negotiation or resolution through the federal court. BMI, on the other hand, terminated the right to play its music as of New Year's Day. CBS asked for an extension of the contracts on an interim basis. At the same time, the networks filed complaints charging both organizations with antitrust violations and asking for the per-use licenses. The defendants were charged with insisting that CBS take the only types of agreements they offered, a blanket or a per-use license.

BMI's dilemma with the networks was further muddied when BMI filed an antitrust suit of its own. Both networks were accused of joining ASCAP in a "contract, combination or conspiracy" to restore the society to its former position of monopoly. BMI charged that the TV chains had discriminated against it by paying ASCAP substantially more than warranted so that the society could continue to "induce" BMI writers and publishers to move to ASCAP.

NBC-TV withdrew its request for the "right to use ASCAP's millions of copyrights" when they were not needed and asking for the right to use only 2,217 specified songs licensed by the society and "certain background music libraries." Network research indicated that performances of ASCAP music and background music on NBC-TV had substantially declined between 1965–66 and 1968–69. An affidavit filed with the petition stated that if any interim fee fixed by Judge Ryan was disproportionate to that paid BMI "it could well be, as BMI contends in its lawsuit, that ASCAP will be able to raid BMI of its members and drive it from business."

A similar application for a limited-repertory license from BMI was filed by NBC, which explained that it intended to deal with program packagers for rights to the background music they controlled, as well as with individual members of both societies for music outside the limited-access arrangements.

Hoping to bring at least ABC-TV into line, BMI terminated its license. A six-million-dollar infringement action was filed against NBC-TV. With both licensing organizations and all three networks at one another's throats, the situation pointedly resembled the ASCAP radio music war of the late 1930s.

The society boldly filed a sweeping counterattack in the New York federal court which asked for the revocation of all owned-and-operated station licenses and charged NBC and CBS with a conspiracy to restrain trade and corner all television programming by creating "a monopoly of program material and access to television facilities." The FCC had recently ruled that major-market stations were barred, beginning in September 1971, from programming more than three hours of network offerings in the 7 to 10 p.m. period, but ASCAP wanted more drastic control. It asked the court to enjoin both CBS and NBC from "hereafter producing, selling, interest in or control over any television program other than news and public affairs programs in connection with the operation of their television networks."

The compromise payment proposed by Judge Ryan to CBS and NBC—$4.3 million for 1970 for access to the ASCAP catalogue while requests for limited licensing were pending—brought BMI once more to the court for relief. Judge Lasker was asked to require CBS to pay 70 percent of the rate the networks agreed on in response to the Ryan suggestion. Though without a BMI license, CBS continued to use its music, lending support to BMI's charge that serious competitive injury followed the loss of some of its principal affiliates to ASCAP.

The escalating bidding for BMI's songwriters and publishers was publicly manifested in a series of trade-paper advertisements. Their message was: The difference between ASCAP and BMI is simple—publishers and writers own ASCAP; broadcasters own BMI. The more money you get, the less money the broadcasters keep, and vice versa.

The two systems of performance payment had been at the heart of the fight since the early 1940s, *Broadcasting* said, in February 1972:

> ASCAP officials have derided what they say is a system that pays one writer in a particular category more than another writer in the same category. BMI claims that ASCAP makes it very hard for writers who are on ASCAP's "four fund plan"—a system of payments that takes membership and recognized works and spreads payments over a longer period than would straight current-performance payments—to leave because they forfeit money they would have received from the fund in the future.
>
> ASCAP claims that it splits payments equally between publishers and writers while BMI gives more to publishers. In fact, BMI pays equally to publisher and writer except in one category: publishers get more for performances on radio stations which pay more than $3,000 in licensing fees. BMI accuses ASCAP of offering large advances and guarantees to prominent BMI writers to woo them away, a practice BMI had freely admitted using in the past.
>
> Both organizations have systems of payment that grant bonuses for hits and take into account length of membership. Both will offer a prominent writer an advance to sign. A writer cannot get out of either society without losing payments on copyrights from the organization he leaves. Both now have "open door policies" concerning membership, as BMI has had from the beginning. And both, under consent decrees, allow writers and publishers to make independent licensing agreements for their works.

It was the last provision CBS intended to manipulate. In a lengthy memorandum, filed in opposition to BMI's request for an annual fee representing 70 percent as much as it paid ASCAP, CBS raised the possibility of negotiating directly with BMI musicians and publishers on the basis of BMI's royalty schedule. On such a basis, CBS would have paid $800,000 for BMI music in 1970, half that called for by its contract.

NBC's bid for a limited ASCAP license was rejected by Judge Ryan, who ruled that the consent decrees did not require the society to issue anything but a blanket license to a television network. Their language called for licenses for "any, all or some" of the ASCAP repertory, and involved "mechanics for setting appropriate license fees, not for establishing the scope of a license. The real issue [is] not one of broad antitrust policy, but the much more limited inquiry as to whether a license to the largest and most influential network in the U.S. of a limited number of ASCAP selections is required by terms of the consent decree. We find that it is not."

BMI was having slightly better luck with the independent TV stations. The permanent All-Industry Committee pushed through a new BMI contract, whose terms substantially reduced the company's present nine-million-dollar income from all independent TV stations. If BMI did not accept the proposition, the committee claimed, its constituents would lose over $50 million in the next ten years.

A compromise settlement tied BMI's rate to the pending ASCAP contract with independent TV stations, and simultaneously brought the two organizations closer to economic parity. In an agreement that would expire with the ASCAP contract at the end of 1978, each station paid BMI fees equal to 58 percent of its payments to the society. The combined saving meant that stations actually paid $60 million less than they would have under previous agreements, but it did represent an increase for BMI. BMI's first victory in the network disputes came at the same time—an undisclosed fixed-dollar contract with ABC-TV.

All the public contentiousness between the networks and the music societies was thought merely to make a record for future litigation. Peace would require realistic compromises. The networks never actually intended to destroy the performance societies. The NBC limited-access proposal, for example, would bring the society $785,000 for 2,217 selections, whereas under the flat-fee blanket contract ASCAP received nearly $6 million from the network in 1969. The networks were conscious of the possibility of an antitrust conspiracy action brought by both societies, which would certainly lead to government intervention and their possible dissolution.

Faced with mounting music fees, the networks demanded limited licenses. If a long-term cap could be placed on music costs, the networks would abandon their suits to change the licensing formula. BMI and ASCAP then might agree to more "reasonable" fees from the networks, provided that blanket licensing remained in effect, and that BMI's share of the market was properly reflected in its income.

Part of such a scenario was about to be played out. Both CBS and NBC accepted the Ryan proposal for a compromise $4.3 million interim annual ASCAP fee while retaining their suits for limited licenses. CBS continued, however, antagonistically to demand that its BMI fees be based on a per-use formula, paid at BMI's published rates. The network budged only after Judge Lasker ordered it to pay BMI an annual $1.6 million for blanket access to its music. Retroactive to the beginning of the year, the sum was the same as had been paid for 1969, the largest BMI had ever received from CBS. BMI's copyright-infringement suits against the network were dismissed with prejudice until the per-use issue was resolved.

Meanwhile, many music industry members unrealistically believed that record sales would pass the one-billion-dollar mark. When they did not, they complained of their having been the victims of a "profitless prosperity." A study of 1964's figures revealed that the combined net profit from sales of $758 million had been 1.7 percent, and the consolidated return on net worth, 3.8 percent. When the business passed the billion-dollar mark in 1967, grumbling focused on competition from prerecorded tapes. Stereo and tape sales far exceeded those of disks, which failed in record numbers. Some 70 percent of the singles and 60 percent of LPs released in 1967 did not recover initial costs.

Rack jobbers, and the department and discount stores they serviced, drove down the standard price of a popular stereo LP to as little as two dollars by 1967. Only twenty-three singles had sold over a million copies that year, and only fifty-eight LPs had sales in excess of one million dollars, qualifying them for the RIAA's Gold Record award.

The monaural record headed for oblivion. To get it there, the price of all monaural records was raised in mid-1967 to that of its stereo counterparts, in the hope that consumers would switch to the more modern technology. However, consumers did not respond as expected. Many companies issued "compatible stereo" recordings, which could be played on either monaural or stereophonic equipment, or "electronically reprocessed" or "enhanced" stereo disks, which actually were from slightly doctored monaural masters and of vastly inferior sound quality. An increase in stereo LP prices inevitably came at the start of 1969. Then, removal of excise taxes obliged the record companies to make cuts.

Having observed a steady near 10 percent annual increase in retail sales, major investment advisers began to court the recording business in the last years of the decade. Five major manufacturers controlled more than half the market—Columbia, Warner-Seven Arts, RCA Victor, Capitol-EMI, and MGM. Thirty-five percent was divided among nearly 100 smaller competitors, and the remaining 10 percent or so was shared by hundreds of others.

Most investors did not recognize that record companies failed more frequently than not. A single 45 had to sell 11,200 copies to break even. This happened, according to Henry Brief, the RIAA's executive director, to 74 percent of all 45s in 1968-69, and 61 percent of all popular LPs, which had to sell a minimum of 7,800 copies to recoup their initial cost.

The Warner-Seven Arts group was among the first to expand in the music business on a large scale, adding Atlantic Records in 1967. With this purchase came its owned-and-operated music firms—Cotillion,

Pronto, and Walden—but not its first publishing operation, Progressive Music, which had been sold to Hill & Range Songs in 1964.

Atlantic was formed in 1948 by Ahmet Ertegun and Herb Abramson, later to be joined by his brother Neshui and Jerry Wexler, who brought out Abramson. The firm's position evolved in the late 1950s when young producers Jerry Lieber and Mike Stoller produced crossover hits by the Drifters, the Coasters, Bobby Darin, and others. At the time of the sale to Warner, Atlantic was doing an annual gross of $20 million. The Erteguns and Wexler signed multi-year contracts with Warner and operated Atlantic as an independent division, with its own distribution, artists, and international licensees.

Steven J. Ross, head of Kinney National Service, brought the option to purchase Warner-Seven Arts and began to acquire Warner common stock. In 1970 Elektra Records joined the roster, and in 1973 Asylum Records was added and eventually amalgamated with Elektra. Warner Communications was formed in the early 1970s as the parent company for this entertainment empire. Within two years, the record group owned almost one-quarter of *Billboard*'s best-seller charts, with Atlantic contributing almost the same share of all profits.

Transamerica, a San Francisco insurance company, purchased United Artists Pictures and its subsidiary music interests, including United Artists and Unart Records in 1967. A year later, it acquired Liberty Records along with its subsidiaries, including the jazz-oriented Blue Note Records, a tape-duplicating plant, and several music companies.

Transcontinental Investing Company diversified by forming the $12-million Transcontinental Music Corporation. The company's head, Robert Lifton, recognizing early that most record companies were incapable of sustaining a nationwide distribution chain, bought several record wholesalers and formed Transcontinental Distributing Corp. It became the largest single-record and tape distribution operation in the United States in 1967. With control of 20 percent of all rack-jobber business, representing the same percentage of retail sales, Transcontinental Distributing amassed control of more than half of the entire record-distribution business.

Electrical & Musical Industries, whose economic fortunes in the record business had continued to skyrocket with Beatlemania, was slowly losing its head in the industry to American Columbia Records. Without the Beatles' knowledge, Brian Epstein had extended their contract with EMI, which expired at the end of 1967, for nine years. Epstein's NEMS Agency now collected its own 25 percent share of Beatles' royalties directly from the manufacturers. The Beatles had already reaped twenty-two Gold Records from the RIAA in America, and it was estimated that they had sold 180 million units in total world sales. The new British 50 percent purchase tax placed a burden on English buyers, which, added to EMI's worldwide switch to stereo and the expansion and automation of manufacturing and engineering facilities, was responsible for a general reduction of EMI dividends.

Capitol had taken over one of the major tape manufacturers, Audio Devices, in which it had long held a minority interest. A program of corporate reorganization put Audio Devices and Capitol Records in control of a new entity, Capitol Industries, in which EMI owned majority control. One of the largest rack jobbers was purchased, but it quickly lost money. Alan Livingston, Capitol's president since 1961, resigned and was followed by two presidents, one of whom, Stanley M. Gortikov, became president of the RIAA following his departure from Capitol in 1971. When the Beatles broke up that year, the company's fortunes reversed completely. A new president, Bhaskar Menon, pulled Capitol into the black by 1973, with prerecorded AudioPak tapes accounting for more than half of all volume.

The European Common Market came into force on January 1, 1968, and a new element of competition entered the European record business. EMI watched its international competitors, Philips Phonographic Industries and Siemens, acquire one of the oldest and most important catalogues of American and British musical-theater and popular-music copyrights: the British and American Chappell companies.

Philips and Siemens formed a partnership in 1963, but they continued to maintain separate manufacturing and sales divisions for all their products, including those of their recording companies, Polydor and Deutsche Grammophon. They signed a contract with the Robert Stigwood Organization to promote some of their popular artists. A rock manager, Stigwood introduced leased taping in England and paid all costs for the master tapes of groups he managed and then sold them for a royalty of 15 percent or more. With a partner, David Shaw, Stigwood obtained from Brian Epstein an option to buy a 51 percent interest in his NEMS artist-management agency, later renamed Nemperor Holdings, of which the Beatles owned 10 percent. Stigwood never exercised his option to buy NEMS because of insufficient funds and the death of Epstein, whose mother inherited 70 percent of the company, and his brother, Clive, got 20 percent. To help pay the large death duties, Clive sold his mother's share to the British Triumph Investment Trust. In a financially intricate turnover of Nemporer, Triumph got back its money and a share of several millions of EMI royalties, frozen in a legal fight that pitted Paul McCartney against John Lennon and the other Beatles, plus 5 percent of all future earnings by the group through 1976. An offshoot of the differences that divided the Beatles and led to their final separation was the 1969 purchase of Northern Songs and Maclen Music in the United States by Lew Grade and ATV, despite all McCartney's and Lennon's attempts to retain the companies.

In 1968, the Beatles announced the sketchy details of their new tax-saving dodge, Apple Corps, Ltd.: a corporation set up to deal in films, music, electronics, and merchandising, each related to their primary business, and thus not suspect to tax men. The company operated for several years and then was sold to the public. The Apple label, on which all future Beatles releases appeared, gave world rights to EMI and American manufacture and distribution to Capitol Records. Royalties were on the terms negotiated by Epstein in 1967.

A new manager, Allen Klein, handled Apple's business affairs. He shrewdly rewrote the Beatles' EMI contract in 1969, which raised LP royalties from thirty-nine cents to fifty-eight cents through 1972, and then to seventy-two cents until the end of 1975 when the agreement expired. The right to release old Beatles material, unavailable under the existing contract, was leased to EMI, and a tidal wave of repackaged albums followed.

RCA's profit margins during the mid-1960s generally ran below those of the parent organization. The record company continued to offer a balanced and homogenized repertory largely ignorant of most contemporary progressive rock. The man who had carried RCA into the modern era in spite of itself, Elvis Presley, had become a film idol, whose soundtrack LPs attracted new members to the company's record club. Then in 1969 his songs "Suspicious Minds" and "In the Ghetto" reminded the public why Presley still should be known as the King of Rock 'n' Roll.

Columbia Records' sales of rock records jumped from 15 percent of the total market to 60 percent in the five-year period ending in 1969, and was due in great measure to the head of its record division, Clive Davis. His presence at the 1967 Monterey Rock Festival led to the signing of a number of major acts.

Davis signed new talent representing what he believed was the best of the revolution in music. Beginning in early 1968, Columbia's steady releases of its new contemporary rock stars set new sales records—Blood, Sweat & Tears' second album sold 3.8 million; Santana's *Abraxas* sold 3.5 million; each new album of Chicago sold well into seven figures.

The Warner/Reprise/Atlantic group worked hard to match Columbia's success, sharing a major portion of each week's *Billboard* Hot 100 chart, and had well over 100 LPs that sold more than a million copies each in the 1968–70 period.

Radio stations developed an immutable affinity for Top 40 programming and the undeniable viability of rock-'n'-roll music, which provided stations with top ratings. The brief period in 1961 when frequency modulation was seen as the best antidote to rock 'n' roll had passed. The approval of FM stereo broadcasting by the FCC resulted only in the duplication of the programming of the AM facilities, which were connected to 60 percent of the 990 FM stations.

Simultaneous AM/FM programming came to an end on October 15, 1965, when all jointly owned AM/FM stations in markets of more than 10,000 listeners were required by the FCC to program non-duplicated music for at least half their airtime. The most prosperous AM broadcasters put their FM operations on a non-commercial basis, giving listeners the kind of music they appeared to want. The FCC order hoped to create a diversity of programming, but those 10,000-watt FM stations that adopted a stereo rock-'n'-roll format selected their music only from the top of the trade papers' pop-music charts.

The FM audience was growing faster than advertising revenues. At the end of 1966, FM advertising revenues were at the $32 million mark, far from local AM time sales of $580 million and a national spot-advertising business of $285 million, all aimed at the eighteen- to forty-five-year-old audience AM had staked out. Much to the chagrin of a business that issued 11,000 new records a year, AM played only from a list of thirty hit singles each week, with a few "specials" thrown in.

Tom Donahue, a disk jockey who was program director of KMPX-FM in 1967 and was later known as the Father of FM Radio, believed that Top 40 radio was dead. Donahue moved to California in 1960 and worked for KYA, San Francisco, when a new owner gave him and other disk jockeys wider latitude over repertoire. KYA, stripped of jingles by Donahue's order, became "Boss of the Bay," using a highly varied "Boss Top 40" format. After leaving, Donahue went to KMPX-FM, San Francisco, in April 1967, and in six months turned its music policy around. He experimented with programming by playing all manner of musical forms, and alternative, free-form radio was born.

The revolution in music propelled the world of FM radio to a 12 percent share of the 100 top national facilities. Free-form radio concentrated on more music, fewer hard-sell commercials, and used no gimmicks. The majority of the time was bought by record companies, enabling stations to emphasize the fresh morality and stark reality of music aimed at the fifteen- to twenty-year-old male market. In the early 1970s, however, concerned practitioners of the new medium, Donahue among them, correctly perceived that because progressive radio was no longer unique, it would disappear or be coopted by commercial AM programming.

11 The Record Business in the Early 1970s: Huge Contracts and Low Profits

THE election of Leslie Arries of WBEN-TV of Buffalo, New York, as chairman of the All-Industry Television Committee in 1971 introduced a new player who would rattle the structure of music licensing and undermine the permanency of ASCAP and BMI income. He established an operating fund to effect a real decrease in broadcasters' payments to the licensing agencies.

The networks' determination to reduce their license fees had little effect on ASCAP. Raids on BMI writers and publishers accelerated after the ASCAP board allowed members to be paid for their collaborations with BMI writers. When negotiations began for a new contract to replace the eight-year-old ASCAP agreement with independent radio stations, the Radio Committee demanded a substantial reduction. It offered statistical evidence that although ASCAP radio fees rose 70 percent since 1963, its share of radio airplay dropped 36 percent. A new contract, approved by Judge Ryan, which set fees of 1.725 percent over a five-year period, brought ASCAP's radio rates nearer BMI's.

In the television arena, CBS refused to terminate its per-use suit and accept blanket licensing. Motions by NBC and ABC for summary dismissal of BMI's two-year-old conspiracy, breach-of-contract, and antitrust suit against ASCAP and the networks were denied by Judge Lasker as was a petition to dismiss BMI's fraud charges. ABC-TV had been made a codefendant in both matters, once it settled a back-payment problem with ASCAP. However, Lasker did grant the ABC motion to sever the antitrust claim from other BMI charges and agreed that the antitrust case be tried first.

The 1972 civil antitrust suit filed by the Justice Department against the three networks sought to prohibit them from engaging in the production of any television programs or backing programs they purchased. The suit asked that nothing more than "first-run right of exhibition" could be acquired, stripping the networks of the right to spread initial costs over reruns.

The matter had originated in 1963, when the FCC barred as anticompetitive those segments of a station's programming over which affiliated stations had conceded to the networks. In 1965, the Justice Department and the FCC examined the covert ownership by network officials of companies from which they purchased television programs. This led first to the prime-time access rule, in 1970, as a holding action until the completed document was filed. A consent decree was finally accepted by the three networks in 1980. It limited the amount of prime-time programming each could produce in the next ten years.

Peace between NBC and BMI came in 1973, when the network accepted a three-year blanket contract. With the signing, NBC was removed as a defendant in the various BMI suits, and its application for a limited BMI copyright license was rescinded.

The All-Industry Radio Committee began to hear complaints about BMI from broadcasters. In the last ten years, BMI's local radio income had increased fivefold, and broadcasters wanted to stop or at least slow down this constant escalation. The committee believed there was no justification for BMI's radio fees to continue at a rate 85 percent as high as that of ASCAP, when that for television was only 59 percent as high.

BMI hoped to pacify the broadcasters by maintaining the current fee in a four-year license extension, but promising lower rates in its final year, 1977. An experimental "incremental" formula would go into place for 1977, using non-network time sales in 1974 as the base on which the final year's rates would be established. Anything earned above the figure would be taxed at a 0.85 percent rate, that below at the 1.7 percent rate, a plan the committee accepted.

CBS v. ASCAP, BMI et al. was resumed in late 1973. Counsel for CBS emphasized the issue of restraint of trade and the lack of a mechanism for direct licensing of music copyrights. ASCAP and BMI attorneys argued that no network had ever attempted to deal directly with writers and publishers since 1929, when CBS first accepted a blanket license from ASCAP, or 1946, when its first blanket television license was signed.

Judge Lasker discussed the CBS complaint, finding that the network had not proved that ASCAP and BMI were guilty of restraint of trade, and that CBS had failed to demonstrate that there were indeed "significant obstacles to direct licensing" negotiations with both societies.

While CBS attorneys believed they could appeal the Lasker ruling, others felt that since the countersuits filed by ASCAP and BMI had not yet been tried, there could be no final judgment. It would be at the Appeal Court's discretion whether the decision was reversed. CBS quickly filed a notice of appeal.

With the threat of termination of all blanket licensing removed, the vendettas of the past were forgotten. As far as BMI and ASCAP were concerned, when one was wounded the other often bled. The CBS application for per-use licensing, NBC's demand for limited access to copyrighted music, and the ensuing legal defenses that followed had freed the societies to work together. The Aiken case, in which the owner of a carry-out restaurant threatened to overthrow the law on which all non-broadcast licensing activities were based, also served as an important unifying agent.

The 1931 Jewell-LaSalle case had established the legal precedent for the licensing of all places where music was played over multiple speakers. Attorneys for Aiken who failed to secure a five-dollar-a-month license for the speakers in his establishment that relayed radio music, unsuccessfully argued in the lower court that radio music was already cleared for performance under the 1909 Copyright Act. The Muzak Corporation, professional supplier of wired music, took a more than incidental interest in Aiken and supported the restaurant owner, hoping that the case would be appealed. It was, and was reversed, the court finding a similarity to *Fortnightly* and *Teleprompter*, in which the Supreme Court had ruled that signals transmitted by cable TV merely extended audience reach and therefore did not constitute a performance.

However, both ASCAP and Muzak sustained a serious economic loss when the Supreme Court declared in *Aiken* that radio music was free. Jewell-LaSalle had been "sidestepped." The court majority found that to require small businessmen owning radio and television sets to take out music licenses "would result in a regime of copyright law that would both be wholly unenforceable and highly inequitable."

The decision underscored the need for omnibus copyright revision. The music industry and the societies realized that a coalition of forces was necessary to effect the desired revisions. Otherwise they would have to learn to live under the 1909 act for years to come.

Emnity between the songwriters organized in the American Guild of Authors and Composers and the publishers represented by the National Music Publishers Association mended, too. In 1967, AGAC president Burton Lane stepped down for Edward Eliscu. Membership in the guild was at an all-time high, and collections in the royalty plan Lane had instituted in 1959 hit a high of three million dollars.

Eliscu encouraged a more collegial relationship with BMI writers and opened the guild's door to all writers and publishers. A new system for dues payments, based on royalty income, was instituted in 1972. Eliscu was succeeded as president in 1973 by Ervin Drake. He struggled to win proper rewards for those who wrote words in English for foreign songs and persuaded ASCAP in 1974 to count foreign performances on an itemized song-by-song, country-by-country basis.

The Composers and Lyricists Guild, next to the AFM the most important musical labor union, first felt shock waves from the CBS per-use suit near the end of 1972. The CLGA still had been unable to obtain a renewal of its original two-year contract with the picture studios and the Association of Motion Picture and Television Producers. The agreement had established base pay, guaranteed health and welfare benefits, and affirmed the right of members to collect performance royalties through the societies. Early in 1972, a $300 million antitrust suit was filed in the New York federal court against the three TV networks and eight major studios. It charged a conspiracy in restraint of trade, because the defendants had signed a new contract that would not allow the plaintiffs to retain ownership of music they had written for the defendants. They also complained that they were precluded from publicly performing their recent TV feature and movie music. The suit asked that the defendants be stripped of the right to assign music to their own publishing companies. CBS severed itself from the lawsuit by returning to those plaintiffs the rights to music written for past programs and offering to include that right in future contracts. In 1974 the suit was dismissed and judged to involve a labor dispute, more properly within the jurisdiction of the National Labor Relations Board, which had given the CLGA its status as a union.

In 1976, motion-picture and television-feature soundtracks once again appeared on the Top 10 LPs of the Year charts when *American Graffiti*, a collection of rock-'n'-roll oldies, made the number-one position. The soundtrack of *A Star Is Born*, featuring Barbra Streisand and Kris Kristofferson, accumulated sales in excess of three million dollars. More than 100 cover versions of the picture's theme song, "The Love Theme . . . Evergreen," were recorded.

The success of film scores like Bill Conti's *Rocky* and John Williams's *Jaws* resulted in greater composer participation in production. Fees for some CLGA members now climbed to as much as $50,000 a film. They also got a share in publishing through their own company; a producer's percentage of the LP list price; and a 2 to 3 percent royalty for re-recording the music. The price for doing what had once been hackwork scores for television films rose, too, as did the quality of the music. Top film composers insisted on clauses in their contracts that required a 50 percent reuse charge when a production was sold overseas for theatrical exhibition. Screen music by older composers had found its way into the catalogues of Mills Music in late 1969, when it merged with Belwin, who published massive compilations of public-domain and original music for the screen musician.

Music publishers now used a small group of well-established companies with first-class printing and distributing facilities for the printed-music portion of their business. E. B. Marks contracted, in 1973, with Belwin-Mills for the latter to be its exclusive representative for all printed products. MCA-Leeds/Dutchess entered into a joint venture around the same time with Mills Music, by then a Belwin-Mills division.

Only those music houses owned by record companies with major foreign publishing and subpublishing connections were able speedily to transfer copyrights in the following years. The Polyder Records division of Phonogram International, the recently activated combination of Philips and Deutsche Grammophen interests, made its U.S. bow in 1970. Polydor arranged for distribution through United Artists, and two years later, Mercury Records Productions and its associated labels, including American Philips (Norelco), were transferred to the Phonogram group. The transfer included Norelco's 49 percent interest in Chappell Music, thus giving Phonogram complete control of the giant music operation. Philips's North American holdings were made public in 1974 in connection with the Dutch combine's projected takeover of Magnavox in order for Philips to move into the American video and audio components business.

The record companies were taking over. As Stanley Gortikov of the RIAA told a copyright revision hearing in 1975, in recent years music publishers had become "heavily administrative and clerical; they are largely service entities, conduits for the processing of income and paper transactions. They don't promote as they used to. They don't employ field representatives as they used to. They don't create demand as they used to. These functions have necessarily been taken over by the recording companies."

Record companies now offered superstars excessive advances. Such bidding had drawbacks for the artists. In the early 1970s, standard artists' royalties averaged around 7.5 percent. There was little risk for Columbia in the $50,000 Clive Davis paid to sign the Electric Flag in 1967. Half of it was immediately written off toward the cost of recording, which was exacted from the artists' royalties. With a profit of around seventy-five cents from each LP unit, the entire $50,000 was more than recouped from the sale of 70,000 copies, with promotion and advertising costs still to be absorbed. However, as Joe Smith of Warner said in *Rolling Stone* in 1971, "the group still does not have its advance money back because they're recovering it at a lesser rate than we [the record company] are. . . . At 100,000 albums they've paid back their advance and recording costs, and from then on they're making money. But only 10 to 15 percent of the albums sell that well."

Artist guarantees skyrocketed. In 1974 when MCA renewed Elton John's contract, all six of his LPs had sold more than a million units. A new Rocket Records label had already been created for John and his protégeś, so MCA offered him the highest recording-artist guarantee yet known, in excess of eight million. It called for one million dollars each for half a dozen new LPs during the five-year life of the contract, as well as $1.40 royalty for every $6.98 album.

The energy crisis in late 1973 had seriously affected record manufacturers, for there was a shortage of polyvinyl chloride, made from petroleum, from which both records and tapes were manufactured. Some small firms were driven out of business, and new releases by independents declined. To cut costs, new talent was offered the release of one single. If album deals were offered, an advance against earnings was paid only after delivery of the finished tapes.

Superstars' guarantees continued to set new levels of privilege. Eight months before his contracts with EMI and Capitol-EMI as a member of the Beatles were to expire, at the end of 1975, Paul McCartney renewed agreements with both for his exclusive services until 1978 for an undisclosed amount. The agreement called for one album a year by McCartney and his new group, Wings.

For much of the 1970s, about two-thirds of the income of ATV Music came from Lennon-McCartney songs. Sir Lew Grade's music company, built on the rocks of Northern Songs and Maclen Music, lost all McCartney songs written after his ATV contract expired in 1973. When Wings began recording, and the ATV contracts were still in effect, many of the songs were credited to Mr. and Mrs. McCartney, which would reduce worldwide royalties on the new material. Grade unsuccessfully instituted a court action for breach of contract, claiming that Linda McCartney was incapable of writing music. All subsequent McCartney songs were published by MPL Communications, owned by Paul and his wife.

Stevie Wonder established a new ceiling on record-company guarantees with a $14-million offer from Motown Records. The new seven-year contract was to become effective as soon as Wonder delivered the two-LP *Songs in the Key of Life* and Motown guaranteed a royalty of 20 percent on the retail price of all future albums as well as ownership of all his compositions.

The superstars now insisted that their music go to their own publishing houses, which generally were administered or operated on a co-publishing basis by well-established firms. One significant exception was Paul McCartney, whose MPL Communications, formed in 1973, was directed by his father-in-law, entertainment-business attorney Lee Eastman. Using royalties paid by ATV for old Lennon-McCartney songs and those paid to MPL for subsequent McCartney music, Eastman created an entity of major proportions. All Buddy Holly copyrights were acquired, followed by other rock-'n'-roll and popular material. McCartney's quarter-interest in Apple Corps was another MPL asset. The only Beatle not represented by Allen Klein, McCartney won court appointment of a receiver for his share of Apple. When the other Beatles sued Klein, in 1973, charging mishandling of their interests, Klein countersued for breach of contract, and also instituted a $42-million action against McCartney, claiming that he had induced the other three to bring the action. The matter ended in McCartney's favor three years later, but Lennon, Ringo Starr, and George Harrison paid $4.2 million to settle the affair. Apple Records closed its doors in 1975.

The splintering of a highly concentrated business into hundreds of new entities had given the NMPA new importance. The association was doing exemplary work in Washington, fighting for copyright revision.

The photocopying machine had provided a new tool for the printed-music pirate's workshop, and the NMPA began to establish "fair use" guidelines for educators and libraries.

The NMPA also sent representatives to the 1971 UCC revision conference and secured the amendment of tax laws that discriminated against music publishers, and the extension of preferential postal rates to printed music and records. The NMPA initiated the first "official" survey of the retail volume of printed-music sales in 1971. During the ensuing four years, sales rose at an average of nearly 20 percent, while the average retail price of sheet music increased by one-third, from $1.00 to $1.50, and the wholesale price from forty-two to sixty-seven cents. Songwriter royalties remained constant.

The NMPA's figures accurately mirrored the decline and fall of the full-service music house. Music-publishing revenues had, in four years, dropped below those of the year before, and a further fall of between 12 and 15 percent was anticipated for 1976.

The publishing of sheet music and the folio, containing from twelve to fourteen songs, had become the monopoly of several large companies: Screen Gems-Columbia Publications, Charles Hansen Publications, Warner Brothers Music, and the oldest, Big Three Music. Screen Gems and Hansen did about two-thirds of their industry's $140-million annual retail gross in 1975. Distribution was handled in Screen Gems' case by regional jobbers who dealt with the estimated 4,000 music stores that carried printed items, while mail-order business was handled by the printing companies themselves.

As the 1960s became more politically volatile, the gap between the generations widened, and was nowhere more apparent than in the recorded music played on FM radio. While the electronic media relied on censorship to alleviate adult pressure on their programming, FM broadcasters made a Freedom Hall of the air, claiming it to be open to any recording.

In response to many protests about irresponsible rock-'n'-roll songs, the FCC was powerless, due in part to recent Supreme Court free-speech decisions. However, the FCC was forced by the Nixon White House to warn broadcasters that they were responsible for the meaning of the words they broadcast. Almost immediately, such hits as the Beatles' "With a Little Help from My Friends" were removed from play lists. The commissioners emphasized that they did not intend to ban the playing of drug-oriented music; instead, they expected broadcasters to police themselves.

The "seven dirty words" case began its slow march to the Supreme Court in October 1973, when a New York City parent complained to the FCC that his son had been exposed to foul language when a George Carlin monologue was broadcast over publicly supported FM station WBAI. The seven words, four nouns and three verbs, were terms for "sexual and excretory activities and organs." Ignoring the statute forbidding it to censor radio communications, the FCC censured WBAI for irresponsibility and banned the seven words from the air.

Four years later the U.S. Court of Appeals in Washington overturned the FCC's "dirty word ruling," finding it too broad and carrying the commission "beyond protection of the public interest into the forbidden realms of censorship." The Supreme Court did so in 1978, saying that the FCC had properly exercised its right to regulate and punish a broadcast of indecent material.

Many radio operators continued to look to the program counselors and the taped-show producers to perform all programming functions. Bill Drake, the best known and most successful of the former, refined and modernized Todd Storz's now stale Top 40 formula by creating a format with more music, less talk, and fewer commercials. In 1965 Drake reduced commercial spots to twelve an hour. He also created a twenty-twenty news policy (twenty minutes past and before the hour), and held his disk jockeys to a tight eight seconds for introducing each record. His play lists included only thirty-three current hit singles, based on *Billboard's* Hot 100 and reports from local dealers, as well as three "hit-bound singles" chosen by Drake, and later by assistants.

With radio-station owner Gene Chenault as his partner, Drake-Chenault Enterprises was formed in 1968. Their regular fee became $100,000 plus a monthly retainer. Drake-Chenault's major division, American Independent Radio, was formed to supply taped programming and did much to eliminate the use of local deejays. AIR's initial package was "Hit Parade," made available only to fully automated stations, followed by other programs that appealed to various constituencies. The packages cost from $550 to $5,000 a month, depending on the size of the market.

The syndicated taped-program business would not have been feasible without the automated facilities installed in many of the country's 2,000 FM stations in 1970. An automated broadcast used only pretaped music, commercials, station identifications, telephone calls, disk-jockey banter, time announcements, and other business, all of which appeared to give the image of spontaneity.

Those stations that declined the services of a programming consultant had the trade-paper charts and mimeographed newsletters to help revise their play lists. The first trade paper to print a music chart, *Billboard* was held in the highest regard by station managements. Second only to *Billboard* in esteem was Bill Gavin's *Record Report,* begun in 1958, which gave predictions based on the editor's expert judgment. Gavin was instrumental in turning stations to formats in addition to Top 40. He and others searched for a format that would hold the twenty-five- to forty-year-old audience, one that had grown up on rock music and was ready for more "adult" fare. In the early 1970s, middle-of-the-road music grabbed that share of the audience and, with Top 40, rock, and album-oriented rock, held more than half of all listeners.

As radio formats provided the public's major access to music, improved prerecorded music tape accelerated interest in magnetic recording tape and players. Professional audio equipment now operated with nearly perfect fidelity at several speeds and could record up to twenty-four tracks on two-inch tape. Equipment for the home encompassed all configurations: four- and eight-track cartridges, reel-to-reel tapes, and the Philips cassette.

The cassette's mechanical drawbacks precluded its immediate acceptance by consumers, who favored the Lear eight-track system well into the early 1970s. The eight-track offered complete compatability in both home and automobile, with all the features that duplicated the sound quality of a high-fidelity record machine.

Eight-track technology improved, and in May 1970 RCA Records introduced the revolutionary discrete four-channel, eight-track compatible sound system for car and home use. High-fidelity aficionados were the first to buy the four-channel equipment to play the QuadraDisc, the name RCA bestowed on its new offspring. Other manufacturers offered new quad conformations.

The record business anticipated an all-quadraphonic industry within five to seven years, but the public failed to share the enthusiasm of hi-fi devotees. Quadraphonic represented just too much electronic gear for most average consumers, and both discrete and matrix disks disappeared, one of the business's greatest fiascos.

American physicist Ray Dolby made the cassette the winner in the battle of configurations that took place during most of the 1970s by his revolutionary miniaturized B-type noise-reduction unit. His A-type system with low-level differential noise reduction made a successful debut in America in 1966, after which Dolby worked on the application of his invention to the eight-track cartridge and the cassette. A new B-type circuit, demonstrated in 1970–71, permitted recording engineers to make favorable contrasts between the cassette and the master tape, leading to its adoption by Decca in England and then by RCA in the United States, all of whose cassette production was completely Dolbyized.

In 1975, sales of tapes of all types accounted for 29 percent of all pre-recorded music sales. In nine years cassette sales jumped from $6 million to a high of $102 million. The future of cassettes was highly predictable after the sale in 1975 of 150 million units of blank tape, an indication of the growing practice of taping music off the air or from borrowed albums.

Disk sales declined in 1971, yet the consumer's growing interest in taped music increased total industry revenues. The record and music business had a number of inherent problems. All sales were on a consignment basis, with a ninety-day collection cycle, and a return factor, for either exchange or cash, that could be as high as 100 percent. The product's obsolence began the first day of its exposure for sale. Some 81 percent of all single releases and 77 percent of all popular albums were failures. Yet new LPs were being released at the rate of 100 a week, and the number of new taped cartridges and cassettes was up by more than 50 percent.

Integrated rack jobbers had gradually expanded their role, adding wholesale distribution and the operation of one-stop retail and self-service discount locations, bringing them control of nearly 80 percent of the business by the early 1970s. They once but no longer serviced the independent labels by delivering copies of releases to local radio stations. The advent of free-form FM gave birth to a new kind of promotion man, one who dealt only with LPs and particular cuts on them.

RCA Records in 1969 inaugurated "dual" or multiple" distribution by restoring to its merchandising chain the company-owned or -franchised distribution facilities that had prevailed in the pre-rack-jobber era. In the next several years, the other eight companies followed. Only A & M and Motown continued to work exclusively through independent distributors.

Dual distribution was a blow to rack jobbers. It could be compensated for only by an increase in suggested retail prices, which did come in 1973, when the medium LP price was raised to $5.98. Raising their prices to accommodate an increase, rack suppliers paid $3 and charged $3.40 for a $5.98 middle-line album, which was sold for $3.98. The manufacturers' branches paid 5 percent less wholesale than the independent distributors and sold the middle-line LP for $3.35.

Recording costs were up by 200 percent. Studio charges had risen for $70 to $120 an hour. The tape that used to be $50 a carton was now $90. A 10 percent royalty was general. In some cases an artist got as much as 14 percent; a 9 percent deal was looked upon as a triumph.

An incipient payola scandal was heralded by the surprise announcement from CBS in 1973 that Clive Davis, president of the reconstituted CBS Records Group since 1971, had been summarily dismissed and a civil action charging him with improper use of company funds was filed. Goddard Lieberson, the once powerful Columbia executive who had been shunted to a senior vice presidency, took his place.

Overnight, the record business became the subject of concentrated critical attention. The FCC investigated dozens of payola complaints. In response, the RIAA announced a code of self-regulatory business practices. It called for company action on the part of the fifty-five-member manufacturing firms as well as by non-RIAA firms; established standards of conduct for all employees; and urged broadcasters to initiate remedial actions.

Another matter meriting examination was the racism inherent in the charges. Payola, the *New York Times* said parenthetically, was "said to be in force mainly among disk jockeys whose music is aimed mainly at the black community. Most of these stations are white-owned. Their disk jockeys, most of whom are black, traditionally are paid far less than their white counterparts."

As recently as 1958, some seventy radio stations in the United States aimed their programming exclusively at a black audience. With the increase in civil-rights action, more sophisticated black audiences emerged, staying tuned to black stations programmed to meet their needs. The word *soul* came to signify what was considered best in black ghetto life. R & B or soul artists soon drew sold-out white audiences, but it was not until their records hit the Top 10 that they appeared on the white Top 40 radio station play lists.

By 1972 three out of every ten records on Top 40 radio were R & B. The success of Atlantic, to some extent Motown, and the Memphis-based group Stax made it possible for them to broaden their lines and move into the lucrative pastures where CBS dominated—pop, middle-of-the-road, and, sometimes, musical theater.

A study of the "soul music environment," made for CBS Records, stated that the company was "perceived as an ultra-rich, ultra-white giant which for the most part had chosen to snub blacks in the business." CBS responded by embarking on a new soul-music division. Black artists were signed; manufacturing and distribution agreements were initiated with existing independent soul-music firms; and a public-relations effort was made to improve the corporation's image in the black community. Six million dollars was advanced against royalties to Stax Records when it turned over its distribution to CBS. The Philadelphia song-writing and record-production team of Kenny Gamble and Leon Huff, together with their Philadelphia International and Gamble Records operation, was added to the CBS Records Group. By 1973, CBS had scored a series of R & B hits made by Gamble and Huff and others and acquired the syndication rights for *Soul Train* and a 15 percent share of the black music business.

Clive Davis in 1974 was hired as a consultant to the recording and music operation of Columbia Pictures Industries. Davis's initial function was to supervise the operation of one of its subsidiaries, Bell Records, which Columbia purchased in 1969. Bell merchandised only records made by independent producers. Its owner and founder, Larry Uttal, built up a network of distributors throughout the United States that matched those of the major companies. The Bell name was replaced with a new Arista logo, and by 1975, Arista showed a increase of three million dollars in profits. It also garnered eighth place on *Billboard*'s annual chart of corporate shares of single and LP sales.

On June 14, 1975, the Justice Department announced the indictment of nineteen persons, four of them record-company presidents: Gamble and Huff, of Philadelphia International Records, and the president of Brunswick and Dakar Records, Nat Tarnapol, all of whose soul music was manufactured and distributed by CBS Records. All three were charged with conspiracy, interstate travel to commit bribery, mail fraud, wire fraud, income tax evasion, and failure to file income tax returns.

Clive Davis, the fourth president to be indicted, was charged with personal tax evasion.

Few remarked on the racial bias running through the indictments, which called to justice only people dealing with soul music. Davis was found guilty of only one of six counts of evasion, involving $2,700 in taxes and $8,000 of undeclared income received for traveling expenses. He paid a $10,000 fine. Tarnapol and four other Brunswick-Dakar employees were found guilty, a verdict that was overturned by the Court of Appeals in 1977. The Philadelphia International defendants pleaded nolo contendere to all charges and were fined $45,000.

A government investigation in mid-1975 of the financial affairs of Capitol Records and its owner, EMI, was precipitated by a class-action antitrust suit involving securities fraud. A Los Angeles grand jury subpoenaed Capitol's financial records. Soon after, a tender offer by EMI to acquire additional stock in the company raised its share of ownership from 70 to 97 percent. Combined with a fall in earnings, the lawsuit and grand jury probe seriously damaged Capitol's financial standing. Its shares fell from fifty-six dollars in November 1969 to twelve the next August, and six a year later.

The plaintiffs in the stockholders' suit contended that Capitol foolishly had bet on the "wrong performers." It lost close to eight million dollars and found itself with almost 20 million unsalable records and tapes. To push these, Capitol gave millions of dollars in discounts to record dealers in the form of cooperative advertising rebates never used for advertising. The company also appealed to dealers not to return records before key financial reporting dates.

The presiding judge accepted the Capitol-EMI defense that the transfer of "interim reserve" funds into sales columns followed generally accepted accounting standards. The Securities and Exchange Commission testified to the contrary. In 1978, the court exonerated Capitol Industries-EMI and concluded that those minority stockholders who had brought the action were not entitled to financial relief.

The American public was ignorant of the payola trials and the virtually exclusive attention to soul music. Black music was actually thriving, having won over many major manufacturers, because of its power to affect Top 40 programming and the major boost it got with the emergence of disco music.

Disco records were cheap to manufacture, for most of the performers worked for scale. Producers usually wrote both words and music and owned the copyright. Computerized rhythm machines laid down the initial bass drum and bass guitar tracks, while the other tracks were made by masses of conventional and electronic instruments, piled over one another and mixed by the producer, the disk's ultimate creator, before the final vocal track was recorded.

When the number-one disco hits stretched beyond their three-minute length into quarter-hour performances, through the use of tape loops, a Latin record manufacturer, the Cayre Corporation, introduced twelve-inch recordings on its Salsoul label. Long after disco had run its course, the twelve-inch single maintained a place in the business. New versions of popular hits, or "mixes," were made by dubbing, echoing, tape manipulation, and the use of as many as four kinds of "reverb."

The majority labels slowly developed new business practices. As Walter Yetnikoff, now president of the CBS Records Group, told *Business Week:* "Things are considerably different . . . today. Now every album that goes out has a complete marketing plan—with details on advertising, displays, discounts for the trade, personal appearances by the artists, sales targets, and national and regional breakdowns."

Still, the record business faced considerable problems. W. T. Grant Stores, a major outlet for the biggest labels, closed its doors, because of the vast bills unpaid by the giant chain-store operation. In 1974, ABC Records surprised the industry with the first step in a major growth program—the purchase of all Gulf & Western's records operations. The sale included the Paramount label—which would be phased out because G & W still owned the name—and Dot, Neighborhood, and Blue Thumb records as well as Word, a major religious communications business that produced white gospel records and associated products.

The industry had grown at a steady 6 to 8 percent rate annually from 1960 through 1965. Then, in an accelerating economy, the introduction and growth of tape led to an increase of 17 to 18 percent. Yet the financial community felt the quality of corporate management had not improved. Business failures, the inability of too many companies to show consistent growth patterns, and the lack of verifiable industry sales figures soured Wall Street. The RIAA's annual survey of record revenues came under serious question in 1975 following the release of *Survey of Financial Reporting and Accounting Developments in the Entertainment Business,* made by the accounting firm of Price, Waterhouse.

The study indicated the revenues of the eight major companies should produce a figure higher than that reported by the RIAA, a difference resulting from the inclusion of income from music publishing, manufacturing, retailing, distribution, and other sources. The RIAA's $2.2 billion for 1974 was based on list figures. But, as *Cash Box* reported in February 1976, when calculated "at the manufacturers' selling price to their distributors (both independent and company-owned) the record business appears to be an $850–950 million industry, less than half the $2.2 billion published figure." The RIAA's explanation was that the $2.2 billion was based on figures submitted by the 85 to 90 percent of the industry affiliated with it; the balance had been projected by a committee. Compiled on a calendar-year basis, the total figure was reached by deducting returns from the sales/shipped figures.

Cash Box continued: "Which figures are more accurate is hard to determine. The RIAA reflects industry sales at an unrealistic list price level, presumably to make the number larger than it should be. And yet when record retailing and manufacturing, music publishing, and domestic revenue from international deals are added in, the overall record industry is a multi-billion dollar industry. But how many multis will remain is hard to determine until more accurate sales statistics are made available."

To create "better trust in the eyes of the investing public," as a WEA executive explained, Warner Communications and CBS disclosed more information in annual corporate reports. CBS reported an income of $484.3 million and Warner $313.8 million. On the basis of other, now more detailed annual reports, expressed in millions of dollars, these two were followed by ABC, with $157.4; MCA, $137.0; Capitol Industries-EMI, $120.2; Transamerica-UA/MGM, $98.7; Columbia Pictures Industries, $30.1; and 20th Century-Fox Film Corporation, $17.9. As had been its practice for years, RCA, the thirty-fourth largest industrial entity in *Fortune*'s 500 list of top corporations, did not separate record/music earnings from its $4.8 billion income in 1975.

As the investigations into their balance sheets troubled the record companies, the counterfeiting of printed music remained one of the music publishers' paramount problems. The first federal anticounterfeiting legislation was passed in 1962; it established a fine of not more than $10,000 and imprisonment for no more than a year for any persons who "transports, receives, sells or offers for sale in interstate or foreign commerce, knowingly and with fradulent intent counterfeited records." Apparently, however, no action was ever filed under this law.

In 1972, amendments to the 1962 legislation provided uniform federal protection for the first time, including both civil and criminal remedies against record piracy, and established federal copyright in sound recordings made following ratification on February 15, 1972, to be effective for the succeeding two-year period. The Criminal Division of the Justice Department was empowered to bring action, but the bill reduced piracy to a misdemeanor, with penalties of one year in prison and/or fines of between $100 and $10,000. The ineffective revised provisions did not strengthen America's position at a convention in Geneva later in the year, at which an international treaty was ratified outlawing counterfeited and pirated sound recordings. The United States became a full party to it in March 1974.

Omnibus copyright revision and its scheduled increase in mechanical rates for record companies gave the RIAA an opportunity to work for stronger antipiracy legislation and use that already on the books to better advantage by seeking retroactive protection for recordings made before February 15, 1972.

By 1974, the NMPA had won affirmative decisions in three more appeals court cases that declared that pirates could not hide behind the existing copyright laws by paying mechanical royalties to music publishers. An amendment to the 1909 laws signed by the President at the end of 1974 stiffened earlier federal penalties: the first offense was made a misdemeanor; fines were raised to $25,000 and to as much as $50,000 for a second offense.

There were many roadblocks to a copyright revision package. The federal court ruling that dismissed the CBS case against TelePrompter for infringement of copyrighted programs over its CATV system was intended to put the propriety of using distant signals on cable into the lap of Congress. The CATV lobby had agreed to accept liability; the FCC had produced regulations governing cable TV acceptable to all sides; intra- and inter-industry differences between cable and commercial television satisfied all parties; and the way to passage of revision of the law, at least as it affected cable, was assured. But then Judge Constance Baker Motley ruled in *CBS v. Teleprompter* that a cable pickup was only an extension of the ability to view the original broadcast, a ruling with which the Supreme Court concurred in 1974, finding cable systems not liable when carrying any broadcasting signals, under the meaning of the 1909 Copyright Act.

Hearings on the newest Senate revision bill continued. The educational and library interests agreed to exempt "musical works" in an amendment extending permission for photocopying scholarly materials. A Copyright Royalty Tribunal was suggested to serve as a forum for statutory rate-change petitions. Some 15 percent of the royalties being paid by cable TV for all copyright materials was allocated for distribution to the music-copyright interests. Following the Supreme Court's decision in *Teleprompter*, this faced serious opposition. A ceiling of three cents on mechanical royalties was approved.

A major bone of contention was the amendment calling for a record-performance royalty first introduced in 1967. Owing to masterly lobbying by the broadcasters, the bill's terms were reduced by half from the initial 2 percent of all advertising receipts by stations making more than $200,000 annually.

The Senate's version of copyright revision was passed in 1974 without the record-performance provision. But the 93rd Congress ended with no action by the House. The several-times-rejected performers'-rights amendment was again introduced in both houses, but with more moderate terms, to assuage the broadcasters' complaints. Chances of the amendment's passage were regarded as better than ever.

An amendment introduced by Senator Charles Mathias, one of the copyright interests' best advocates, called for the application of the statutory licensing principle to public broadcasting in connection with all copyrighted materials. A new proposal raised jukebox license fees to $19.70 per machine from the eight dollars proposed by the Library of Congress in 1967. ASCAP was ready to stay with the eight-dollar fee, while BMI wanted to keep both Congress and the Tribunal out of the matter and rely on the marketplace to originate an appropriate rate.

The decisive power of organized broadcasters was demonstrated at a

Senate committee hearing on the performance-royalty bill. Although she supported the concept wholeheartedly, Register of Copyrights Barbara Ringer urged that the "killer provision" be dropped, leaving it for study as a separate piece of legislation, chiefly because of potential powerful opposition from the NAB. Nevertheless, CBS President Walter Gortnikov attacked the broadcasters, demonstrating with charts that, although 75 percent of their programming time was devoted to recorded music, they provided "zero sales benefit." The vast majority of new releases never got any airplay, and the Top 40 stations added only five or six new songs a week to their play lists. Record manufacturers, he said, paid far more money to advertise their product on the air than they asked the industry to pay.

The war over rates began. The suggested increase from two cents to three cents per side would, in Gortikov's words, impose "the burden of an additional $50 million in annual royalty fees." Economic studies introduced by the RIAA showed an increase in mechanical fees in the seven-year period ending with 1972 from $41 million to $78.2 million. The reduction proposed in record-performance royalties from broadcasters, to 2 to 1 percent, would, he complained, deprive the recording industry of "a meaningful source of new income . . . while suffering" the added burden of a 50 percent increase in mechanical fees.

Citing the risk of losses taken by the manufacturers, the RIAA illustrated the problems currently affecting its constituents. Using figures for 1972, it demonstrated that, to break even, a 45 single had to sell about 46,000 units, but that four out of every five singles failed to do so. The break-even point for LPs in 1972 was about 61,000 units. Some 23 percent of all releases did that or better; the remainder sold 20,000 copies or less.

"The record-by-record odds against success are especially difficult for the smaller or newer company, which can produce only a few releases a year," the RIAA study concluded. "An increase in statutory rates, if not passed on, would raise the break-even point and the odds against success for all record companies still higher. . . . There is not, and has not been, any significant amount of bargaining or *real* negotiation about these rates. Any statutory rates would become the norm."

The executive vice president of the NMPA, Leonard Feist, said that the all-too-familiar RIAA presentation overlooked two facts. The proposed royalty was only a ceiling which increased the range for bargaining, and not the rate actually paid. At hearings ten years earlier, the RIAA predicted that a one-cent raise in the ceiling would push the manufacturers to raise the price of a $3.98 LP with its twelve songs by twelve cents. Since then, without any change in the law, the price of LPs had increased by three dollars or more, to $6.98 and up, and the number of songs had dropped to ten. What the NMPA and the AGAC wanted, Feist said, was to create a ceiling with the same purchasing power as the 2½ cents suggested by the Register of Copyrights in 1964. Since then the

Consumer Price Index had gone up by more than 70 percent. Therefore, the NMPA and the AGAC asked for a new ceiling of four cents.

In the interest of achieving unanimity with the Senate, the House committee began to match, word for word, its draft with that passed by the Senate in the last days of the 93rd Congress. The performers' royalty had been put over until the 95th Congress. Responding to advice from the Justice Department that the Senate-created Copyright Royalty Tribunal represented a violation of the separation of powers, the committee made it a regulatory commission, funded by Congress, with executive-appointed members.

Long before these and other changes had been agreed upon, the Senate Judiciary Committee on Copyrights completed an updated version of the bill that had been passed in 1973. Subjected to intense RIAA pressure, a majority of the Senate Committee turned back the clock for songwriters and publishers to 1967 insofar as record royalties were involved. When the bill emerged for a vote, all provisions were unchanged except the performers' royalty provision, which was excised. The "no surprise" legislation passed unanimously.

The House bill would be ready for a vote in September. Anticipated alterations failed to materialize. Among the provisions dealing with music in the final House bill were: elimination of the "for profit" restriction of public performance, with specific exemptions for educational and other nonprofit uses; copyright protection extended to life plus fifty years; "fair use" guidelines for the duplication of copyrighted materials, including the exemption of music from the new right of public libraries to make copies under certain circumstances; overthrow of the Supreme Court's *Fortnightly* and *Teleprompter* decisions, with cable television now liable for compulsory licensing; the creation of a new regulatory body, the Copyright Royalty Tribunal, to review royalty rates regularly in connection with compulsary licensing of recordings, jukeboxes, public broadcasting, and cable TV; and an eight-dollar jukebox tax per machine.

Immediately after the House voted in favor of the revised bill, both copyright committees began to reconcile the differences between their versions. The result was essentially the bill passed by the House, with only a few minor changes, one of them, a 2.75-cent mechanical rate, or a limitation of 5 cents per minute of playing time.

This struggle to change U.S. copyright law ended, after a twenty-year fight, on October 19, 1976, when President Gerald Ford signed the omnibus revision bill, which would become effective on January 1, 1978.

12 The Late 1970s: Betting on the Million-Seller

THE temporary pause in the CBS legal actions gave BMI an opportunity to plead "undue hardship." The company's reserves eroded, while CBS enjoyed record profits. BMI pleaded for expeditious retroactive adjustment of its interim fees and expected additional income in excess of $10 million from the network. Judge Lasker rejected the petition, because the matter was now before the Court of Appeals.

Meanwhile, the use of BMI repertory grew. It licensed sixty of the one hundred most played songs on American radio in 1975, music in 68 percent of the leading film box-office earners, and thirty-four of the fifty-three RIAA Gold Singles during its fiscal year, ending June 30, 1976.

On the basis of a new sampling system made possible through access to material provided by *TV Guide,* BMI promised a new bonus-payment system, beginning in 1977, which acknowledged all music used on local syndicated television shows. This plan doubled the minimum half-cent rate for popular songs on local FM and that for concert music from four cents a minute; increased royalties from four to six cents for popular music played on AM stations; instituted a new bonus song plan; and vastly increased television music rates. Payments for popular songs performed on Group A television shows (those on the air between 7 and 11 p.m. or on variety shows made expressly for showing before 1 a.m.), were increased from thirty-six cents to $2.25 per station hooked to a TV network; and from twenty-five cents to $1.25 on Group B stations.

The new plan shifted BMI policy to an emphasis on performances. Songs with more than 25,000 feature performances were paid at 1½ times the basic rate; those with from 100,000 to 499,000 performances, twice the basic rates; from 500,000 to 999,000, 3.8 times. Simultaneously BMI also reduced its 10 percent collection fee on foreign income to 5 percent.

ASCAP's new and more realistic competitive policy toward BMI was bearing fruit. The society's share of *Billboard*'s Hot 100 charts jumped from 12 percent in 1972 to nearly half. Revisions in the payment system had particular attraction to young writers who could be paid on a current-performance basis or have their earnings spread out by accepting the four-fund royalties that had been established for the benefit of older members. The amendment provided that only new members could start with the more attractive current-performance payoff and switch to the four-fund system after three years.

A four-million-dollar windfall in license income was released to ASCAP by the court and distributed in early 1976, covering network play between 1964 and 1970. While music on the owned-and-operated stations virtually vanished, many advertisers moved over to radio, where rates were substantially more attractive. With the changes in radio content as leverage, the networks drove down their ASCAP fees as much as 70 percent.

This reduction raised hopes of a further cut in their ASCAP payments among independent radio-station owners. Citing a greater use of the BMI repertory, they had already won an estimated $2.4 million reduction to the society. Their annual 1.72 percent ASCAP rate was just a bit higher than BMI's 1.7 percent fee. Now the representative of 2,200 stations, the All-Industry Radio Committee bargained for a new 1.3 percent rate, a reduction of nearly 25 percent. Aware of the request, ASCAP proposed to increase rates by 16 percent.

Eighteen months later, the court approved a new five-year contract between ASCAP and independent radio. The 1.725 percent rate remained in force, but concessions by the society, including an increase from 5 to 15 percent of standard optional deductions, put the stations in a position to save $6.5 million to $8 million during the life of the agreement.

Similarly deadlocked negotiations between BMI and the Radio Committee for their new five-year contract took twenty-seven months to conclude. The agreement lifted the rate back to 1.7 percent for stations with income over $100,000 a year. Like ASCAP, BMI lifted the optional standard deduction to the 17 to 18 percent range and simplified some reporting procedures.

The CBS petition asking the Court of Appeals to reverse the Lasker decision was reaffirmed in August 1977 and resulted in bad news for BMI and ASCAP. The per-program option offered by both societies was found to be merely another form of blanket license, Appeal Court Judge Murray Gurfein wrote, because neither allowed "a licensee to pay only for those compositions it actually uses." Gurfein held blanket licensing to be price fixing, and with respect to the television networks could not serve as a "market necessity" defense.

The Gurfein decision was hailed by the All-Industry Television Committee and particularly its chairman, Leslie Arries, who continuously fought to reduce the money being paid to the licensing bodies. Arries's subsequent presentations at NAB conventions emphasized the discrepancy between money paid for music by the networks and by independent stations. The 1976 figures, for example, showed the differences to be $18.5 million from the networks, much of which was paid by affiliates because of the twilight-zone bookkeeping reimbursement, and $48 million from local TV stations.

Briefs filed in 1978 by ASCAP and BMI appealed to the Supreme Court that blanket licensing did not constitute an unlawful practice. "The time is now," ASCAP pleaded, "not years from now after federal courts all over the country have been inundated with plenary lawsuits and counterclaims in infringement actions, in which users assert that the unavailability of an ASCAP 'per use' license devised to meet their particular needs entitles them to use copyrighted music for nothing." Pointing to the intrinsic differences in the structure and operation of the defendants, BMI reminded the court that "every governmental body which has looked carefully into the facts in recent years has concluded that blanket licensing is a reasonable and lawful response to the unique problems of licensing music performance rights."

In December 1978 a class antitrust action was filed in the New York federal court by five local TV entities on behalf of the 700 independent U.S. television stations. It alleged that ASCAP and BMI unfairly set prices for performing rights, and required the plaintiff licensees to pay for all compositions represented by the two organizations whether or not they wanted to use them. As a result, local TV stations were proportionately paying more to ASCAP and BMI than any other group of music users.

Although the networks provided about two-thirds of their affiliates' programming, the plaintiffs were paying two and a half times as much as the networks. Arries submitted an affidavit to substantiate the claim. Asking for an injunction against continuance of blanket licensing, the plaintiffs also sought a halt to the "practice of 'splitting' performance and synchronization rights for pre-recorded television programs," which would make feasible the "clearance at the source licensing by TV packagers." The court issued a temporary restraining order placing in escrow 20 percent of all plaintiffs' payments to ASCAP and BMI as well as temporarily extending all licenses.

The unexpected reversal on the part of independent TV station owners and managers toward BMI was due to changes in the company's station-relations policy. As the men Carl Haverlin had recruited for BMI's station-relations department retired, their younger replacements became little more than bill collectors, and fraternization between them and broadcasters was discouraged. As a consequence, a fire such as the "Buffalo station case," which would have been quickly extinguished in Haverlin's regime, had burst into a major conflagration.

Arguments were heard by the Supreme Court in January 1979, including a Justice Department brief that suggested that the legality of blanket licenses could be tested under a "rule of reason." The deputy solicitor general stated that blanket licensing lowered prices and provided real economic benefits. No one else could sell what ASCAP and

BMI did, and to accept the CBS arguments would turn the antitrust laws into a redundancy.

ASCAP's attorney pointed out that CBS actually had an extraordinary range of options: it could deal with individual copyright owners, negotiate with ASCAP for a per-program blanket license, or have the federal court that supervised the society's negotiations with its licensees establish a reasonable fee. The CBS attorney responded that the network had no obligation to deal with individuals for a license.

BMI's attorney said that the issue was whether offering a blanket license really constituted a violation of the Sherman Act. Judge Gurfein incorrectly had assumed that copyright holders would prefer the blanket license to individual bargaining, yet testimony clearly demonstrated that individual licenses could be had for the asking.

The eight-to-one Supreme Court ruling overturned Gurfein, holding that the laws against price fixing were not automatically violated by blanket licensing. Justice Byron R. White wrote for the majority that "[because] the blanket license has provided an acceptable mechanism for at least a large part of the market for performing rights to copyrighted musical compositions, we cannot agree that it should automatically be declared illegal in all of its many manifestations. Rather, when attacked, it should be subjected to a more discriminating examination under the rule of reason. It may not ultimately survive the attacks, but that is not the issue before us today."

CBS unsuccessfully in August 1980 petitioned the Supreme Court for a rehearing. The network stopped all payments to ASCAP when Gurfein announced his decision, but resumed them earlier in the year, paying nearly nine million dollars in back fees plus interest. A retroactive adjustment waited for final disposition of the case, as did that for BMI, to which CBS had been paying the ordered $1.7 million annually, and, in 1979, had added a temporary additional adjustment of $900,000 a year, ordered by the Lasker court.

Two other significant developments had taken place at the licensing agencies during 1980. Stanley Adams was replaced as ASCAP President by Hal David, a lyricist and former leader of the "Young Turks," who wished to make the society's distribution system more equitable. At BMI stockholders put candidates of their choice, and not the usual self-perpetuating slate, on the company's board. Leslie Arries collected sufficient proxies to elect two representatives of the Buffalo case plaintiffs.

The completely revised Copyright Act required that new rules be drawn up and clarified, regulations interpreted, and a system set up for licensing jukeboxes. The RIAA and the NMPA actively lobbied for new mechanical rates, and an attempt was being made to legitimize performance royalty, in connection with which the Copyright Office had been instructed by Congress to prepare an economic study.

The Copyright Royalty Tribunal was activated in 1977 and took up a number of issues. Copyright owners and users each explained their structure, functions, and position on public-broadcasting rates and jukebox regulations. Compulsory licensing of music by public broadcasters, fees that had been privately negotiated by the licensing societies and those dealing with synchronization and mechanical rights by the Harry Fox Agency, were also discussed.

A slightly modified licensing bill was introduced in 1977. In addition to exciting provisions, including a 1 percent tax on radio stations, it would require jukebox operators to add an additional dollar to the eight dollars per machine they paid. Broadcasters continued to argue that if the recording artist is entitled to more money, record companies should pay it. The money performers get is directly owing to radio-station airplay, they held.

The Copyright Office found that broadcasters could pay a recorded-music license "without any significant impact" and would pass fees along to sponsors. It concluded that only a fraction of union musicians made a living wage, a third earning $7000 or less. "Clearly," the report stated, "musicians who do receive sales and performance royalties are in the minority, and in some cases the extreme minority. Most performers are not in a strong enough position to bargain with record companies for a sales royalty." For those who were, all production, studio, and other costs had to be repaid before any royalty was received.

Two new disk performance bills were introduced during the 1979–80 Congress. Concerned broadcasters offered a new survey of stations, estimating the commercial time value of free airplay for recordings at between $150,000 and $490,000 a week. In response, Stan Cornyn, of Warner Communications, rhetorically asked what would happen "the day radio died," to which he responded, "if it weren't for radio, half of us in the record companies would have to give up our Mercedes Benz leases."

When asked if only big recording stars would benefit, the manufacturers promised that all royalties would be shared equally by all participants on a recording. Still, opposition was based in part on the ground that the record companies should not share in any way. A series of changes introduced in May 1980 would exempt many of the smaller-earning stations, public broadcasters, other nonprofit bodies, and small businessmen who used their own recordings. The manufacturers, on the other hand, were confronted with the problem of securing protection against the growing business of illegal home taping.

The multi-million-dollar receipts from eight-dollar jukebox licenses which, with cable television fears, were expected to be between $10 and $13 million in the first year, and did not materialize. In 1978, the Copyright Office collected about $1.1 million from 144,468 coin machines, and in the next year, $107 million from about a thousand fewer. The money represented license fees from fewer than half the machines known to be in operation. The only compulsory-license royalties collected by the Copyright Office, those from cable and jukeboxes, were invested at interest to await distribution by the Copyright Royalty Tribunal.

The Tribunal conducted hearings at the end of 1980 to determine royalty-rate adjustments. No increase was ordered in jukebox fees for 1981, but beginning in 1982 the $8 tax was increased to $25, to $50 for 1984, and subjected to a cost-of-living adjustment in 1987. The Amusement and Music Operators Association argued that the industry was going downhill. Annual jukebox manufacture had fallen from 700,000 to 25,000 units. In 1980 only 300,000 units were operating and only between 3,000 and 5,000 jukebox operators were still in business. However, they possessed enough political influence to induce Congressmen to introduce bills to eliminate the annual royalty and substitute a single one-time license fee of $50 for each new machine and $25 for each jukebox already in use. As a result, ASCAP and BMI organized the Action Committee for the Arts, to lobby on Capitol Hill.

The 1975–76 "penny war" over mechanical rates escalated in the spring of 1980 when the Tribunal determined adjustments for the next seven years. During hearings held by the Copyright Office on jukebox licenses, the NMPA unsettled the RIAA by seeking a 6 percent increase in the suggested price of disks and tapes, because the current royalty "served only to maximize the revenues of the record companies, [allowing them] to buy cheap and sell dear."

A new battle of economic studies began. The publishers maintained that the two-cents-per-song royalty had dropped from 6 percent of the suggested $3.98 retail price to 3.4 percent of the $8.98 LP now standard. In support of the NMPA demand, the AGAC found that nearly 70 percent of its members had averaged income from music-related sources just above the poverty line in the past five years. Consequently the guild wanted not a 6 but an 8 percent mechanical royalty.

Once again the RIAA painted a starkly bleak picture. The year 1977 was the industry's worst one in history, with estimated pretax losses of $208.7 million. High costs, Stan Cornyn said, were effected when record manufacturers had taken over promotional and marketing functions from music publishers. Warner Brothers Records had spent nearly $12 million on national promotion, artist development, and advertising in 1979, a 289 percent increase over 1975. Album costs had increased correspondingly: those for a recording session up from $50,000–$65,000 to $125,000–$150,000; cover artwork up from $2,000 to $3,000–$3,500. The cost of vinyl was up 36 percent, shrink-wrapping had increased by 71 percent, and board jackets were up 53 percent.

Cornyn further testified that Warner had released 136 albums in 1979, of which ninety-one were by singer songwriters who received 81 percent of the company's total mechanical royalties for the year. Sixteen of them got $16 million, or 57 percent of the total. If the rate was increased to 6 percent of retail list, Cornyn said, "this extra income will go to the titans of the business, not the little songwriter starving in the attic."

When the Tribunal recessed in a tangled morass of contradictory figures, it asked the NMPA to gather further financial data, including 1977–79 domestic and foreign income from mechanical licenses, public-performance royalties, print income and other revenue, as well as operating expenses and general administrative costs. It indicated that the music publishers' profits had increased by 5.17 percent in 1977, 7.04 percent in 1978, and 8.41 percent in 1979. Dollar profits for those years were $9.6 million, $15.9 million, and $18.8 million. Total mechanical revenues fell from $20.5 million in 1978 to $12.9 million in 1979; total operating costs rose from $63 million in 1977 to $77.9 million in 1979.

In conclusion, the RIAA suggested that no adjustment be made immediately. Rather, the RIAA board wished to use the average recommended list price of the 200 top LPs on *Billboard, Cash Box*, and *Record World* charts during the previous year to make adjustments in 1982 and again in 1985.

The Copyright Royalty Tribunal increased the mechanical rate to four cents, or ¾ cents per minute, whichever was larger, as well as made annual inflation rate adjustments. The decision, the Tribunal explained in its report, reflected the fact "that between 1973 and 1979 sales of recorded music in the U.S. almost doubled, from $2 billion to nearly $4 billion. In our opinion, based on the evidence in this proceeding, the fortunes of the record companies, the copyright users, have been enhanced in the last decade. The evidence shows that at the same time, the fortunes of songwriters and music publishers, the copyright owners, subject to a price-fixed mechanical royalty in a period of great inflation, have dwindled."

In preparation for a decision before year-end affecting cable television, ASCAP and BMI joined the Motion Picture Association, the NAB, and several sports organizations to petition the Tribunal for a 15.4 percent increase and the institution of a system that would adjust cable rates regularly on the basis of inflation. The licensing bodies, including SESAC, had been alloted a 4.5 percent share of all cable royalties, which was divided in shares approved by the Tribunal. CATV rates were also steeply raised to 21 percent, reflecting the period from October 1976 to January 1, 1980, and were to remain in effect until hearings in 1985. The decision brought down the wrath of both the NAB and the National Cable Television Association, the former accusing the Tribunal of subsidizing cable by setting fees that were too low, the latter regarding it as an enemy.

As such arguments indicated, the weakness of the nation's copyright protection in relation to that of other countries was long evident. Beginning in 1922, a drive was started to bring the copyright law into closer line with the Berne Convention, particularly in the protection of authors and composers for the period of their lives and fifty years thereafter. Measures calling for revision of the 1909 Copyright Act and ratifying the Berne Convention resulted in revision legislation that finally emerged in 1976. It did bring the country closer to Berne standards, but this did not mean participation in that international copyright agreement would occur quickly. However, American music publishers had long before devised a back-door method of enjoying the benefits of the Berne Convention.

The NMPA was instrumental in early 1978 in the formation of the International Federation of Popular Music Publishers which would serve as spokesman for its members in their industry's world forums. The fifteen publishers' associations that founded the IFPMP represented roughly 150 music companies throughout the world. They were united by concern over the illegal duplication of recordings, as well as by the fight for increased mechanical royalties. The NMPA embarked on a public-relations campaign to celebrate the sixtieth anniversary of its founding. In association with the Copyright Office, it held a series of national workshop forums, to explain the intricacies of the new copyright act.

A full-time publisher now was required to do everything, from production of demonstration cassettes of new material, to securing recorded cover versions of hit songs or ones dug out of the catalogue, to marketing elaborate printed folio collections of popular music. Recording artists frequently accepted a tremendous advance on royalties from all sources and then insisted on merely a short-term relationship with a music firm. A publisher's administrative and promotional services entitled them to ask for and get up to half of a song's income, including that from the original release. Those who agreed only to adminster income usually offered 10 percent of all revenues.

The future of printed music looked rosy. Sales of $211 million in 1979 represented a 193 percent cumulative increase over the $72 million of 1967. Superstar folios and songbooks were in favor. Columbia's new LP-sized AlbuMusic songbook-folios attracted buyers as much for the supersized poster insert of their composer as for the music. Publishers found it was more profitable to give up publications rights to their music on an open basis to the several specialists in printed-music marketing than to finance their own print operation.

As printed music sales rose, piracy and infringement followed. The traffic in illegal "fake books," compilations of the melody and chord lines of old and new hit songs, had been curbed by the production of licensed collections. When the print pirates turned to the high-priced songbooks and folios, the NMPA sought new statutes to strengthen property rights. Two actions brought by the NMPA established precedents that served as warnings against future infringement by organized religious groups and educational institutions, which had long considered themselves beyond the law.

A hymnal, *Songs for Worship and Fellowship,* published by the Unification Church led by Korean evangelist Sun Myung Moon, was alleged to contain forty songs copyrighted by twenty-two publisher members of the association. After two years of legal maneuvering, the matter was settled in favor of the plaintiffs, and the church paid $40,000 and legal fees.

The other action, the first against a college for infringement of music copyrights, was filed in 1980. Longwood College, a small, state-funded institution in Virginia, was charged with "willing and intentional" violation of the copyrights of five musical works by allowing them to be photocopied. The case was settled with acknowledgment by the defendants of innocent infringement and payment of damages and attorneys' fees.

Record companies continued to dictate policy and practices in the music business, although not without the existence of certain irregularities. M. William Drasilowsky, co-author of the standard industry reference work, *The Business of Music,* alerted *Billboard* readers to such irregularities as cut-ins by recording artists and record producers, the "Chinese bookkeeping" that short-changed songwriters and recording artists, the filtering of funds from international licensing by music publishers, record-plant overruns that robbed almost all copyright owners, consumer mail fraud in regard to record-club subscriptions, kickbacks to recording artists from independent studio owners, cross-collateralized advances between artist-songwriter accounts by the record companies, off-the-book sales of phantom seats for rock concerts, disk-jockey payola, and laundering of funds. He omitted to mention the growing tendency on the part of the public to make tapes instead of spending money for an LP.

While the major labels complained that home taping was eating at their profits, their control of artists and the year's 100 best-selling LPs and singles was nowhere better shown that in the pages of *Billboard*'s 1975 *Talent in Action* supplement. Major conglomerates had an 81 percent share, leaving the balance to the few remaining independent labels. With record sales maintaining a steady growth to $2.4 billion in 1975 and $2.73 billion the next year, the major manufacturers were emphasizing the short-term success from the sales of their superstars. Walter Yetnikoff, of the CBS Records Group, said, "If an artist can only sell 100,000 records . . . that this company is not interested in pursuing that artist. We're looking for the major, major breakthroughs."

Supersales made for quicker recouping of investment in new performers. Studio costs for a typical rock album of more than $100,000 could be spread out more profitably over big sales. WEA's lead in raising the price of its front-running popular LPs to $7.98 began another game of follow-the-leader. The industry's excuse for the price increase was the high royalty factor, which had risen to an average of 15 percent, and as much as 17 or 18 percent for superstars. The jump from $6.98

to $7.98 meant thirty-five cents more in artist's royalties on a 15 percent contract, and with the average twelve-cent increase in mechanical royalties that was to go into effect in 1978, forty-seven of the forty-eight-cent profit was gone.

While CBS and its major rival Warner Communications had combined sales of $1.21 billion in 1976, the Dutch PolyGram Company along with its American Subsidiaries—Mercury—Phonogram, Philips, MGM, and United Artists Records—become sixth among the leading American record companies. Its international operations did not match this success. In the ten years since Philips and Siemens had merged, in 1963, to form it, PolyGram had become an international giant with subsidiaries in thirty-five countries. The transfer of control of Chappell Music to PolyGram had strengthened its position in music publishing. So, too, did the addition to Chappell Intersong of a major portion of the Aberbach family's Hill & Range-controlled catalogues, not including the two Presley firms, which Chappell would continue to administer, but including many early hits from country-music firms. Yet it remained for an alliance with Robert Stigwood in 1975 to propel PolyGram into its position in 1978 as the first company to enjoy worldwide music-and-entertainment sales of $1.2 billion.

After Brian Epstein's death, which effectively froze Stigwood out of any further participation in the sale of NEMS/Nemperor, he diversified his activities under the name Robert Stigwood Organization. Resuming personal management of a number of major rock stars, he operated a music-publishing company that owned the copyrights of all music written by his artists, and produced a number of successful stage musicals, including Tim Rice and Andrew Lloyd Webber's *Joseph and the Technicolor Dreamcoat* and *Jesus Christ Superstar*, features for the American "Movie of the Week" series, and the original English TV productions of *Sanford and Son* and *All in the Family*. RSO went public in 1970 and was sold to PolyGram for one dollar a share, much above market value. The transaction cost PolyGram eight million dollars, from which Stigwood got one and a half million.

Early in Atlantic Records' licensing relationship with Stigwood, it formed RSO Records to handle his artists in the United States. The Polygram Group acquired the recently re-formed Robert Stigwood Organization Ltd. by offering Stigwood a place on its board, a guaranteed income of $10 million from the various Stigwood music companies, and other benefits, including the investment of five million dollars in each of the following five years for acquisition and development of screen properties. While the final papers were being prepared, Stigwood chose as his first film production an adaptation of a magazine article about the rising addiction to disco dancing.

Stigwood offered Paramount Pictures $2.5 million against profit to make *Saturday Night Fever*, starring John Travolta. Paramount returned the $2.5 million to Stigwood Productions for distribution rights for just under a 60 percent share of the profits. Music for the film was written by the Bee Gees and published by Stigwood. To enforce his philosophy that one "built on a musical vehicle so that it got better known than the play or movie of which it was part," Stigwood hired Al Coury, Capitol's vice president for national promotion, and made him president of RSO Records. Coury's first move was to build a seven-man promotion staff to handle RSO exclusively.

Coury's promotion strategy, "cross-over marketing," began in 1977. In December, the $12.98 double-LP soundtrack album of *Saturday Night Fever* was "shipped gold," with half a million copies distributed all at once. In the four months before the LP's release, Coury released four singles from the album, an unprecedented departure from customary practice in dealing with soundtrack music. Some 8,000 major retailers and radio stations were sent a full-color giant poster showing John Travolta in the disco-dance position that was used on the LP cover and on all film advertising. A thirty-second trailer with the Bee Gees singing the picture's theme song, "Staying Alive," was shown simultaneously in 1,500 theaters, coinciding with the release of another Bee Gees single from the movie, "How Deep Is Your Love." Immediately after Thanksgiving, a new three-minute trailer was exhibited nationally, featuring excerpts from all four of the RSO *Saturday Night Fever* singles. Soon all four singles were in the top ten, the first time such a feat had been accomplished since the Beatles' days. Coury projected a sale of between $100 and $129 million for the package.

To expand its U.S. operation, PolyGram also invested in Casablanca Records, founded and owned by Neil Bogart, the erstwhile "King of Bubble Gum Music," who had been one of the earliest to switch to disco, with its queen, Donna Summer. Bogart recruited groups who depended for their success as much on their physical appearance as on their music. The most popular was Kiss, whose members dressed in Kabuki costume and featured a fire-breathing bass player, and the six-man Village People, earliest of the "gay to straight cross-over" disco groups.

RSO and Casablanca sold $300 million worth of records in 1978. RSO released ten LPs during the year, five of which attained Gold or Platinum awards. Three soundtrack albums were responsible for two out of every three PolyGram sales: *Saturday Night Fever,* which sold 15 million LPs in the United States; *Grease,* starring Travolta and Olivia Newton-John, 22 million units around the world; and *Sgt. Pepper's Lonely Hearts Club Band,* based on a London stage musical, which sold three million copies at a list price of $15.98.

Not since the unexpected success of *Hair,* the folk-rock musical, had a soundtrack or musical-score LP enjoyed such mass marketing. Rivaling the success of *My Fair Lady* on recordings, it had three cast albums and fourteen songs covered by major artists in the United States alone. But not until "Aquarius/Let the Sunshine In," recorded by the MOR group the Fifth Dimension, began to become a hit did *Hair*'s music and records start to sell, and the original cast recording remained on the charts for 151 weeks.

Although record companies had lost the habit of courting the Broadway stage for soundtrack material, the success of *A Chorus Line, Hair* and other rock-oriented musicals like *The Wiz,* based on *The Wizard of Oz,* attracted them, as did the sales of those shows' original cast recordings. However, record companies were ready only to pay for recording costs and hoped that the music publisher would promote the LP.

Despite a still-pending stockholders' class action, Capitol Records was generally regarded as responsible for a major contribution to the financial well-being of EMI, much of whose international record profits still came from Beatles albums. Not only did EMI and its international subsidiaries manufacture and distribute one-fourth of all records sold in thirty-two countries, but EMI had further diversified its interests in all branches of the entertainment and communications business. To expand its worldwide publishing operations, EMI bought almost the entire music-publishing division of Columbia Pictures Industries in 1976. Columbia's music print division, Columbia Pictures Publications, was not involved in the transfer.

Now owner of the world's largest music-publishing operation, EMI was enjoying a 43 percent rise in sales of music and records. In 1979, EMI picked up United Artists Records for three million dollars. In a restructuring of U.S.-Canadian operations, EMI grouped all its record holdings in the United States under a Capitol/EMI group flag.

There had been a 21 percent increase in profits from the worldwide EMI Music in 1978, but due to unexpected losses, a drop in profits was expected in the second half of fiscal 1978–79. Lord Bernard Delfont, who had recently been installed as chief executive of EMI Ltd., began talks with Charles Bludhorn, president of Gulf & Western, about the possibility of a merger. Bludhorn had been watching massive profits from music copyrights used in the successful Paramount productions *Saturday Night Fever* and *Grease* flow into the pockets of Robert Stigwood and PolyGram and wished to increase Paramount Pictures' role in the entertainment industry.

The EMI-Paramount merger would involve the latter's investment of $154 million to buy a half-interest in EMI's music holdings, with which the film company's fifty-year-old division Famous Music would be integrated. In 1979, the merger was called off, while two months later, Thorn Electrical Industries, a British conglomerate involved in consumer electronics, white goods, and television rentals, acquired EMI Ltd. for $348 million. It was one of the largest mergers in British history and was expected to put Thorn-EMI in a very strong position in the major domestic appliance market of the 1980s—videodisks, video recorders, and video cassettes.

The pre-recorded music business had now climbed a new Mount Everest: a $4.1 billion retail gross. The figure represented $1 billion from 137 million pre-recorded cassettes; $1 billion from 133.6 million eight-track tapes; and $2.1 billion from 531.3 million recorded disk units.

Payola remained a constant and invidious force, and in early 1977 the FCC resumed hearings as a result of complaints that disk jockeys on an influential black music station in Washington had received payments in excess of $14,000. Reports simultaneously indicated that some manufacturers had talked about the application of pressure on discount record stores to end the fiction of "list price," as well as tie-in sales involving recordings by superstars and price fixing. It was also alleged that some executives had engaged in discussions to put a stop to escalating royalty rates being paid to artists.

The need for standardized methods of artist-royalty accounting was emphasized by a Price, Waterhouse study, which commented that there were many methods of payment and that bookkeeping methods for compensating artists were "unique to the record industry." The report found that "accounting procedures for royalty advances paid to artists which can be recouped out of future income are not uniform."

Contracts with superstars lacked uniformity and often involved unheard-of advances, clauses, stipulations, guarantees, and rising royalty rates. For example, for a $20 million guarantee, Paul McCartney renewed his agreement with EMI for all countries in the world except the United States and Canada, leaving him free to stay or not with Capitol Records. He left, signing a contract with CBS Records that guaranteed him two million dollars on each new LP. When this was recouped, a royalty of 22 percent on each additional sale of his $8.98 LPs would begin. Under this deal, at least 250,000 units had to be sold before any profit was realized. But although all five of the LPs McCartney made for CBS prior to the return of his North American rights to EMI-Capitol, in late 1985, were Gold and three also Platinum, only one, *Tug of War* (1982), actually climbed to the number-one position on the trade-paper charts.

Seymour L. Gartenberg, CBS senior vice president of finance, candidly stated that all artists, especially superstars, were not equal. With a national 50 to 75 percent defective rate of superstar records, CBS instituted a 20 percent exchange limit on most albums. This replaced the troublesome and chaotic 100 percent return privilege that had been introduced to interest the rack jobbers but resulted in returns of 40 percent of all shipped merchandise.

CBS offered new talent only one-year contracts, with options for four more years, which if picked up required delivery of two new LPs a year. The new artists' royalty was generally between 18 and 24 percent "all in," meaning inclusion of a producer's 6 to 8 percent. Mechanical royalties were limited by CBS to 27.5 cents on a single LP, but the rock groups' propensity for longer tracks often required the company to swallow the difference and drain the mechanical-royalties budget.

In June 1977, an FCC administrative judge made it clear that, although the Washington black-station disk jockeys did accept gratuities, they were exonerated of the charges made against them earlier in the year. However, black radio was under more guns than one. The NAB found the lyrics of a number of disco hits suggestive and warned members to screen all material. Some manufacturers did take action. CBS revised the original suggestive lyrics of a projected release and discouraged artists "from delivering offensive material by telling them we can't get it exposed." None of these gestures effected the success of black music in 1977. Its sales represented approximately two-thirds of the $3.5 billion retail gross that year.

Black-owned or -controlled business in America had a combined business volume for the year of $896 million. But there were few black owners of broadcasting properties. Of the more than 300 black-oriented stations in the United States, only sixty were owned by blacks. In 1971, the number had been six. With the narrowing of space bands on the broadcasting frequencies in order to allow 125 new AM stations on the air, it was expected that conditions would improve once the economic barrier against black stations was overcome.

The FCC moved the payola hearings to the West Coast in mid-1977, where it learned that bribery in the record business was racially undiscriminatory. Back in Washington, the commission turned from the takers to the givers, independent promotion men, whose targets were not as widespread as an industry with 8,600 AM and FM stations might indicate. A few years later, a former promotion man said privately that only 2,700 stations were important to the record companies.

After five years of complete inactivity on the payola problem, a House subcommittee looked into it in 1984. After discussions with broadcasters, promotion men, and record-company executives, the investigation turned up the payment of millions of dollars for promotion, which the Communications Act did not forbid, and the whole business was dropped.

Despite these legal entanglements, the record business once again was prospering. According to 1978 figures, PolyGram and EMI had topped one billion dollars in worldwide sales. The CBS Records Group was in a strong third position: $946.5 million for the year. WEA had reached $617 million, almost half of Warner Communications' total sales. RCA Records enjoyed an estimated 25 percent increase in one year, to $500 million in sales; and MCA's record and music-publishing unit achieved $131.5 million in sales, a 32 percent increase. Stripped of revenues from manufacturing and excluding foreign operations, CBS Records had sales of around $500 million; Warner Brothers Group, $416 million; RCA Records, an estimated $200 million; Capitol-EMI, $100 million; and MCA Distributing, about $70 million.

Suddenly confronted in 1979 with the beginning of another recession, the major companies stared possible disaster in the face. CBS and Warner had spent in excess of $20 million to erect new manufacturing facilities in order to meet expected large sales. Finding themselves, instead, with millions of LP returns, they lowered the boom on the 100 percent return policy. There was also the problem of counterfeit LPs among the returns. Between 20 and 40 percent of the returns of smash-hit LPs of all labels were illegal duplicates.

After $80 million in losses over four years, the American Broadcasting Company put its record-and-music operation up for sale in 1978. Though a record sales boom was in progress, there were no bidders. MCA was able to get the initial $55-million asking price lowered, and it absorbed the properties in early 1979. The possibility of establishing ABC Records, as a third independent MCA line was soon abandoned, and most of its employees were discharged. So, too, were some of the staff of ABC Music Publishing, whose 20,000 copyrights were transferred to MCA Music. By the end of 1979, the MCA record-and-music unit reported an operating loss of $6.1 million, a 163 percent drop since 1978.

The decline was regarded as almost matter of course by an industry whose leaders coped with the problem of falling demand by raising prices. More inventory risk than they cared to assume fell on the shoulders of retailers, who coped by reducing orders. The price increase meant only fifty cents, after discounts, to many consumers, but it represented an industrywide certainty that the public would accept without protest a more than 10 percent jump in list prices for superstar recordings.

Living on a small profit margin and hoping for large sales, the largest independent record companies were the victims of traditional distribution practices. They were forced to raise wholesale prices, while their manufacturer competitors maintained theirs, to $3.60 wholesale for a $7.98 LP, which was resold to retailers for $4.30. The manufacturers' branches, meanwhile, paid $3.47 for a similar product and passed it on to retailers for $4.09. It was a fact of life for the independent record companies in 1978 that it was more profitable for them to use an independent distributor, but it was also true that rack jobbers, subdistributors, and retailers went to the manufacturer-owned branch operations for their stock.

To resolve the cash-flow problems that began to plague them in 1979, most independent record companies transferred their business to the manufacturers. They did receive less money per wholesale unit, but there was a significant cash advance against royalties, as well as a most-favored-customer pressing charge, and problems with slow-paying customers ended.

Elton John's Rocket Records, 20th Century-Fox Records, and A & M Records signed distribution agreements with RCA Records, which had increased its sales in the past few years and strengthened its worldwide record business. In 1976, RCA had produced more than half of its revenues from that source for the first time. The next year, it had more than forty international licensees to handle its records. It had bought German Teldec Records, entered into a joint-venture project with the Japanese Victor Company to market records and tapes, and increased production of its SelectaVision videodisk player and library in association with Matsushita of Japan. RCA's relationship with A & M alone was particularly successful and expected to yield a 50 percent increase in sales volume, to an overall 13 percent share of the American market and 35 percent of the world market.

In 1983, RCA acquired a 50 percent interest in Arista, which had lost about $54 million that year, buying it from Bertelsmann, with which it entered into a worldwide merger in 1985. Seventy-five percent of the joint company was owned by RCA in the United States and some other countries, 51 percent by Bertelsmann in Germany, Switzerland, and Australia. There was a branch-distribution agreement with CBS in 1982, and Motown became an MCA client in 1983, giving the six manufacturers distribution rights to virtually all recorded music produced in the United States.

Although CBS's sales increased to one billion dollars in 1979, its operating profits actually fell 46 percent. Many attributed the general fall in record sales for fifteen years to blank tapes. A CBS study of home tapes in 1980 indicated that they were responsible for an annual industry loss of $700 to $800 million, or 20 percent. There were some 40 million buyers of blank tape. Any possibility that home taping might be reduced in the future was thwarted by Sony's introduction of its "silent disco player," the Walkman, a hand-held playback-only unit with a full-size cassette.

The threat of home taping should not have come as a surprise to the industry, many of whose members engaged in the profitable production not only of pre-recorded tapes but also of blank reel-to-reel tapes and cassettes. Sales of pre-recorded tapes in all formulations had reached the one-billion-dollar mark within ten years of their introduction, and blank cassette sales grew from 150 million units in 1975 to a dollar volume of $170 million in 1978, during which 220 million Dolbyized cassettes were sold.

The copyright owners of motion pictures, television programs, and phonograph recordings made their first concentrated effort to legislate on the home-taping problem in the late summer of 1977. There was no mention of the matter in the new copyright act. A House subcommittee report on antipiracy laws in 1971 had failed to show any concern about home taping where no commercial gain was attached.

Technologically varying versions of the videotape and the first videotape recorder—introduced by Bing Crosby Enterprises in 1951—had been tested and marketed by a number of major electronics manufacturers. This was before the Betamax, introduced by Sony in 1975, precipitated the legal action on which relief from home taping might hinge.

In their 1978 complaint in the "Betamax case"—*Universal City Studios—Walt Disney v. Sony of America*—the plaintiffs held that the use of videocassette recorders in the home infringed on their motion-picture copyrights, and they asked that either the sale of VCRs be outlawed or the machines be mechanically adjusted to limit their recording capabilities. In October 1979, the lower court found for the defendants, eliciting an immediate promise to appeal from the plaintiffs.

The surveys of potential losses of record and tape sales due to home taping were entered into the Copyright Royalty Tribunal's records at the end of 1979. Both showed that an average of 23 percent of all respondents had taped music in the past year, but that the biggest tapers were also the largest purchasers of music, and that the higher the family income, the higher the incidence of taping.

It appeared that little legally or technically could be done to stop taping. With LP prices rising, which had caused unit sales to fall 20 percent across the country, it was cheaper to tape than to buy. One out of every two U.S. households owned a cassette unit, and Americas were buying the latest best-selling item in audio and retail-record stores—premium blank cassettes priced around $4.98.

The most significant change since electrical recording and the switch from shellac 78s to vinyl 45s and LPs—computerized or digitally mastered tape-recording decks—had been developed in the United States prior to 1979 by Thomas Stockman, Jr., a Salt Lake City engineer; the 3M Company; Battelle Northwest of Columbia-Ohio; and some smaller companies. A catalogue of several hundred hybrid LPs recorded by digital methods was available in the United States by 1980, and they won critical raves for their sound and presence.

During the late 1960s, Philips had embarked on experiments in optical electronics, and it had produced its compact disk system by early 1979, offering an hour of sound, free of surface noise, on a dime-thin 4½ inch reflective minidisk, whose information was played inside out by an optical laser. Music was transferred through early-stage Philips compact-disk players into a standard home high-fidelity reproducing system.

Soon after the introduction of the Philips CDs, Sony joined the enterprise as a full partner, contributing enhancements to the system that led to the final Sony compact-disk player, introduced in 1983, after several years of deals, during which licensing arrangements and special mechanical-royalty rates were effected throughout the world.

13 The 1980s: The *Buffalo* Case, MTV, and Further Fractionalizing of the Marketplace

BUFFALO v. ASCAP, BMI, et al. was tried in 1981 in the U.S. Southern District Court in New York, before Judge Lee Gagliardi. It unveiled the workings of Hollywood television-film production as well as music licensing. Eight companies controlled the distribution of off-the-screen and syndicated film packages: Viacom Enterprises, MCA/Universal, Columbia Pictures Television, Paramount Television Distribution, Warner Brothers Television Distribution, United Artists Television, 20th Century-Fox Television, and MGM Television. These companies distributed 82 percent of all syndicated off-network reruns and 52 percent of all programs made for syndication. Almost all of the ASCAP and BMI music used in them was assigned to publishing firms owned by the corporations. In 1979, only 13 percent of all ASCAP and BMI publishers had received any television royalty distributions, and fewer than 8 percent enjoyed more than 75 percent of all TV performance royalties.

The music for a syndicated program was generally written on a work-for-hire basis. The producer cleared it through either an ASCAP or a BMI publisher but granted the composer his share of performance royalties. Producers were unable to assign the TV performing rights in any manner that would revoke the composer's right to collect his royalties. Existing copyrighted music was licensed by the Harry Fox Agency. In this case, too, the performance rights were separate, having been already assigned to ASCAP or BMI. It was this splitting of performing rights from the other rights on which the Buffalo attorneys concentrated, asking the court to enjoin the blanket license and allow their clients to obtain the music rights directly from syndicators, who would obtain them "at the source" from composers and publishers.

Judge Gagliardi's 1982 decision found that the local-television blanket license "unreasonably" restrained trade in violation of the antitrust laws. Referring to the Supreme Court decision in *BMI v. CBS,* he wrote:

> Unlike CBS local television stations could not by virtue of their market power effect the transition to a reasonably practical, centralized system of direct licensing. . . . Those with the incentive to change the system lack the power, those with the power lack the incentive. The local station has no choice but to purchase access to the entire repertoires of ASCAP and BMI although the station's needs could certainly be satisfied with a far more limited selection. . . . The court accordingly holds that for the plaintiffs, direct licensing is not a realistically available marketing alternative to the blanket license.

Leslie Arries and his colleagues saw their victory as the end of blanket licensing. Music costs could be spread across the entire marketplace. The music business regarded the Gagliardi decision as a complete misinterpretation of the Supreme Court ruling in the per-use case and felt that it would be overthrown by the appellate court.

Nearly $20 million, representing the 20 percent of TV-station license fees being held in escrow, had been collected by 1980. Gagliardi deferred his final judgment until at least February 1984 and ordered the plaintiff stations to continue paying their ASCAP and BMI fees until then, but at the 1980 levels. The disposition of all fees would be determined by the judge in 1984, at which time he was also expected to issue an order barring both societies from collecting television performance royalties from the local stations under the blanket-licensing arrangement.

ASCAP President Hal David reported that ASCAP was headed for another banner year. He changed the voting procedure to permit nonvoters to control the fate of new proposals. Performances became ASCAP's criterion for popular- and country-music awards. The society aggressively campaigned to win Nashville writers away from BMI. However, faced with the imposition of severe economies due to the Gagliardi decision, ASCAP and BMI suspended all cash advances and guarantees.

Blanket music licensing was once again debated in 1983, when a panel of three judges heard arguments in the Buffalo case. The plaintiffs filed a brief complaining that ASCAP and BMI sought to obscure the fact that the TV stations were paying a "super-competitive price—two and one-half times the network rate—for no visible reason other than the defendants' monopoly power to discriminate." The Buffalo counsel pointed out that 200 new stations had gone on the air since the Supreme Court ruling, and they could not be denied a hearing. BMI's attorney urged that, because the issues and the form of license were the same as in the CBS case, they should be dealt with the same way. Buffalo's chief argument was based on the plaintiffs' inability to deal directly with composers and publishers so long as blanket licensing existed.

Belligerent relations between CBS and BMI continued. New licenses for both the ABC and the NBC owned-and-operated television stations had been negotiated, but CBS had, according to BMI, "reneged" in early

1983 on a deal that provided it would pay BMI "either $1.85 million under one set of circumstances or $2 million under another set." Once again BMI filed a damage suit for infringement of copyrights by the CBS-owned stations, and the threat of a preliminary injunction to prevent the network's unlicensed use of BMI music was sufficient for CBS to accept a new blanket license agreement.

Hal David was re-elected as ASCAP president in 1984. He negotiated a new contract with the All-Industry Radio Committee in 1983 which simplified reporting procedures but did nothing to reduce the cumulative $80 million being paid annually to ASCAP and BMI. It was understood that ASCAP would not ask for an increase provided BMI was not given one.

ASCAP had another setback when a New York district judge refused to modify the 1960 consent order and allow the society to reject the ABC-TV demand for a per-program license. ASCAP maintained that to give ABC a per-program license while it still had a blanket license from BMI would be "anticompetitive." Its motion pictured BMI as an arm of the broadcasting industry and alleged that an ASCAP per-program license was used to discriminate in favor of music licensed by BMI.

In opposing the ASCAP motion, the Justice Department stated that "if BMI acted as the tool of the networks, the networks would face the risk that many composers would respond by joining a union or guild. Indeed, many other creative artists are actively represented by such organizations. The networks potentially could view a strong composers guild as less desirable than the status quo." The federal court in New York rejected the ASCAP petition.

The fierce antagonism between BMI and ASCAP colored the negotiation with the All-Industry Radio Committee. "BMI, Radio Stations on Verge of War," a *Variety* headline read on June 27, 1984. When Edward Cramer, BMI's president, had been asked for yet another contract extension, he said he would grant it for a small increase. The suggestion was rejected, as was Cramer's proposed arbitration. New five-year contracts were mailed, which raised rates by about 10 percent and gave BMI the right to examine a station's program logs.

The All-Industry Committee's attorney indicated that the contract would raise rates by an overall 15 percent, and in some cases as much as 30 to 40 percent, and threatened legal action. A truce was arranged, which reduced the rate proposed by 1985 by 8 percent, and specified negotiations "from scratch" to determine the terms of 1986 and after. It was also agreed that in case BMI and the broadcasters could not agree on new contract terms, a "formal rate-fixing procedure" would be created.

Meanwhile, the appeals court upheld the blanket TV license. Options did exist, the court declared, for, "since the blanket license restrains no one from bargaining over the purchase and sale of music performance rights, it is not a restraint unless it were proven that there are no alternatives." Those that did exist included per-program licensing, direct licensing from composers and publishers, and "source licensing," passed on to local stations by the producers of the syndicated programs as part of the cost of the package.

Attorneys promised an appeal for another reversal would be made to the Supreme Court. Leslie Arries added that while he expected ASCAP and BMI to seek retroactive compensation for revenues lost, "the industry will not voluntarily relinquish one penny of those savings." They were significant. In addition to the $20 million in escrow, they had been at least $25 million over and above the $57 million paid by the TV stations in 1983.

With Judge Gagliardi's freeze due to expire on November 1, 1984, BMI notified all stations that it intended to seek a "retroactive upward adjustment" of fees, and that all existing interim arrangements would be canceled unless the stations resumed payment at the pre-1980 rates. The committee asked the court to postpone any action to raise rates until a month after the appeals process was exhausted and final order was issued.

In 1984, the Second Circuit Court of Appeals denied the plaintiffs' application for a rehearing, and the next February the Supreme Court denied a review of that reversal. Immediately after the Copyright Royalty Tribunal raised the basic mechanical royalty rate to four cents, the RIAA challenged the ruling. A panel of three judges upheld the new mechanical rate, but it remanded the proposal to adjust rates annually. It suggested that a "reasonable mechanism [be devised] for automatic rate changes in interim years."

Months passed before the Tribunal announced, in December 1982, that it would adjust the rate in steps: on January 1, 1983, to 4¼ cents, or 0.08 cents a minute; on July 1, 1984, to 4.5 cents or 0.085 cents; and on January 1, 1986, to 5 cents, or 0.09 cents per minute.

The absence of further opposition from the RIAA to the Tribunal's regulations was the result of a sudden "friendship" between copyright holders and users. United by the need for government intervention in the hometaping problem, the ancient antagonists—RIAA, NMPA, AGAC, and the Nashville Songwriters Associations—had been forced into an alliance.

At a 1981 Senate hearing, the new register of copyrights, David Ladd, called for repeal of compulsory licensing in the case of cable television only, a stand seconded by the MPPA, the NAB, and sports organizations. The National Cable Television Association pleaded for retention of the status quo, to which the Tribunal agreed, adding that "to consider restricting or eliminating the compulsory license with respect to cable is inconsistent if the same is not considered for the other three compulsory licenses under the statute: those for phonorecords, jukeboxes and public broadcasters."

Accelerating technological change indicated a clear need for copyright revision. One of the many issues that was not anticipated in the 1976 act, home taping of video programs, was currently before the Supreme Court. A new omnibus copyright revision bill introduced in 1984 proposed the exemption of home video and audio taping from copyright liability; amended the "first sale" principle of copyright law, thus giving copyright holders control over subsequent sales of their recorded video and audio works; and overrode the Copyright Royalty Tribunal's distant-signal ruling and restored the former rates.

The cable provision immediately coalesced opposition from the Motion Picture Association, broadcasters, video retailers, and the consumer electronics industry, leaving only the cable interests to support the measure. The Supreme Court's inaction on the use of videotape recorders in the home to record television programs temporarily defused the entire taping issue.

The registrar of copyrights warned that the international piracy of copyrighted materials was a matter of urgency. However, Ladd cautioned, "excessive penalties may lead to a shift of public attitudes away from basic respect for authors' rights to commercial regulation of copyright products. And it may also polarize pro- and anti-copyright positions in areas not directly related to piracy, such as off-the-air recording, lending rights and private copying."

While the recording industry dominated the music business, established music publishing daily was becoming more complicated. There was no longer a standard publishing contract. The more commercially attractive a song, catalogue, or songwriter was, the more enticing the terms offered. Bidding was as heated as that for recording stars, and half of all songs that appeared on *Billboard*'s Hot 100 list in 1982 were co-published.

The growing breach between songwriters and publishers was illustrated by the "Who's Sorry Now" affair. The new copyright act enabled the original owners of songs copyrighted before January 1, 1978, and or their heirs to terminate all original ownership and publication rights to a work and transfer them, provided the original fifty-six-year term had expired. The newly granted nineteen-year term was still in effect. Only "derivative" works, arrangements, sound recordings, and motion pictures based on the original work were exempt. The issue was central to the pending Harry Fox Agency suit on behalf of Mills Music, original publisher of "Who's Sorry Now," and Ted Snyder Music, formed by his heirs after the songwriter's death, to which the rights of the song had been transferred. Once again involved in the destiny of the music business, Judge Edward Weinfeld found for Fox and Mills in July 1982, giving Mills the right to license new releases of recordings made prior to 1978. Snyder Music and the AGAC filed appeals.

An appeals court decision had reversed a lower court's finding in the Betamax case, ruling that home video taping was an infringement of copyright. The verdict mobilized the Coalition to Save America's Music, a group of music-business organizations ranging from record manufacturers to music-trade and songwriters' associations.

Bills had been introduced in both houses of Congress to exempt video taping when it was not for commercial use, as well as to establish it as a fair use of copyrights. Both would decriminalize the ownership of videocassette recorders.

A 1 percent increase in printed-music sales during 1982 emphasized the weak position of the music-publisher members of the NMPA, now 307 in number. They believed that an additional $25 million might have been earned had it not been for illegal photocopying. The loss to publishers and songwriters of additional millions was the result of money going in and out of the "suspense accounts" of European mechanical societies not policed by the Harry Fox Agency. Because the Fox Agency did not collect on behalf of composers, it did not turn over the unclaimed royalties to Fox for distribution, being privileged under terms of the International Copyright Convention to return the money to the user of the composition.

In 1981, six of the major film studios changed ownership or management. Various entertainment conglomerates lumbered into an alliance to try to wrest away some control of home and cable-TV business: Warner Communications and American Express; Fox Films, Coca-Cola/Columbia Pictures, and CBS; Coca-Cola/Columbia Pictures and RCA; and Walt Disney and Westinghouse. They, too, fought to establish a royalty for home video taping. Their lobbyists talked about a fifty-dollar fee on the sale of VCRs, and a two-dollar charge on blank videotape.

The Sony Betamax case came to an end in January 1984, following oral argument before the Supreme Court. The chief issues had been isolated by Sony's attorneys to "whether all Americans, broadcasters and audiences alike, are to be denied the benefit of time-shift home television [taping for delayed viewing] because a few program owners object," and by counsel for the film industry to "unauthorized copying of motion pictures, that has never been permitted by copyright laws and the fact that it is now being done at home makes no difference."

The latest advances in technology threatened to change the character of how, when, and where Americans amused themselves. Time-shifting had increased the available motion-picture viewing audience, and the sales of feature movies to the networks and cable television were booming. The high cost of videocassettes of full-length movies had created a thriving rental market of as many as 30 million persons.

The study *Prerecorded Home Entertainment Industry*, prepared late in 1983, explained the recent upturn in record-industry profits by pointing to the "five mega hit albums." The study found that the "longterm growth outlook for traditional records and tapes is severely limited by home taping." It suggested that future profits lay in music videos, the fastest-growing record-business product, from $40 million in sales for 1983 to a projected $1.25 billion by 1988.

It all appeared both relevant and irrelevant when the Supreme Court, dealing only with the issue of time-shifting, found that home TV taping did not constitute a violation of copyright. The five-to-four decision directed the motion-picture-company plaintiffs to seek redress from Congress. The public, now the owners of nearly four million VCRs, showed little interest in any future legislation.

The Motion Picture Producers Association continued to call for compensation to performing artists and producers whose works were copied from radio, television, and recordings. When Congress adjourned, only the record-rental bill had finally worked its way to the President's desk and his signature. Passage of the record-rental bill, which ended the proliferation of stores engaged in the business, with consequent losses to songwriters and music publishers, provided a note of joy. So, too, did the appellate court decision in the Buffalo case, which reversed the lower court and affirmed the validity of blanket licensing, on whose revenues most publishers depended.

The piracy of printed music had moved to more impressive environs, among them some of Chicago's Catholic churches and the University of Texas. F.E.L. Publications, of Los Angeles, owned by Dennis J. Fitzpatrick, learned in 1976 that the archdiocese of Chicago was responsible for continued illegal photocopying of F.E.L. and Fitzpatrick copyrights. He took his case to the federal district court, where, after eight years, a grand jury granted F.E.L. $3.1 million in damages. In the field of higher education, the music department of the University of Texas, Austin, was discovered to be photocopying masses of copyrighted concert and choral works and forced to acknowledge the practice as well as destroy any existing copies.

The five-to-four Supreme Court decision, announced in 1985, favoring Mills Music in the "Who's Sorry Now" case was a clear setback for the Songwriters Guild and all authors and composers, and a victory for the NMPA and the publishers. It restored to the original publisher the right to collect a share of income for the sale of the original and all later recordings of a song, even if the royalties were paid years later and the copyright had been transferred. Legal circles had been surprised when the Supreme Court agreed to hear the publishers' appeal, particularly because the Second Circuit Court in New York, which had ruled in favor of the Snyder heirs, was the first federal court ever to consider the question.

The major record companies did begin to make profits again late in 1980 by raising prices for the fifth consecutive year. The new 20 percent limit on returns for credit also helped but was responsible for the closing of uncounted small retail operators and the demise of several hundred independent distributors. And there was the surprise CBS release of a budget line of LPs reduced from the old catalog price of $7.98 to $5.98.

With millions of LPs on their hands, the small independent retailers and distributors found their salvation in rack jobbers, discount houses, and chain operations who offered a 100 percent return policy on excess stock and old merchandise. The bargains were displayed as "cheap stuff," "inflation fighter Lps," or "budget special values," and sold for $3.98, two for $5, or five for $10. They reportedly accounted for as much as 85 percent of retail sales in some major markets.

One-stop and retail stores were victims of a general "silent deletion" policy that removed hundreds of titles from the catalog without notice. Because there were so few "hot" new LPs, records stayed on the trade-paper charts for a much longer time. By the end of 1981, Top 50 LP chart life had grown from the fifteen weeks of 1979–80 to twenty-one weeks and more, and some LPs reached their first anniversary on the *Billboard* charts.

At the same time a thriving business was being done by independent record companies who dealt with music in which the major labels showed little interest—punk rock, salsa, reggae, disco, ethnic music, electronic concert works, classical-musical esoterica, old-time folk music, forgotten rock-'n'-roll, and new age. It was on the last that Windham Hill Records concentrated, doubling sales every month during late 1982. Independent record operations needed the sale of only 5,000 to 10,000 disks to recoup recording and initial distribution costs as well as to earn a small profit.

"Is Rock on the Rocks?" *Time* asked, pointing out that Top 40 was disappearing from the major markets. WABC-AM, king of that roost, had recently changed to all talk. Ratings were slipping everywhere for AOR stations, whose mixture of old best-selling rock LPs, timidly dosed with applications of new rock music, was losing its core audience of white adult working males. Stations with adult-contemporary formats were calling on programming consultants whose advice, according to *Time*, was that they rely on "passive research" and offer "not what listeners actually like but what they find least offensive," or "play it safe, whatever you do."

There was gnashing of teeth in the Big Six executive suites. The sale of pre-recorded cassettes was below two billion dollars a year for the first time since 1977, though combined shipments of all pre-recorded music exceeded the revenue of any other form of entertainment. Superstars continued to fail to meet the expectations of high-powered executives.

Looking for allies in the fight for a tax on blank cassettes and tape recorders, the RIAA and its members allied themselves with their opponents of long standing—the publishers and songwriters, who had recently received a mechanical rate hike to four cents a selection—and joined the Coalition to Save America's Music. Printed-music sales had leveled off at around $240 million. The publishers benefited from this increase because their royalty from the specialty printers was generally about 20 percent of the list price. The organized songwriters remained fixed to the terms of the standard AGAC contract—10 percent of the wholesale price on the first 200,000 sales, sliding up to 15 percent for sales in excess of half a million copies.

While the record industry waited on Congress to give a levy on blank cassettes, it instituted economies that seriously affected the smaller retailer. Pressing plants and regional offices and warehouses were closed, leaving only central facilities for direct-to-factory orders and slowing down deliveries. The imposition of merchandise quotas of major labels terminated many small retail operations.

The best-selling LPs of 1982, *Asia* and John Couger Mellencamp's *American Fool,* had neared the three-million sales mark, giving executives the feeling that their future and that of the industry lay in new artists. Warner Brothers experimented with untested talents, issuing only six cuts of their music on mini-LPs that sold at reduced prices and were promoted on Music Television (MTV), the Warner Express cable channel with twenty-four hours a day of music videos.

The music short made for television was not new. In 1951, Bob Horn of Philadelphia captured a teenage audience with shorts made by the Snader Company featuring Capitol Records artists. Beginning in the mid-1950s, the 200 TV disk jockeys were provided, without charge, a variety of three-minute "tele-records." The medium changed very quickly from film to tape and then to the earliest videodisks. The French musical-video jukebox Scopitone, which featured three-minute performances, and ColorSonics, a similar machine, were introduced. The programming of these musical shorts declined, and by 1967 only about seventy stations featured them.

The success of the Beatles' *A Hard Day's Night* and other films featuring British rock performers as well as television shows like *Top of the Pops* created the first interest in short "concept" musical films. In 1970, the quixotic popular-music visionary Van Dyke Parks was made director of Warner Brothers' Television Film Company and charged with the production of ten-minute sixteen-millimeter promotional films starring new Warner artists. However, when expenses for Parks's department climbed to over half a million dollars, his services were dispensed with, and the project was reduced in scope.

Other video departments continued to produce less costly promotional film clips, though their use on commercial television dropped off dramatically following the emergence of late-night rock-music shows. ABC's monthly ninety-minute "In Concert" series and Don Kirshner's syndicated New York Concert and his annual Rock Awards telecasts all presented rock stars in a concert format.

MTV, the twenty-four-hour all-music cable service which went on line on August 1, 1981, for Warner-Ammex Satellite Entertainment, was the child of a former disk jockey, Bob Pittman, and targeted an audience of twelve- to thirty-four-year-olds. Videos were often played four to five times a day, introduced with taped announcements with video-jockeys who filled in gaps on the screen with gossip about rock stars. Within a year, MTV had signed 125 sponsors, yet it showed a $15-million pretax loss. The record companies had at last found the break that might turn the economic tide. *Billboard* found that new acts who made their bow on MTV enjoyed an immediate 10 to 15 percent increase in sales. However, performers often were obliged to pay up to half the production costs of videos out of their royalties.

American radio's failure to program black music other than that which hit the top of *Billboard*'s pop charts was also true on MTV. Album-oriented rock ruled and all other forms of popular music were shunned. Seven of the thirty-five best-selling albums on *Billboard*'s charts in November 1982 were by black artists, but Bob Pittman had little to do with them, pointing out that "You can't be all things to all people."

The absence of black music not only on MTV but also on almost all AOR radio was regarded by many as "rock racism." In the early 1970s, prior to the advent of disco music, soul music generally had lived alongside white pop music on Top 40 stations. But when disco knocked off Top 40 and by 1978 accounted for the major portion of all *Billboard* LP and singles charts, a backlash against the eight-billion-dollar business and black music in general followed.

Black radio stations in major markets began to call their formats "urban contemporary," and the management and backers of one of them, WBLS-FM in New York, unsuccessfully tried to adapt its programs to cable television, featuring only black performers. Another cable service, Washington-based Black Entertainment Television, showed only two hours a week of *Video Soul* in 1982.

The Home Recording Act of 1983, introduced shortly after the 98th Congress convened, called for negotiations between copyright holders and the tape-equipment and software manufacturers and importers, though it left the distribution of royalties or taxes to the Tribunal. Simultaneously bills to exempt home recording for liability were brought to the floor, as were bills calling for repeal for the "first-sale" doctrine.

Congressional reluctance to co-sponsor these bills was attributed by the music and record business to the effectiveness of the opposition's lobbying front, the Coalition for Home Recording Rights, whose chief spokesman was a former FCC chairman, Charles D. Ferris. His financial backers and chief clients in this matter included all the biggest producers of blank tape and hardware.

The spectacular sales in the first quarter of 1982 of $421 million worth of Berzerk, Defender, and, particularly, Model 2600 Pac Man, as well as other products of Warner Communications' Atari Videogame Division, represented two-thirds of the conglomerate's total profits, and more than compensated for its record- and music-publishing subsidiaries' losses of $101 million in the same period.

In a five-year period, the manufacture of arcade and home video games had become a $7-billion industry, with sales of the latter expected to bring in $1.7 billion for 1982 and $3 billion the following year, by which time 15 million American homes would own video-game modules.

A few months later, Warner Communications stock fell 16.75 points on the New York Stock Exchange, in anticipation of a severe loss in Atari's earnings. Almost overnight, fierce competition and consumer negativism toward new video-game products resulted in a $1.1-billion loss in Warner common stock and by the second quarter of 1983 a deficit of $283.3 million. By autumn, the Warner music-publishing and record division's operating income was up 80 percent over 1982, to $14.7 million, and in time for the Christmas season when LP prices rose once again.

A 1982 study sponsored by opponents of taxes on blank tape and recording decks found that 52 percent of the tapes made by 1,018 individuals involved in a telephone survey were not of music; tapers owned more pre-recorded music than non-tapers; and the $8.98 price for an LP was greatly responsible for home taping.

In early 1982, Warner Communications added between $10 million and $12 million in anticipated annual income with the acquisition of the 20th Century-Fox music operation. In May, Warner negotiated to bring in the United Artists music catalogue and the Big Three music-printing operating, both owned by Transamerica, to guarantee an additional income of $25 million to $30 million for its music division. The transaction would double the size of its music catalog and represent the largest single sale since North American Philips and others purchased Chappell Music. The sale, however, required Justice Department approval.

Many record company executives opposed the purchase, pointing to its inherent antitrust implications because Warner already owned the largest ASCAP house, Music Publishers Holding Corporation, and other major catalogues. When it was clear that United Artists would not deliver music rights to future MGM/UA productions slated for release on videocassette, Warner called off the negotiations.

Confronted with growing deficits from an expansion of its book- and record-club empire into the United States, as well as unexpectedly poor sales by its American record labels, Arista and Ariola, Bertelsmann, a German communications conglomerate, sold its four-year-old Innerworld Music Group catalogue to Chappell Music in the summer of 1982.

Now that Warner Communications was out of the picture, CBS Records resumed talks with the MGM/UA Entertainment Company. By the end of 1982, CBS acquired UA's 50,000 music copyrights. The contract included a five-year co-publishing arrangement with MGM/UA for all music written for its future film productions and first rights to its soundtrack recordings. In September 1983, the CBS Catalog Partnership was formed to administer CBS songs and subsidize the corporation's purchase of UA Music.

The upturn of record sales in 1983 reflected sudden overnight hits as purchases of LPs and pre-recorded cassettes by teenagers rose. The latter now accounted for at least half of all sales and were at their lowest in years, due to drastic price cutting by most companies. In all of 1982, only three albums had sold more than two million units. In the first quarter of 1983, four LPs were expected to do the same or better—two of them by Michael Jackson, with his *Thriller* album moving 60,000 units a day.

Emphasis remained on deal-making. CBS gave the Rolling Stones a four-LP contract that added up to a guaranteed $28 million and included reissue rights to their fifteen most recent albums. After signing multi-million-dollar long-term contracts with Diana Ross and Kenny Rogers, the former "sleeping giant" of the industry, RCA Records did business with Bertelsmann, the German colleague which handled U.S. sales and distribution of its Arista Records. It bought a half-interest in the label, with an option to buy the remainder.

In the summer of 1983, Warner Communications and Philips-Siemens combine began discussions looking toward a merger, contingent on the purchase by Warner of Siemen's half-interest in PolyGram. Such a combined operation showed a potential of $1.5 billion in annual sales. The plan called for Warner Communications to transfer its record business to Warner Records, and PolyGram to Chappell. The companies would then merge into Warner-PolyGram and would form the world's largest record operation. It would also serve to take some of the impact on the American industry of the imminent CBS compact-disk venture with Japanese Sony, pending Justice Department approval.

The erratic nature of the album market was confirmed by an analysis of *Billboard*'s popular-album charts for 1983, Only six LPs had hit the number-one spot, and only three of them had remained there for more than two weeks, the fewest since 1979. Five albums had sold a combined 28.5 million units in the United States, all of them doing better than four million.

MTV was making small tentative gestures to remove its color line by featuring clips by Michael Jackson, Tina Turner, Donna Summer, and Prince. The Black Entertainment Channel stepped up its *Video Soul* segment to fifteen hours a week, while ABC-TV introduced a black-oriented after-prime-time TV production, *New York Hot Tracks*, and NBC-TV scheduled an integrated weekly *Friday Night Videos*.

In late 1981, FM stations found a new format, which coincidentally also had the appeal of MTV's video jukebox. Beginning in major markets, they rediscovered the Top 40 format, called it contemporary hits radio, or CHR, and played hits all the time. These shows were targeted for baby-boomers, the very audience CHR wanted, while men in the eighteen to twenty-four age bracket found CHR too soft.

With *Billboard*'s charts expanding to cover changing styles and tastes, CHR FM stereo was on the top of hits as they broke. In less than a year, the formula was the hottest new trend in radio. It was the third most popular format, after contemporary and country music.

CBS's Walter Yetnikoff voiced unqualified optimism about the record industry and its future. He cited the video revolution as a source of promotion. His company was the principal beneficiary of MTV's almost mystical ability to sell both the records and persona of the most popular CBS singer of all time, Michael Jackson, whose cross-over LP *Thriller* was already the best-selling solo album in history, with worldwide sales in excess of 25 million so far. It provided CBS Records with $120 million in a twelve-month period and lifted its net income by one-quarter over 1982, to $187 million. It was anticipated that Michael Jackson Inc., a holding company, would pay its board chairman in excess of $50 million by the end of 1984.

The proposed Warner-PolyGram merger had touched off consideration of a similar move by RCA, which, like Warner, Philips, Sony, and its European partner Bertelsmann, had been suffering the indigestion of overdiversification. In early 1984 some unprofitable units were sold, electronic military business was expanded, and the annual $100 million loss from videodisks was about to be terminated. RCA Records, which had sales of around $600 million, was in a position to seek a global tie with Bertelsmann, which itself was in the process of concluding negotiations with MGM/UA, MCA, and Paramount for joint pay-cable projects in Germany.

Following World War II, Bertelsmann had grown into a multi-national media giant with holdings in book- and record-club operations and ownership of leading magazines and paperback-publishing concerns. Its Ariola-American record company and the Arista label produced losses, however. Though RCA and Ariola did not share distribution in Europe, Bertelsmann's computer software division, Sonopress, made RCA records for sale on the continent. When RCA expressed an interest in buying the remaining share of Arista Records, and talk about a Warner-PolyGram merger became public, negotiations began for the formation of Ariola-RCA Music in Germany, and RCA-Ariola International elsewhere.

The West German Cartel Office and the United States Federal Trade Commission both opposed the Warner-PolyGram marriage. The FTC maintained that the concentration levels "in this case exceed levels which courts have in the past held to establish a prima facie case of illegality under antitrust laws." Nonetheless, plans for the joint RCA-Bertelsmann venture continued throughout the year, culminating with approval by the Cartel Office in early 1985.

Warner Communications was not as fortunate in the outcome of its plans to join with PolyGram. The FTC held that the resulting reduced competition would not be cured by "new competitors," because "the volume necessary to break even in national distribution exceeds $125 million in annual sales." Independent distributors could not respond to the majors' price increases by distributing more hit records, since the best-selling artists were under contract to the majors. "Moreover, the small volume of the remaining independent distributors makes them high-cost competitors for the small amount of mainstream music they continue to distribute."

Warner/Reprise-Elektra/Asylum-Atlantic Records had brought in profits in 1983 of $60.7 million, from sales of $775 million, but the parent corporation was in perilous straits, having lost $535 million on the Atari unit alone in the first three quarters of the year. The $875-million credit line for its cable ventures, including MTV, had barely escaped a hostile takeover.

A federal judge had refused in April to block the merger, but after hearings in September, an appeals court agreed to stop it, pending a hearing of arguments. An additional roadblock to the partnership with Philips-PolyGram was placed by Siemens's refusal to advance any further funds after learning that PolyGram had lost in excess of $200 million in the United States recently and faced a $15-million loss for the year. Siemens wanted to withdraw from the ownership of Deutsche Grammophon it had shared with Philips since 1963, 90 percent of which it returned to its partner in early 1985. After introducing the world's first VCR for home use in 1972, Philips had lost its hold on that market to the Japanese while continental sales of Philips's home electronics and industrial computer systems started to fall off because of American competition.

Only after an attempt to demonstrate that competition from the increasing incidence of home taping mitigated against the FTC charge that a Warner-PolyGram entity would represent a monopoly in restraint of trade did Warner Communications and Philips concede defeat. Their argument had not dissuaded the district court judge, nor the commission, whose attorneys suggested that, rather than become part of a monopoly that would control 26 percent of the U.S. record market, PolyGram, which sustained losses of $255 million in the past six years and was currently losing $300,000 a day, should seek an alliance with MCA. The increase in MCA's revenues was misleading, as it was actually the contribution of Motown Records, which had switched its distribution to MCA in 1983. In fact, the company's record-and-music division showed a decline for the first quarter of 1984. An alliance with PolyGram would be a more than proper move for the company.

The Warner Music Group's increased sales in 1984 were reinforced by the sale of 10 million units of Prince's *Purple Rain* soundtrack, the year's soul-music runner-up to all-time champion Michael Jackson. However, the races continued to remain mostly separate in the music business and on radio. Of the twenty-two artists enjoying much play on MTV in early 1984, none was black. Separate sales charts in *Billboard* and other music-trade publications confirmed the situation. Black bands had difficulty obtaining record-company financing for their videos, a crucial competitive factor.

The problem of rising video-clip production costs appeared to be solved by three of the Big Six in June 1984. No longer sending videos to MTV without charge, CBS, MCA, and RCA signed contracts with the music channel for exclusive "light rotation" rights for predetermined limited periods of selected videos. The reported $16 million paid by MTV to guarantee exclusivity raised once more the question of exactly who owned the clips. There was also the possibility that these agreements violated antitrust laws. Well aware of their rights in the matter, ASCAP and BMI negotiated contracts, based on income, with MTV. The fees were related to the latest financial information on MTV's progress, which showed that its revenue in the second quarter of 1984 was $6 million. In the first half of the year, cable operators had paid $13.741 million to carry MTV. The sum allotted in the prospectus for video-clip exclusivity was reported as $4.585 million, of which $925,000 would be offset by advertising time.

A suit filed in Los Angeles in September by Discovery Music Network raised the issue of illegal monopoly and "exclusive and coercive arrangements" into which MTV had "intimidated" CBS, MCA, RCA, and PolyGram to enter into an exclusivity agreement. The announcement of plans for a twenty-four-hour Discovery music channel and those for a similar service operated by Turner Broadcasting Cable Music Channel was followed by an announcement from MTV that it would mount a similar operation, VH-1, for twenty-five- to fifty-four-year-olds. While Discovery's suit was bogged down, Turner sold MTV its music channel for one million dollars and $500,000 of advertising time.

After its introduction in Japan in the fall of 1983, the compact-disk player and the compact disk went on general sale in America, at between $800 and $1,200 for the player. CBS and Sony took over and jointly reopened the Terre Haute plant for annual production of 10.5 million CDs. Within a year, during which 2,000 CD releases were put on the market and 4.3 million were sold, the price of a CD stabilized at around twelve dollars, and a player could be had for just below $300. A U.S. representative of PolyGram, which staked much of its future on the new technology, predicted that there would be 645,000 CD units in use by the end of 1985.

When it was suffering the problem of maintaining an adequate cash flow with which to develop the compact-disk market, and unable to raise sufficient funds to buy out Siemens in early 1983, PolyGram considered the sale for approximately $150 million of Chappell-Intersong and associated catalogues. Even earlier in the 1980s, when the top management team of the PolyGram Group was switched from Dutch executives to a four-man German team, it was reported that Philips was considering a withdrawal from the entertainment business. Nevertheless, rumors in the United States and Britain that its publishing empire, which had annual revenues of $60 million, might be sold were regularly denied by PolyGram.

When the sale took place in 1984, the buyers of Chappell-Intersong were headed by Freddy Bienstock and his Anglo-American Music Publishing Corporation. Financial details were not made public, but it was believed that the purchase of price had been in the $100 billion range. The dozens of publishing operations functioning under the Chappell banner included all of the Dreyfus companies, the Aberbach brothers' music houses, other companies acquired through the years, and many firms administered by Chappell on behalf of their owners.

Bienstock's first major acquisition since his purchase of New York Times Music in 1977 had been, in 1983, one of the two remaining family-owned old-line music companies, Edward B. Marks Music, bought in partnership with Williamson Music. His catalogues plus Chappell's represented perhaps the largest repository of popular-music copyrights in the world.

Estimates by CBS Records, which had shown a high degree of accuracy in the past, indicated that the record industry had made another comeback, a new high of $4.464 billion in 1984, 17 percent over the previous year and 8 percent over the 1978 previous all-time peak. Official RIAA figures, representing only the sales of its members and none of those of small independent labels, were released in April 1985. Prerecorded cassette sales were the key to the industry's $4.37 billion gross. They were up 32 percent over the previous year, to $2.38 billion, more than half of all sales revenue. Compact-disk shipments had jumped by 625 percent in a year, representing a total dollar volume of $103.3 million, up 500 percent from 1983's $17.2 million. Total unit shipments of 697.8 million were still below those in 1978, 1979, and 1980. So, too, was the number of new album releases—about 2,000 in 1984 and 2,300 in 1983, considerably lower than the 4,056 in 1972—indicating a steady diminution of new recorded music.

The success in 1984 of Tina Turner, Prince, and Michael and the other Jacksons had lifted some of the burden from black artists and had added more than a touch of black gospel and soul music to white rock, the Hot 100, and MTV. Singles plus music video still made an album a bit hit. In the first quarter of 1985, the RIAA certified twenty Platinum albums, thirty-one Gold LPs, and five Gold singles. One non-theatrical video was certified as both Gold and Platinum in the same month, signifying sales of more than 40,000 units and cumulative retail sales in excess of $1.6 million. One was certified Gold, representing the sale of 20,000 units and $800,00 in retail sales.

CBS was the first, in 1985, to raise the list price of selected superstar albums to $9.98. If the other labels followed, it might induce consumer resistance once more and put an end to the recovery. The world that lay ahead for the recorded-music business, and, in a sense, for the printed-music business as well, was essentially in the hands of the small group that had shaped its destiny for years and had firmer control than ever of the distribution apparatus: RCA Records, now RCA Ariola International; Philips-PolyGram, until its conglomerate masters sold off their entertainment holdings, having already sustained a $220-million loss in the previous five years; Warner Communications, with whatever new partner it might take on in the interest of greater worldwide distribution; MCA Records, which needed a new alliance in the business; Thorn-EMI/Capitol; and the CBS Records Group, at whose heels the hounds of takeover were baying, promising to dispose of it in order to pay for the CBS Television Network.

Conclusion

THE corporate control of the musical marketplace has only increased since 1984 and monopolizes which music is recorded and which artists are promoted. The retailing of music has also undergone centralized control. Owner-operated "mom and pop" record stores virtually have disappeared and been replaced by such chains and "superstores" as the Record Bar and Tower Records. At the same time, while record companies and retailers exert a virtual stranglehold on the music business, consumers have begun to purchase an increasingly diverse body of musical forms, which has led to the meteoric rise of black rap artists and heavy metal bands, whose records dominated the charts in this period in an unprecedented manner. Record executives now realize that rap music appeals to more than just inner-city teenagers, and hard-rock bands can expand their audience beyond its ordinarily hard-core male constituency. Motion-picture producers, too, tapped into the youth market and adopted the crossover strategy initiated by Al Coury's merchandising of *Saturday Night Fever*. Soundtrack albums recently have dominated the charts, adding to the coffers of both record companies and movie studios. And yet the mainstream marketplace has not been alone in its promotion of diverse musical forms. In part that diversity has been due to a boom in "fringe" products available from the "indie" or independent labels and distributors. Even if it is estimated that the "indies" control 5 or 10 percent at most of market sales, they have enabled the inquisitive consumer to sample forms of music the majors have avoided or denigrated. Furthermore, the A & R staffs of the major labels appear to treat the "indies" as farm teams, awaiting to see which of their artists have found a constituency and then offering them more lucrative contracts, to which a number have succumbed. The result of these many forces has been expanding markets offering an ever-widening range of product. Even if consumer choices continue to be dictated by a small number of companies, the diversity of music one purchases and the format on which it is recorded, what with the innovation of compact disks and digital audio technology, can only continue to proliferate.

If any musical forms can be said to have dominated the charts in the last several years, they would be movie soundtracks, R & B, and heavy metal. In 1984, ten soundtrack albums went platinum and two, *Footloose* and *Purple Rain*, held the number-one chart position for more than half a year. The former generated six Top 40 singles, while the album by Prince sold in excess of eight million copies, a sales record topped only by that of *Saturday Night Fever*. Both albums benefitted from artist participation in the film and the music: Prince not only starred in and wrote but also scored *Purple Rain*, while Dean Pitchford wrote the screenplay and lyrics for *Footloose*. However, in too many cases, soundtracks have suffered from a lack of such input. Performers and writers have often done little more than contribute a song to a film score and then been given too little time to prepare the material. The result has been soundtrack albums glutted with mediocre filler or work by prominent artists that was not up to their usual high standards. Despite the often hasty lack of coordination, other soundtracks have dominated the charts, including 1986's bestselling *Top Gun*, which yielded three chart-topping singles, and 1987's *Dirty Dancing* and *La Bamba*. The *Dirty Dancing* album was followed by yet another with music from the film and even led to a tour, in the summer of 1988, by singers featured on both recordings as well as dancers performing the slow grind gyrations of the early 1960s. *La Bamba*'s success proved that soundtracks need not include contemporary material, as Los Lobos's re-recording of the Ritchie Valens's 1959 hit soared to the top of the charts, bought in many cases by the children and grandchildren of its original purchasers. This was also the case with the 1983 *The Big Chill*, which resulted in two collections of Motown-dominated R & B of the 1960s. Still, for all the successes, many rock-dominated soundtracks failed, as did the films they promoted. For a soundtrack to be successful, the music must support the film's storyline as well as stand on its own as a coherent composition. Also, it was often forgotten that the best musical marketing tool for a film was not an album but an identifiable hit single accompanied by a frequently aired video, preferably one including enticing clips from the film. When a successful marketing strategy, good music, and a crowd-pleasing film can be combined, the results are more than marketable.

Rhythm and blues performers have garnered substantial sales in the last several years. The market's acceptance of black music can be attributed to the meteoric success of Michael Jackson's *Thriller*. It is now the all-time best-selling album, having racked up over 30 million sales worldwide, and held the number-one spot on the album charts for two years in a row. Even if *Thriller*'s success caused many to assert that Jackson's follow-up album, *Bad*, could not possibly equal its predecessor, it accumulated 2.75 million advance orders, the largest number in CBS's history. Prince, too, achieved substantial sales, the *Purple Rain* soundtrack selling 1.3 million copies on the first day of release. A song from that score, "When Doves Cry," was the top pop and black single of 1984, the first such record to win both categories since Bobby Lewis's

"Tossin' and Turnin' " in 1961. Other black performers in this period who gained wide constituencies and placed records on more than one chart include the Pointer Sisters, Tina Turner, Kool and the Gang, Whitney Houston, Sade, and Lionel Ritchie, formerly of the Commodores group, who, in addition to contributing to soundtracks and scoring dance hits, was the top adult contemporary artist for two years, 1983–84.

However, the most influential R & B format has been rap music. Any close study of the black album and singles charts indicates that rap currently moves more units than mainstream black music and has acquired a loyal audience among urban teenagers. Rap began as a chart phenomenon with the 1979 "Rapper's Delight" and was taken up as a commercial vehicle by Hollywood in 1984 with the release of several successful films, including *Breakin'* and *Beat Street*. And yet, it was not until 1986 that the music business recognized the remarkable audience for rap with the release of Run-DMC's second album, *Raising Hell*. It peaked at number three on the pop album charts and sold 3 million copies. The group added to its established urban audience with their successful single "Walk This Way," a remake of the 1975 Aerosmith hit recorded in conjunction with members of that group. In 1987, Run-DMC's sales record was broken by the Beastie Boys' *Licensed to Ill*, which became the first rap album to hit number one on the pop charts. While many still consider rap a musically unsophisticated form, the careful listener can discern a wide range of styles among such artists as the teen-idol LL Cool J, the comic Fat Boys, the overtly political Public Enemy, and the all-woman rap group Salt-'n'-Peppa. Unsophisticated or not, rap continues to be one of the best-selling forms of contemporary popular music.

Heavy metal, too, increased its marketability in this period. It has always possessed a loyal following among young males, going back to the late 1960s with the early recordings of bands like Black Sabbath and Deep Purple. Over the last two decades, heavy-metal groups consistently filled arenas with enthusiastic fans, but it was not until Bon Jovi's 1987 album *Slippery When Wet*, which sold over 8 million copies in the United States alone, logged 38 weeks in the top five on the Top Pop Album charts, and placed three singles in the Top Ten, that the "acceptable face of heavy metal," in *Billboard*'s words, was apparent to the public. Bon Jovi combined a hard-rock sound, memorable anthemic choruses, and stinging guitar solos with their good looks to attract not just the customary leather-clad adolescent male but also young girls. The group's success helped to lay the groundwork for other bands to place albums in the Top 40, including Cinderella, Poison, Whitesnake, Motley Crue, Poison, Guns N' Roses, and Def Leppard.

Other musical forms fared less well during this period and seemed temporarily unable to find their ideal constituency. Country and western in particular appeared locked in an attempt to crossover onto the pop charts and, as a result, lost sight of its roots. The success of the 1980 film *Urban Cowboy* and its soundtrack caused producers unsuccessfully to aim at demographically expanded audiences, and traditional country fans felt something was missing. A backlash started in 1984 as a conservative return to the music's traditional roots spurred the success of acts like the Judds, Ricky Skaggs, Reba McEntire, Dwight Yoakam, and George Straight. The back-to-basics movement found its most successful exemplification in Randy Travis, whose 1986 debut album sold over 600,000 copies. The marketing of country music became more sophisticated and utilized such promotional outlets as the cable channel *The Nashville Network* where fans could watch videos of popular country artists.

Jazz artists, who have perennially struggled to achieve wider commercial acceptance, made some headway in this period, due largely to the popularity of trumpeter Wynton Marsalis, whose Columbia albums contained music reflecting the jazz mainstream that appealed to a wide public. Marsalis also gained listeners by his recordings as a soloist of classical compositions, alerting many people to not only his but many other jazz artists' technical virtuosity. Those included his brother Branford, who also recorded for Columbia, as a member of Wynton's group and as leader of his own. When Branford joined the band accompanying the rock star Sting on tour in 1985, rock fans probably got their first taste of masterful jazz improvisation, and many sought out jazz releases to enlarge their awareness of what was to them a novel musical form. This period also saw an increasing emphasis on reissues as many labels resurrected back catalogs, in the case of Prestige with their original covers and liner notes, in order to appeal to collectors eager to sample long out-of-print material.

The ever-widening rock audience was appealed to by its favorite artists during this period at a number of large-scale public service events, The first, Live Aid, occurred in July of 1985. It was a massive concert staged simultaneously in London and Philadelphia, broadcast over international television, and raised funds to feed starving Africans suffering from a catastrophic famine. The organizer of the event, Bob Geldof, also composed a single, "Do They Know It's Christmas," which was an international hit and whose profits added to Live Aid's coffers. In September of the same year, rocker John Cougar Mellencamp and country star Willie Nelson organized Farm Aid, a live benefit concert again broadcast over national television which raised $10 million dollars to benefit financially strapped family farmers. Benefit records proliferated, with various elements of the music community raising money for a variety of causes. Two of particular note were "We Are the World," written by Lionel Ritchie, whose sponsor group USA for Africa raised $92 million for starving Africans, and Little Steven's "Sun City," which aimed to increase public consciousness of government repression in South Africa.

The televising of these public service events, particularly by MTV, the music video channel, indicates how much the medium continues to be one of the music industry's major marketing tools. A number of artists, including Duran Duran, Culture Club, Cindy Lauper, and Madonna, cannily utilized video as a means of merchandising their music through the construction of a striking visual image. However, it also became increasingly clear that merely projecting a "hot" video image, if not backed up by a substantial repertoire, could easily result in artistic burn-out. Nonetheless, many record companies continued to rely on MTV's services, charging the cost of video production against artist royalties. Companies signed exlusivity contracts with the channel to insure that their artists had the best chance of being in "heavy rotation" and pulling the widest possible audience. At the same time, MTV underwent organizational changes. The management that founded the channel moved on, and in 1985 it was bought by Viacom International. Ted Turner threatened in 1984 to compete with MTV by forming an alternative music video channel, but he failed and shut down the channel in December, selling it to MTV who established a new format, Video Hits One. VH-1, as it is known, was targeted to an older demographic audience of 25-to-49-year-olds; the repertoire was softer and less aggressive, including more middle-of-the-road and easy listening material than MTV. Its success in turn caused MTV to re-examine its programming. A number of the original vee-jays were fired, and the continual accusations of a racist lack of attention to black music began to be remedied as the channel incorporated more dance tracks.

Video copies of motion pictures have become a most marketable item in the last several years. As ownership of VCRs increased and the rental of films became a common pastime, the manufacturers of video tapes realized that exorbitant prices had made consumer purchase of films economically prohibitive. Even if tape manufacturers could make a substantial profit from rental stores, who, to keep up with public demand, would purchase several copies of popular films, they were unable to tap those many people who would be willing to purchase a film if the price was reasonable. The Media Corporation initiated price-cutting in 1984, when they lowered the list price of much of their product to $19.95, and Paramount Pictures, who had mysteriously removed a number of films from circulation the year before, reintroduced them at premium prices with their "25 for $25" campaign. The Disney firm joined the bandwagon by issuing a number of their classic cartoon features for $29.95. These actions resulted in the sale of 50 million units in 1985, with profits of between 1.5 and 2 billion dollars. This did not include the mushrooming public domain market, which issued films no longer protected by copyright claims at $10 or less, many of them sold by chain stores and mass merchandisers. Box-office mega-hits, such as the Eddie Murphy comedy *Beverly Hills Cop*, reaped profits equal to or greater than their initial release when issued on tape at a reasonable price. The result was that in one year's time video sales more than doubled to 5 million dollars from the purchase of 80 million units. Such profits allowed video firms to diversify, and in 1986 Vestron was the first software manufacturer to initiate a commercial film division. It also led firms to cement deals with corporate sponsors, as in the case of Paramount's *Top Gun*, whose video tape began with a commercial for Pepsi-Cola. The commercial's inclusion allowed the company to sell the tape for $26.95 and thereby increase their profits. Video tapes furthermore were no longer sold by "mom and pop" owner-operated stores, for those same chains which had consolidated record sales began to market video tapes in large speciality stores, outselling their competition by increasing the number of available titles.

The consolidation of retail music marketing led to the phasing out of the sale of recorded music by department stores, as was the case with Montgomery Ward in 1984. Now the retail business was controlled by chains like Tower Records and Record Bar, the former of which began to open stores with 18,000 or more square feet packed with discounted product. Such stores could market numerous copies of a wide range of selections, easily attracting consumers who knew they could find current hits at reasonable prices. Large corporations took advantage of the chain stores' success, leading to Transworld's buy-out of the Record Land chain in 1985. Retail chains expanded, either through the construction of new stores or aquisition of other organizations. This was particularly true in 1986 on the West Coast, where, in what became known as the "battle of California," Musicland acquired the Southern California chain Licorice Pizza from Record Bar in addition to most Record Bar units west of the Mississippi. Simultaneously, Warehouse Entertainment bought out the San Francisco chain Record Factory.

The marketing, production, and performance of music during this period was not free of governmental interference. In a series of actions reminiscent of the Pastore and Celler hearings, the Parents Music Resource Council (PMRC) sought to compel record companies to engage in censorship of material the council felt to be pornographic and injurious to the nation's youth. Their activities began in 1985 when a Washington, D.C., parent was aghast at hearing a Prince song, "Sugar Walls," whose lyrics referred to masturbation. The wives of a number of government officials, including Senator Albert Gore and cabinet member Jim Baker, led the group and petitioned record companies to enact several measures to alleviate their concerns: include lyric sheets within all albums, post a warning label on the cover of albums with questionable lyric content, and admonish artists to consider the effect of their music upon impressionable audiences. Capitol Hill hearings on the matter were held in August of 1985, and later that year the RIAA acceded to the request for warning labels. A number of artists, including such disparate figures as Frank Zappa, heavy metal vocalist Dee Snider, and John Denver, vehemently protested what they considered an abridgement of

First Amendment rights and unnecessary censorship. However, the PMRC's actions had an impact upon the music industry in a number of cases. In August of 1986, the Wall Mart chain refused to carry heavy metal product any longer, specifically objecting to the albums' often outrageous covers, and the San Antonio city council enacted a local ordinance in September 1985 outlawing any performers whose onstage behavior they considered offensive. Some musicians were more personally affected, such as lead singer Jello Biafra of the punk rock band the Dead Kennedys. He was charged in 1985 with the distribution of harmful pornographic materials for including a suggestive poster by the Swiss artist H. R. Gieger in the album *Frankenchrist*. The fact that he was exonerated in August 1987 indicates the degree to which the PMRC's influence has waned, but it should be added that as a consequence of the suit the group disbanded and is still many thousands of dollars in debt for legal fees.

Musicians and writers have faced another threat to their economic livelihood from the television networks during this period. While the resolution of the *Buffalo v. ASCAP, BMI, et al.* case by the Supreme Court denies broadcasters the right to collect payments for music "at the source," they have once again engaged in legal actions to secure similar rights with the Television Source Licensing Bill. If the bill is passed, a composer would be required to negotiate an up-front deal with each program producer for the use of his material. That negotiation would occur before the composer could judge the worth of his work. In other words, if the program was a runaway success and eventually went into syndication, he would not receive any further fees than those obtained in the initial negotiation. The broadcasters have incorrectly argued that they are forced to purchase "blanket licenses," which convey the rights to a wide range of material, whereas they may be interested only in a few songs. In truth, broadcasters are not forced to buy any single form of licensing, and all performing rights organizations offer a flexible set of options. Final adjudication of the bill awaits congressional action, and most members of the music business consider the potential ramifications of its passage the most serious to date to the rights of writers and composers.

Like much else in the music business, radio stations were undergoing consolidation in this period. Regional network chains began annexing suburban stations in communities adjacent to major markets. For example, ABC paid $9 million to buy Hicks Communications' KIXX, Denton, Texas, which was then transformed into Dallas's "Kiss-FM KTKS," broadcasting a Top 40 format to the metroplex area. The year 1985 saw an unprecedented number of acquisitions as conglomerates began to swallow up major broadcasting chains: Capital Cities purchasing ABC; United Stations, RKO; and Westwood One, the Mutual Broadcasting System. The Storz Broadcasting chain, associated with the Top

40 format since the 1950s, began disbanding, as outlets were sold to Price Communications in 1985. Now, the manufacture, retail, and playing of recorded music was in the hands of a small body of individuals.

The formats on which music is recorded and played recently underwent their most radical transformation since the innovation of stereo recordings. In the summer of 1983, the compact disc, or CD, entered the American market as an alternate to, or some argue a replacement for, the stereo LP. The new format's principal selling point is its dynamic range: a CD perfectly reproduces recorded music and allows no distortion or noise to interfere with its reception. Listening to a CD requires a digital audio disk player. The small, approximately 4¾-inch, plastic disc rotates within a sealed compartment. No portion of the CD's operation is visible, as all information on the disc is read by a "laser eye." The player can be connected to a standard preamplifier or receiver in the same manner as a cassette deck. At first, prices of both individual CDs and digital players exceeded the means of the average customer, but within three years prices lowered to the point that CDs outsold LPs. By the end of 1987 annual domestic production of the discs swelled to over 100 million. The LP now accounts for as little as 10 percent of many record retailers' sales, as the CD has overtaken their inventory. This has resulted in a transformation of retail marketing: most record stores, in response to demands for space to accommodate CDs, have cut inventory in their "deep catalog" LPs: older but steadily moving titles, including not only classic rock albums but also less fast-selling musical forms such as jazz, reggae, blues, and folk. Many in the music industry believe that as soon as CDs are accepted fully by adolescents and lower-income customers, the LP will be a thing of the past. This dire prediction is corroborated by the fact that little to no research and development is presently being done to improve the manufacture of turntables or the phonograph record itself. However, many record collectors who have an interest in the "deep catalog" items retailers have dismissed recognize the inherently archival nature of the CD format. As the disc will not deteriorate with time, older and more fragile recordings can benefit from being transposed to a CD. Some record labels have responded to this need by investing in the release of reissues of "deep catalog" items, often including alternate takes or tracks unissued on the original recording.

If the CD has sent shockwaves through the music industry, the invention of digital audiotape, or DAT, promises the potential of even more devastating impact. DAT will allow the consumer to reproduce the digital-quality sound heretofore available only on CD. In addition, it will allow them to make unlimited copies of CDs with no loss of fidelity. DAT, like CDs, uses digital coding for sound reproduction. Music is converted to a code of electronic pulses which, when played back, is converted into music. Understandably the music industry is horrified by the possibility of pirating albums with DAT. If a consumer can make or purchase a perfect copy of a CD, why would they buy it? Furthermore, as the record industry has invested heavily in CD technology, they are not ready or willing to see it displaced by an alternate process. Appeals for congressional protection from DAT await resolution, but as the Supreme Court has already decided that home taping is legal, the industry has heavily invested in the Copy-Code system, whose inaudible signal placed on all CDs would impair the "illegal" recording process.

The "deep catalog" which technology threatens increasingly has become the province and livelihood of the independent or "indie" record labels. In all too many cases, the majors are uninterested in signing and promoting an artist who cannot reap *big* profits. As a result, they have not only ignored or at best paid little attention to musical forms like reggae, blues, and folk music but also dropped substantial artists from their roster when they failed to meet the bottom line. Public awareness of this trend was aroused when Warner Brothers in 1984 dropped more than thirty artists, including new-wave performers like the Roches and Tom Verlaine as well as such perennials of the industry as Arlo Guthrie and Van Morrison. Some of them turned to independent artist-owned labels, Guthrie starting Rising Son and singer-songwriter John Prine Oh Boy when he was dropped by Elektra/Asylum. The result has been artists receiving greater profits and being in complete control of their material. Many more, however, turned to the "indie" marketplace, those "little outposts of unpredictable aesthetic principle, pockets of structural resistance in the struggle for fun," as *Village Voice* music critic Robert Christgau has written. If the "indies" continue to share only a small slice of the economic pie, they have undeniably added to the industry by releasing a variety of music and often commiting themselves to their artists through active promotion, tour support, and continuous inclusion of their material in an active catalog. Some labels have specialized in one musical genre: Alligator with blues, SST with hard-core rock, Charly with reissues of classic rock and R & B, and Flying Fish with folk music, among countless others. The very plentitude of available material is overwhelming, leading even the most dedicated listener to abandon the possibility of "keeping up" with innovative music. At the same time, members of the "indie" community have discerned a certain kind of homogenization reminiscent of the majors. The advent of college radio as a marketing tool and the use of new tip sheets like the College Music Journal (CMJ) raise the possibility of appealing to established audiences rather than creating new ones. Also, a number of "indies," including Rhino, Enigma, and Twin/Tone, have signed distribution deals with the majors, leading many to question if those labels, too, will succumb to a bottom-line mentality. The hope remains, as *New York Times* critic Jon Pareles states, "that somewhere on another lower commercial echelon, there's still scruffy strange weird nasty obnoxious stuff, which is what indies were put on this earth to produce."

While it may not be characterized as nasty or obnoxious, certainly one of the most significant current additions to the musical marketplace initiated by the "indies" has been the mushrooming field of international music. For years, folk music from a variety of foreign cultures has been available, but never to the present degree. Also the fusion of foreign musical styles with American popular music has reached new heights, notably as a result of the success of Paul Simon's 1987 *Graceland* album. His inclusion of native south African musicians and musical forms on the record as well as his subsequent world tour alerted many people to the dynamic range of musical expression on the African subcontinent. While some have accused Simon of ransacking a foreign culture for his own benefit, it must be said that without his efforts African music would have remained an indigenous and alien phenomenon except to the musical cognoscenti. And yet, African music has not been alone in this new and ever-expanding field. American consumers can now readily purchase albums of Trinidadian soca, Zairean soukous, and French Caribbean zouk. Perhaps the label most involved in this process has been the New Jersey-based Shanachie. Started in 1973 by Richard Nevins and Dan Collins, they began releasing some 10 records a year, selling an average of 5 to 10,000 copies, Now, their annual list includes some 40 releases, and sales average 30,000 and more. Shanachie has made its name marketing a number of international forms, starting with Irish folk music, then reggae in the early 1980s, African music in 1984, and in 1988 began its World Beat/Ethno-Pop line. These latest records range from Yemenite Israeli dance music to Arabic-rock fusion played by a mix of Arabs and Germans. Randall Grass, the third manager of the label, explains their choice of repertoire: "The first and most important factor is whether we're excited by the music. If we are, we'll try to find a way—unless the economics are just daunting, meaning that people involved want amounts of money that are just out of the realm of reality—to try and release it. We have, on occasion, released records that had little chance of making money just because we thought the music was great." At times, this inundation of novel and exciting music seems a veritable tower of babble, too much product for any one ear to absorb. And yet, as Robert Christgau writes, absorbing this material is "a critical-perceptual project [that] could take decades to bear its own fruit—that is, genuinely international rock and roll. Which as far as I'm concerned is a guarantee that things will stay interesting."

As the music business enters the 1990s, it confronts a multi-billion-dollar marketplace dominated by a small handful of international conglomerates, while conservative forces of censorship threaten the free flow of creation. Many are concerned that between powerful market forces that emphasize the bottom line and a regressive strain in American society that squelches controversial subject matter, the security of adventurous musicians is threatened.

At present, six major labels—PolyGram, CBS, WEA, EMI, BMG, and MCA—are estimated to account for 93 percent of all record sales. The market share controlled by independent labels has shrunken, as several of the most successful privately owned companies have been taken over. In late 1989, PolyGram purchased Chris Blackwell's Island Records for approximately $270 million and shortly afterward added A & M Records, at a cost of $460 million to its roster. The following March, MCA acquired the Geffen label from its owner, David Geffen, at a cost of $550 million in stock, thus making Geffen MCA's largest individual stockholder. However, he possesses restricted voting rights even though his yearly dividends will amount to approximately $7 million. All three labels acquired the benefits of corporate distribution and promotion through their takeover, yet many in the industry wonder whether consolidation in the hands of a small number of conglomerates will benefit progressive or experimental musicians. It is hard to imagine that innovative music necessarily is well served by an increasing bottom-line mentality. At the same time, emergence of several new labels—SBK, a pop-oriented division of Disney, Charisma (run by Virgin), DGC (a Geffen subsidiary), and a Warner-distributed company to be run by former agent and MCA Records President Irving Azoff—bodes well for an expanded marketplace.

If many in the music business are wary of corporate consolidation, many more have failed resolutely to respond to the accelerated censorship of musical expression. The influence of the PMRC has not abated in the last five years as accusations of obscenity and endangerment of the public safety have targeted elements of the music community, particularly members of the rap and heavy-metal contingents. In particular, the rap group Public Enemy has been charged with alleged anti-semitism, while the heavy-metal stars Guns N' Roses have made racist, sexist, and homophobic comments in song lyrics and public statements. In response, a number of conservative organizations and public officials have proposed legislation or put into force existing laws against the sale of certain records to minors and advocated either the stickering of records as a warning to the consumer or the outright banning of certain music for sale. In Florida and Alabama record clerks have been charged with violating obscenity statutes by selling releases by 2 Live Crew, a raunchy rap group whose album *As Nasty As They Wanna Be* already is stickered as well as available in an edited version. The group itself furthermore has been arrested for public adults-only performance in Florida of what a local judge has deemed pornography. Finally, the FBI has threatened legal action against the radical rap group NWA for a song said to advocate violence against the police.

While a number of musicians and corporate executives have spoken out against the endangerment of the First Amendent, the industry response to public pressure has been vague, perhaps in the hope that the matter might blow over in time. However, as that pressure has accelerated and a number of record chains now refuse to sell certain contested records, the industry realized the pressure was a threat not merely to its reputation but moreover to its bottom line. As a result, in early 1990 the fifty-five-member major trade organization the Recording Industry Association of America and the National Association of Independent Record Distributors and Manufacturers capitulated to the pressure and recommended the adoption of warning stickers, which they hope will allay the threats of civic and governmental intervention and avoid passage of restrictive state and national laws. However, it should be added that, using the PMRC's own figures, of the 7500 albums released between January 1986 and August 1989, only 121 contained questionable lyrics and of those 49 already had been stickered.

The rap community in particular has faced another legal dilemma due to its use of a new technological aide: the digital sampler. This electronic keyboard allows the user to convert any musical element into a processable computer code; in effect, if one can type, one can "create" music. Rap writers have used samplers to incorporate portions of recordings as elements in their own work. In virtually all cases the sampled material is credited, but the technology has called into question whose beats, rhythms, riffs these are, anyway? A number of suits have been pursued, including the Turtles' prosecution of De La Soul, the top-selling rap group of 1989, and will force an inevitable re-examination of copyright law. Questions including whether the fair-use law covers sampling and if a performer's characteristics are themselves copyrightable, as, for example, James Brown's scream, one of the most sampled phenomena in the rap field, must be resolved. Such considerations will be effected by the recent successful suits brought by singers Tom Waits and Bette Midler against advertising agencies for employing "copycat" singers, thereby, they argued, stealing their sound. The technological revolution sure to continue in the 1990s will insure a close examination of the processes of creation, duplication, and imitation.

Finally, the reproduction of sound recordings has itself gone through a revolution comparable to the battle of the speeds of the 1950s. The compact disk has taken over the vinyl market. In a number of record stores vinyl occupies a minimal amount of available floor space, and a virtual 90 percent drop in vinyl sales has been observed. Certain formats have virtually disappeared, particularly the 45 rpm single. Admittedly, the cassette tape still dominates the market, as 13 cassettes are sold for every 6 CDs and single album. Vinyl will continue to be a collectors' medium, but it is virtually being marketed out of existence. However, concerns have been raised about the CD format, particularly by environmentalists who view the boxed packaging as wasteful of the diminishing international forest reserves.

Clearly, changes in the music industry will continue to confound one's expectations. When one thinks back to Mr. Edison's speaking into the first phonograph, it seems more innovation has occurred than could ever have been imagined. The future of the music business is uncertain but bright. May a hundred forms of music bloom. May listeners discover sounds that inspire and excite them. And may the major record labels and retail chains allow us to hear and enjoy them.

Bibliography

General

Ackerman, Paul, and Lee Zhito, eds. *The Complete Report of the First International Music Industry Conference*. New York: Billboard Publishing, 1969.

Aldrich, Richard. *Concert Life in New York 1902–1923*. New York: Putnam, 1941.

Allen, Frederick Lewis. *The Big Change: America Transforms Itself, 1900–1959*. New York: Harper & Row, 1969.

———. *Only Yesterday: An Informal History of the 1920s*. New York: Harper & Row, 1964.

———. *Since Yesterday: The 1930s in America*. New York: Harper & Row, 1972.

American Society of Composers, Authors and Publishers (ASCAP). (All published by ASCAP, in New York.) *ASCAP Biographical Dictionary of Composers, Authors and Publishers*. 1948, 1966, 1980.

———. *ASCAP Grew With Music—So Will Your Business*. 1957.

———. *ASCAP in Action*. 1979–.

———. *ASCAP Journal*. 1937–.

———. *The ASCAP Story*. 1961.

———. *Chords and Dischords*. 1941.

———. *How the Public Gets Its Music: A Statement of Some of the Reasons for the Copyright Law, Its Operation and How It Benefits the Public*. 1933.

———. *Minutes of a Conference Held at 56 West 44th Street in the Offices of The ASCAP, September 20, 1922, New York City*.

———. *The Murder of Music*. 1933.

———. *Notes from ASCAP*. 1950–.

———. *Nothing Can Replace Music*. 1933.

———. "President Stanley Adams' Speech to the ASCAP Membership." March 31, 1964.

———. "Statement of ASCAP Before the National Commission on New Technological Uses of Copyrighted Works (CONTU), March 31, 1977." Mimeo.

———. *The Uses of Music and Why*. 1934.

Aptheker, Herbert, ed. *A Documentary History of the Negro People in the United States*. New York: Citadel Press, 1951.

Archer, Gleason L. *Big Business and Radio*. New York: American Historical Company, 1939.

———. *A History of Radio to 1926*. New York: American Historical Company, 1938.

Arnaz, Desi. *A Book*. New York: William Morrow, 1976.

Arnold, Elliott. *Deep in My Heart: A Story Based on the Life of Sigmund Romberg*. New York: Duell, Sloan and Pearce, 1949.

Arnold, Thurman. *The Bottlenecks of Business*. New York: Reynal and Hitchcock, 1940.

Artis, Bob. *Bluegrass*. New York: Hawthorne Books, 1975.

Austin, William. *Susanna, Jeanie, and The Old Folks at Home*. New York: Macmillan, 1975.

Autry, Gene. *The Art of Writing Songs and How to Play a Guitar*. Evanston, IL: Frontier Publishers, 1933.

Autry, Gene, with Mickey Herskowitz. *Back in the Saddle Again*. Garden City, NY: Doubleday, 1978.

Baker, David N., Lida M. Belt Holt, and Herman C. Hudson. *The Black Composer Speaks*. Metuchen, NJ: Scarecrow Books, 1977.

Baker, W. J. *A History of the Marconi Company*. London: Methuen, 1971.

Bakewell, Dennis, ed.: *The Black Experience in the United States*. Northridge, CA: San Fernando State College Foundation, 1970.

Bane, Michael. *The Outlaws: Revolution in Country Music*. New York: Country Music Magazine Press, 1978.

Banning, William Peck. *Commercial Broadcasting Pioneer: The WEAF Experiment, 1922–1925*. Cambridge, MA: Harvard University Press, 1946.

Barnes, Ken. *The Bing Crosby Years*. New York: St. Martin's Press, 1980.

Barnouw, Eric. *The Golden Web: A History of Broadcasting from 1933 to 1953*. New York: Oxford University Press, 1968.

———. *The Image Empire: A History of Broadcasting from 1953*. New York: Oxford University Press, 1970.

———. *A Tower in Babel: A History of Broadcasting to 1933*. New York: Oxford University Press, 1966.

Barzun, Jacques. *Music in American Life*. New York: Doubleday, 1956.

Baskerville, David. *Music Business Handbook*. Denver: Sherwood, 1979.

Bastin, Bruce. *Crying for the Carolines*. London: Studio Books, 1971.

Belz, Carl. *The Story of Rock*. New York: Oxford University Press, 1971.

Bennett, Lerone, Jr. *Before the Mayflower: A History of the Negro in America, 1619–1964*. Baltimore: Penguin Books, 1964.

Berkman, Paul L. *The "Rhythm and Blues" Fad: An Exploratory Study of a Popular Music Trend*. New York: Columbia University Bureau of Applied Social Research, 1955.

Bernheim, Alfred L. *The Business of the Theatre: An Economic History of the American Theatre, 1750–1932*. New York: Benjamin Blom, 1972.

Bierley, Paul E. *John Philip Sousa, American Phenomenon*. Englewood Cliffs, NJ: Prentice-Hall, 1973.

Bigsby, C. W. E., ed. *Superculture: American Popular Culture and Europe*. Bowling Green, OH: Bowling Green State University Press, 1975.

The Billboard Music Year Book. 1944.

The Billboard Encyclopedia of Music. 1945–.

The Billboard International Buyers' Guide. 1980, 1982–83.

Bing Crosby on Record: A Discography. San Francisco: Mellos Music, 1950.

Black Perspective in Music. 1973–.

Blesh, Rudi. *They All Played Ragtime*. New York: Oak Publications, 1971.

Bohn, Thomas W., and Richard Strongren. *Light and Shadows: A History of Motion Pictures*. Sherman Oaks, CA: Alfred Publishing, 1975.

Bond, Carrie Jacobs. *The Roads of Melody*. New York: Appleton, 1927.

Boorstin, Daniel J. *The Americans: The Democratic Experience*. New York: Random House, 1973.

Bordman, Gerald. *The American Musical Theatre*. New York: Oxford University Press, 1978.

———. *Jerome Kern: His Life and Music*. New York: Oxford University Press, 1980.

Bowers, David. *Put Another Nickel In*. New York: Bonanza Books, 1966.

Bowker, Robert Rogers. *Copyright, Its History and the Law*. Boston: Houghton Mifflin, 1912.

Bradford, Perry. *Born with the Blues*. New York: Oak Publications, 1965.

Braun, D. Duane. *Toward a Theory of Popular Culture: The Sociology and History of American Music and Dance, 1920–1968*. Ann Arbor, MI: Ann Arbor Publishers, 1969.

Brawley, Benjamin. *A Social History of the American Negro*. New York: Collier Books, 1970.

Broadcast Music, Inc. (BMI). (All published by BMI, in New York.) *The ABC of BMI*. 1940.

———. "BMI License Rates Have Been Reduced Since 1940: A Word from BMI." 1943.

———. "BMI Memorandum to the Committee on Interstate and Foreign Commerce, United States Senate, in Regard to the Smathers Bill." 1957.

———. "BMI Memorandum Submitted to the Federal Communications Commission Feb. 1960." 1960.

———. *BMI Newsletter*. 1943–1958.

———. *BMI 1940–1960: Twenty Years of Service to Music*. 1960.

———. *The Many Worlds of Music*. 1958–.

———. "Memorandum of the Status of BMI as of July 27, 1940." 1940.

———. *Poor Richard's Almanac*. 1942–1943.

———. "Statement Before the National Commission on New Technological Uses of Copyrighted Works (CONTU), March 31, 1977."

———. *Your Stake in BMI*. 1948.

———. *Facts on Source Licensing*, 1987.

Broadcasting. 1931–.

Broadcasting Yearbook.

Broonzy, William, and Yannick Bruynoghe. *Big Bill's Blues*. New York: Grove Press, 1955.

Broven, John. *Walking to New Orleans: The Story of New Orleans Rhythm and Blues*. Bexhill-on-Sea: Blues Unlimited, 1974.

Brown, Les. *Television: The Business Behind the Box*. New York: Harcourt Brace Jovanovich, 1971.

Brown, Les, ed.: *The New York Times Encyclopedia of Television*. New York: Times Books, 1977.

Brown, Peter, and Steven Gaines. *The Love You Make: An Insider's Story of the Beatles*. New York: McGraw-Hill, 1983.

Brown, Sterling A. *Negro Poetry and Drama and the Negro in American Fiction*. New York: Atheneum, 1969.

Burton, Jack. *The Blue Book of Broadway Musicals*. Watkins Glen, NY: Century House, 1952.

———. *The Blue Book of Hollywood Musicals*. Watkins Glen, NY: Century House, 1953.

———. *The Blue Book of Tin Pan Alley*. Watkins Glen, NY: Century House, 1950.

Butcher, Margaret Just. *The Negro in American Culture*. New York: Knopf, 1967.

Butler, Tobias, and Will D. Cobb. *The Butler-Cobb Method of Successful Songwriting*. New York: Publishers Press, 1921.

Caesar, Irving. "A letter to writer and/or publisher members of ASCAP." Attached to a typescript extension of oral presentation made to the ASCAP Board of Directors May 22, 1969.

Cahn, Sammy. *I Should Care*. New York: Arbor House, 1974.

Calloway, Cab. *Of Minnie the Moocher and Me*. New York: Crowell, 1978.

Carmichael, Hoagy, and Stephen Longstreet. *Sometimes I Wonder*. New York: Farrar, Straus & Giroux, 1966.

Caron, Paul. *The Devil's Son-in-Law: The Story of Peetie Wheatstraw and His Songs*. London: Studio Books, 1971.

Carpenter, Paul S. *Music, an Art and a Business*. Norman: University of Oklahoma Press, 1950.

Carr, Patrick. *The Illustrated History of Country Music*. New York: Doubleday, 1980.

Cashbox. 1942–.

Castle, Irene. *Castles in the Air*. Garden City, NY: Doubleday, 1958.

Castle, Vernon and Irene. *Modern Dancing by Mr. and Mrs. Vernon Castle*. New York: Harper, 1914.

Chapple, Steve, and Reebee Garofalo. *Rock 'n' Roll Is Here to Pay: The History and Politics of the Music Industry*. Chicago: Nelson-Hall, 1977.

Charles, Norman. "Social Values in American Popular Songs." Thesis. University of Pennsylvania. 1958.

Charles, Ray, and David Ritz. *Brother Ray: Ray Charles' Own Story*. New York: Dial, 1978.

Charters, Ann. *Nobody: The Story of Bert Williams*. New York: Macmillan, 1970.

Charters, Samuel. *The Bluesmen: The Story and the Music of the Men Who Made the Blues*. New York: Oak Publications, 1967.

———. *The Poetry of the Blues*. New York: Oak Publications, 1963.

Charters, Samuel, with Leonard Kunstadt. *Jazz: A History of the New York Scene*. New York: Doubleday, 1962.

Chase, Gilbert. *America's Music: From the Pilgrims to the Present*. New York: McGraw-Hill, 1955.

———, ed. *Music in Radio Broadcasting*. McGraw-Hill, New York: 1946.

Chilton, John. *Who's Who of Jazz: Storyville to Swing Street*. Chicago: Time-Life Books, 1978.

Christgau, Robert. *Any Old Way You Choose It: Rock and Roll and Other Pop Music*. Baltimore: Penguin Books, 1973.

Churchill, Allen. *The Great White Way: A Re-creation of Broadway's Golden Days of Theatrical Entertainment*. New York: Dutton, 1962.

———. *Remember When: A Loving Look at Days Gone By, 1900–1942*. New York: Golden Press, 1967.

Claghorn, Charles Eugene. *Biographical Dictionary of American Music*. Nyack, NY: Parker Publishing, 1973.

Clark, Dick, and Richard Robinson. *Rock, Roll and Remember*. New York: Crowell, 1976.

Clarke, Garry E. *Essays on American Music*. Westport, CT: Greenwood Press, 1977.

Clarke, Norman. *The Mighty Hippodrome*. New York: Barnes & Noble, 1968.

Coad, Oral Sumner, and Edwin Mimms. *The American Stage*. New York: U.S. Publishers, 1929.

Cohan, George M. *Twenty Years on Broadway and the Years It Took to Get There*. New York: Harper, 1925.

Cohen, Nathan. "State Regulation of Musical Copyright." *Oregon Law Review*, Summer 1938.

Cohen, Norm. Liner notes for "Minstrels and Tunesmiths: The Commercial Roots of Early Country Music, 1902–1923." John Edwards Memorial Foundation LP 109.

Cohn, David L. *The Good Old Days*. New York: Simon & Schuster, 1940.

Cohn, Nik. *Rock from the Beginning*. New York: Pocket Books, 1969.

Collier, James Lincoln. *The Making of Jazz: A Comprehensive History*. New York: Dell, 1978.

Conant, Michael. *Antitrust in the Motion Picture Industry*. Berkeley: University of California Press, 1960.

Condon, Eddie, and Hank O'Neal. *Eddie Condon's Scrapbook of Jazz*. New York: St. Martin's Press, 1973.

The Congressional Record. Washington D.C. Government Printing Office, 1909–1985.

Cook, Bruce. *Listen to the Blues*. London: Robson Books, 1975.

Coon, Caroline. *The New Wave Punk Rock Explosion*. London: Orbach & Chambers, 1977.

Coon, O. Wayne. *Some Problems with Musical Public-Domain Material Under United States Copyright Law as Illustrated Mainly by the Recent Folk-Song Revival. Copyright Law Symposium Number Nineteen*. New York: Columbia University Press, 1971.

Cooper, Al, with Ben Edmonds. *Backstage Passes: Rock 'n' Roll Life in the 60s*. New York: Stein & Day, 1977.

Copyright Enactments: Laws Passed in the United States Since 1793 Relating to Copyright. Washington, D.C.: Library of Congress, 1978.

Corio, Ann, and Joseph DiNona. *This Was Burlesque*. New York: Grosset & Dunlap, 1968.

Corry, Catherine S. *The Phonograph Record Industry: An Economic Survey*. Washington, D.C.: Library of Congress Legislative Reference Service, 1965.

Coslow, Sam. *Cocktails for Two: The Many Lives of a Great Songwriter*. New Rochelle, NY: Arlington House, 1977.

Crosby, Harry Lillis. *Call Me Lucky: An Autobiography*. New York: Simon & Schuster, 1953.

———. "My Kind of Music Is Coming Back." *This Week*, April 24, 1960.

Crowther, Bosley. *The Lion's Share: The Story of an Entertainment Empire*. New York: Dutton, 1957.

Csida, Joseph, and June Bundy Csida. *American Entertainment: A Unique History of Popular Show Business*. New York: Watson-Guptill, 1978.

Cummings, Tony. *The Sound of Philadelphia*. London: Methuen, 1975.

Cuney-Hare, Maud. *Negro Musicians and Their Music*. Washington, D.C.: Associated Publishers, 1936.

Dachs, David. *Anything Goes: The World of Popular Music*. Indianapolis: Bobbs-Merrill, 1964.

———. "Thunder Over Tin Pan Alley." North American Newspaper Alliance, August 17–21, 1957.

Dalton, David, and Lennie Kaye. *Rock 100*. New York: Grosset & Dunlap, 1977.

Dance, Stanley. *The World of Duke Ellington*. New York: Scribner, 1970.

———. *The World of Earl Hines*. New York: Scribner, 1977.

Dannett, Sylvia, and Frank R. Rachel. *Down Memory Lane: Arthur Murray's Picture History of Social Dancing*. New York: Greenburg, 1954.

Davies, Hunter. *The Beatles*. New York: McGraw-Hill, 1968.

Davis, Clive, with William Willwerth. *Clive: Inside the Record Business*. New York: Morrow, 1975.

Davis, Steven. *Bob Marley*. New York: Doubleday, 1985.

———. *Reggae Bloodlines: In Search of the Music and Culture of Jamaica*. New York: Anchor Books, Doubleday, 1977.

Debas, Allen G. Liner notes for "The Early Victor Herbert, from the Gay Nineties to the First World War." Smithsonian LP 30366.

DeForest, Lee. *Father of Radio: The Autobiography of Lee DeForest*. Chicago: Wilcox & Follet, 1950.

DeKoven, Anna. *A Musician and His Life*. New York: Harper, 1926.

De Long, Thomas A. *The Mighty Music Box: The Golden Age of Musical Radio*. Los Angeles: Amber Crest Books, 1980.

———. *Pops: Paul Whiteman, King of Jazz*. Piscataway, NJ: New Century Publishers, 1985.

Denisoff, R. Serge. *Great Day Coming: Folk Music and the American Left*. Baltimore: Penguin Books, 1973.

———. *Solid Gold: The Popular Record Industry*. New Brunswick, NJ: Transaction Books, 1975.

Denisoff, R. Serge, and Richard A. Peterson, eds. *The Sounds of Social Change: Studies In Popular Culture*. Chicago: Rand McNally, 1972.

Dennison, Sam. *Scandalize My Name: Black Imagery in American Popular Music*. New York: Garland Publishing, 1982.

Dethlefson, Ron, ed. *Edison Blue Amberol Records 1912–1914*. Brooklyn: APM Press, 1980.

———. *Edison Blue Amberol Records 1915–1929*. Brooklyn: APM Press, 1981.

DeTurk, David A., and A. Poulin, Jr., eds. *The Folk Scene: Dimensions of the Folksong Revival*. New York: Dell, 1967.

DeWhitt, Bennie Lee. "The American Society of Composers, Authors and Publishers 1914–1938." Diss. Emory University, 1977.

Dilello, Richard. *The Longest Cocktail Party: A Personal History of Apple*. Chicago: Playboy Press, 1972.

DiMaggio, Paul, Richard A. Peterson, and Jack Esco, Jr. "Country Music: Ballad of the Silent Majority." In *The Sounds of Social Change*. See Denisoff and Peterson.

DiMeglio, John E. *Vaudeville, USA*. Bowling Green, OH: Bowling Green State University Press, 1973.

Dixon, Robert, and John Goodrich. Recording the Blues. London: Studio Books, 1970.

Dowd, Jerome. *The Negro in American Life*. Chicago: Century, 1926.
Dragonette, Jessica. *Faith Is a Song*. New York: David McKay, 1951.
Dranov, Paula. *The Music Publishing Business 1978–1983*. White Plains, NY: Knowledge Industry Publications, 1977.
Drew, Joan. *Singers and Sweethearts: The Women of Country Music*. New York: Doubleday, 1977.
Dubin, Al. *The Art of Songwriting*. New York: Mills Music, 1928.
Duke, Vernon. *Listen Here*. New York: Oblensky, 1963.
Dulles, Foster Rhea. *A History of Recreation: America Learns to Play*. New York: Appleton-Century-Crofts, 1965.
Dunlap, Orrin, E. *The Story of Radio*. New York: Dial, 1935.
Dunleavy, Steven. *Elvis, What Happened?* New York: Ballantine, 1977.
Dunn, Don. *The Making of "No, No Nanette."* Secaucus, NJ: Citadel Press, 1972.
Dunning, John. *Tune in Yesterday: The Ultimate Encyclopedia of Old-Time Radio, 1925–1976*. Englewood Cliffs, NJ: Prentice-Hall, 1976.
Edelman, Jacob M. *The Licensing of Radio Services in the United States 1927–1947*. Urbana: University of Illinois Press, 1950.
Ehrenberg, Lewis Allan. "Urban Night Life and the Decline of Victorianism: New York City's Restaurants and Cabarets 1890–1918." Diss. University of Michigan, 1974.
Eisen, Jonathan, ed. *The Age of Rock: Sounds of the American Cultural Revolution*. New York: Vintage Books, Random House, 1969.
———. *The Age of Rock* 2. New York: Vintage Books, Random House, 1970.
———. *Twenty-Minute Fandangos and Forever Changes*. New York: Vintage Books, Random House, 1971.
Emery, Lynn. *Black Dance in the United States 1619–1970*. Palo Alto, CA: National Press Books, 1972.
Engel, Lehman. *The American Musical Theatre: A Reconsideration*. New York: CBS Records, 1967.
———. *Words With Music*. New York: Macmillan, 1972.
The Entertainment Industry: A Survey of Financial Reporting and Accounting Developments in 1975. New York: Price, Waterhouse, 1976.
Erickson, Don. *Armstrong's Fight for FM Broadcasting: One Man vs. Big Business and Bureaucracy*. University, AL: University of Alabama Press, 1973.
Etzkorn, Klaus Peter. "Musical and Social Patterns of Songwriters: An Exploratory Sociological Study." Diss. Princeton University, 1959.
Escott, Colin, and Michael Hawkins. *Catalyst: The Sun Records Story*. New York: Aquarius Books, 1975.
Ewen, David. *All the Years of American Popular Music: A Comprehensive History*. Englewood Cliffs, NJ: Prentice-Hall, 1978.
———. *The Life and Death of Tin Pan Alley*. New York: Funk & Wagnalls, 1964.
———. *Great Men of Popular Music*. Englewood Cliffs, NJ: Prentice-Hall, 1970.
Federal Communications Commission. *Reports*. Washington, D.C.: GPO, 1935–.
Feist, Leonard. *Introduction to Popular Music Publishing*. New York: National Music Publishers Association, 1980.
Finkelstein, Herman. "The Composer and the Public Interest." *Law and Contemporary Problems*, Spring 1954.

———. "The Copyright Law: A Reappraisal," *University of Pennsylvania Law Review*, June 1956.

———. "Public Performance Rights in Music and Performance Rights Societies." 7 *Copyright Problems Analyzed*. New York: Commercial Clearing House, 1952.

Finnis, Rob. *The Phil Spector Story*. London: Rockton, 1975.

Fisher, William Arms. *One Hundred and Fifty Years of Music Publishing in the United States, 1783–1933*. Boston: Ditson, 1933.

Fleming, Len. *Book of Information for Songwriters and Composers*. New York: Leo Feist, 1913.

———. *Dollars and Sense: A Fortune in Popular Songs*. New York: Leo Feist, 1912.

Fletcher, Tom. *The Tom Fletcher Story: 100 Years of the Negro in Show Business*. New York: Burge, 1954.

Foreman, Ronald Clifford. "Jazz and Jazz Records 1920–1932: Their Origins and the Significance for the Record. Diss. University of Illinois, 1968.

Franklin, John Hope. *From Slavery to Freedom: A History of Negro Americans*. New York: Random House, 1969.

Freedland, Michael. *Irving Berlin*. New York: Stein & Day, 1974.

———. *Jolson*. New York: Stein & Day, 1972.

Freeman, Larry. *The Melodies Linger On: Fifty Years of Popular Song*. Watkins Glen, NY: Century House, 1951.

Frith, Simon. *Sound Effects: Youth, Leisure, and the Politics of Rock 'n' Roll*. New York: Pantheon, 1981.

Fuld, James J. *American Popular Music 1875–1950*. Philadelphia: Musical Americana, 1956.

Furnas, J. C. *The Americans: A Social History of the United States*. New York: Putnam, 1969.

Gabree, John. *The World of Rock*. Greenwich, CT: Fawcett, 1968.

Gaillard, Frye. *Watermelon Wine: The Spirit of Country Music*. New York: St. Martin's, 1978.

Gaisberg, Fred. *The Music Goes Round*. New York: Macmillan, 1942.

Gammond, Peter, ed.: *Duke Ellington, His Life and Music*. New York: Roy Publishers, c. 1958.

Garland, Phyl. *The Sound of Soul*. Chicago: Regnery, 1969.

Garraty, John A. *The American Nation*. New York: Harper & Row, 1971.

Geijerstam, Claes. *Popular Music in Mexico*. Albuquerque: University of New Mexico Press, 1976.

Gelatt, Roland. *The Fabulous Phonograph, 1877–1977*. New York: Macmillan, 1977.

Gershwin, Ira. *Lyrics on Several Occasions*. New York: Knopf, 1959.

Gersohn, Frederic, ed. *Counseling Clients in the Performing Arts*. New York: Practising Law Institute, 1975.

Gilbert, Douglas. *American Vaudeville: Its Life and Times*. New York: Dover, 1963.

———. *Lost Chords: The Diverting Story of American Popular Songs*. New York: Cooper Square, 1970.

Gilbert, L. Wolfe. *Without Rhyme or Reason*. Hollywood: Vantage, 1956.

Gillespie, Dizzy, with Al Fraser. *To Be or Not to Bop: Memoirs*. New York: Doubleday, 1979.

Gillett, Charlie. *Making Tracks: Atlantic Records and the Growth of a Multi-Billion-Dollar Industry*. New York: Dutton, 1974.

———. *The Sound of the City: The Rise of Rock and Roll*. New York: Outerbridge & Dienstfrey, 1970.

Gillett, Charlie, and Simon Frith. *Rock File 3: Sources of British Hit Songs, Writers and American Hits*. London: Panther Books, 1975.

Giovannoni, David. "The Phonograph as a Mass Entertainment Medium: Its Development, Adaptation and Pervasiveness." Thesis. University of Wisconsin-Madison, 1980.

Gleason, Ralph. *Jam Session: An Anthology of Jazz*. New York: Putnam, 1958.

———. *The Jefferson Airplane and the San Francisco Sound*. New York: Ballantine, 1969.

Goldberg, Isaac. *Tin Pan Alley: A Chronicle of American Popular Music*. New York: Frederick Ungar, 1931; paper, 1970.

Goldmark, Peter. *Maverick Inventor: My Turbulent Years at CBS*. New York: Saturday Review Press, 1973.

Goldstein, Richard. *Goldstein's Greatest Hits*. New York: Tower, 1970.

———. *The Poetry of Rock*. New York: Bantam, 1969.

Goodman, Paul. *Growing Up Absurd: Problems of Youth in the Organized Society*. New York: Vintage, 1960.

Goodman, Robert Israel. "Music Copyright Associations and the Antitrust Law." *Indiana Law Journal*, Fall 1950.

Goodrich, John, and Robert M. W. Dixon. *Blues and Gospel Records, 1902–1942*. London: Storyville, 1969.

Goss, Madeleine. *Modern Music Makers: Contemporary American Composers*. New York: Dutton, 1952.

Gottlieb, Polly Rose. *The Nine Lives of Billy Rose*. New York: Crown, 1968.

Grau, Robert. *The Theatre of Science: A Volume of Progress and Achievement in the Motion Picture Art*. New York: Broadway Publishing, 1914.

Green, Abel. *Inside Stuff on How to Write Popular Songs*. New York: Paul Whiteman Publications, 1927.

Green, Abel, with Joe Laurie, Jr. *Show Biz: From Vaude to Video*. New York: Henry Holt, 1951.

Green, Archie. *Only a Miner: Studies in Recorded Coal-Mining Songs*. Urbana: University of Illinois Press, 1972.

Green, Douglas B. *Country Roots: The Origins of Country Music*. New York: Hawthorn Books, 1976.

Green, Stanley. Liner notes for "Ziegfeld Follies of 1919." Smithsonian LP 14272.

———. *The World of Musical Comedy*. New York: A. S. Barnes, 1980.

Greenfield, Jeff. *No Peace, No Place*. Gordon City, NY: Doubleday, 1973.

Grissim, John. *Country Music: White Man's Blues*. New York: Coronet, 1970.

Groia, Philip. *They All Sang on the Corner: New York City's Rhythm & Blues Vocal Groups of the 1950s*. New York: Edmond, 1973.

Groom, Bob. *The Blues Revival*. London: Studio Books, 1971.

Guralnik, Peter. *Feel Like Going Home: Portraits in Blues and Rock 'n' Roll*. New York: Outerbridge & Dienstfrey, 1971.

Guthrie, Woody, and Robert Shelton. *Born to Win*. New York: Macmillan, 1965.

Gutman, Herbert G. *The Black Family in Slavery and Freedom, 1750–1925*. New York: Pantheon, 1976.

Hackett, Alice Payne. *Sixty Years of Best Sellers*. New York: R. R. Bowker, 1956.

Hadlock, Richard. *Jazz Masters of the 20s*. New York: Macmillan, 1965.

Hamm, Charles. *Yesterdays: Popular Songs in America*. New York: Norton, 1979.

Hammond, John. *John Hammond on Record*. New York: Ridge Press, 1977.

Hampton, Benjamin A. *A History of the American Film Industry*. New York: Convici Friede, 1931.

Hansen, Barret Eugene. "Negro Popular Music 1945–1953." Thesis. University of California at Los Angeles, 1967.

Haralambos, Michael. *Right On: From Blues to Soul in Black America*. London: Eddison Press, 1974.

Harbach, Otto A. "Affidavit and Supplementary Affidavits Regarding a Conspiracy Between Broadcasters and Broadcast Music Incorporated, 1951, 1953." Typescript.

Harlow, Alvin F. *Old Wires and New Waves: The History of the Telegraph, Telephone and Wireless*. New York: Appleton-Century, 1936.

Harris, Hebert, *American Labor*. New Haven, CT: Yale University Press, 1939.

Harris, Herby, and Lucien Farrar. *How to Make Money in Music*. New York: Arco Publishing, 1978.

Harris, Sheldon. *Blues Who's Who: A Biographical Dictionary of Blues Singers*. New Rochelle, NY: Arlington House, 1979.

Harrison, Hank. *The Dead Book: A Social History of the Grateful Dead*. New York: Links Books, 1973.

Hart, Dorothy. *Thou Swell, Thou Witty*. New York: Harper & Row, 1976.

Hart, Philip. *Orpheus in the New World: The Symphony Orchestra as an American Cultural Institution*. New York: Norton, 1973.

Haskins, James. *The Cotton Club: A Pictorial and Social History of the Most Famous Symbol of the Jazz Era*. New York: Random House, 1977.

———. *Scott Joplin*. New York: Doubleday, 1978.

Hatch, James, and Omanii Abdullah. *Black Playwrights 1823–1977*. New York: R. R. Bowker, 1977.

Haverlin, Carl. "Affidavit . . . in Opposition to ASCAP's Motion for an Order Modifying Amended Final Judgment of March 14, 1950." Typescript.

Hayakawa, S. I. "Popular Songs vs. The Facts of Life." In Rosenberg, Bernard, and David Manning White, eds. *Mass Culture: The Popular Arts in America*. New York: Free Press, 1957.

Hays, Will H. *See and Hear: A Brief History of Motion Pictures and the Development of Sound*. New York: Motion Picture Producers and Distributors of America, 1929.

Heilbut, Tony. *The Gospel Sound: Good News and Bad Times*. New York: Simon & Schuster, 1971.

———. Liner notes for "Precious Lord: New Recordings of the Great Songs of Thomas A. Dorsey." Columbia LP KG 32151.

Hemphill, Paul. *Bright Lights and Country Music: The Nashville Sound*. New York: Simon & Schuster, 1970.

Henn, Harry. "The Compulsory License Provision of the U.S. Copyright Law." *Copyright Law Revision,* Washington, D.C.: Library of Congress, 1960.

Hentoff, Nat. *The Jazz Life*. New York: Dial, 1961.

Hentoff, Nat, with Albert McCarthy, eds. *Jazz: New Perspectives on the History of Jazz by 12 of the World's Foremost Critics and Scholars*. New York: Rinehart, 1959.

Herman, Pinky. *Showbiz and Me*. Lauderdale Lakes, FL: Manor Music, 1977.

Herndon, Booton. *The Sweetest Music This Side of Heaven*. New York: McGraw-Hill, 1964.

Higham, Charles. *Ziegfeld*. Chicago: Regnery, 1972.

Hirsch, Paul. *The Structure of the Popular Music Industry: The Filtering Process by Which Records Are Presented for Public Consumption*. Ann Arbor: Institute for Social Research, University of Michigan, 1970.

Hitchcock, H. Wiley. *Music in the United States: An Historical Introduction*. Englewood Cliffs, NJ: Prentice-Hall, 1969.

Hodier, Andre. *Jazz: Its Evolution and Essence*. New York: Grove Press, 1956.

Hoffman, Charles. *Sounds for Silents*. New York: Drama Book Specialists Publications, 1970.

Hopkins, Jerry. *Elvis: A Biography*. New York: Simon & Schuster, 1971.

———. *Elvis: The Final Years*. New York: St. Martin's Press, 1980.

Hornblow, Arthur. *A History of the Theatre in America, from Its Beginnings to the Present Time*. 2 vols. New York: Benjamin Blom, 1965. Reprint of 1919 ed.

Horstman, Dorothy. *Sing Your Heart Out, Country Boy*. New York: Dutton, 1975.

Howard, John Tasker. *Our American Music: A Comprehensive History from 1620 to the Present*. 4th ed. New York: Crowell, 1965.

Hubbell, Raymond. *From Nothing to Five Million a Year: The Story of ASCAP by a Founder*. Washington, D.C.: Library of Congress. Mimeo.

Hughes, Langston, and Milton Meltzer. *Black Magic: A Pictorial History of Black Entertainers in America*. New York: Bonanza, 1967.

Hurst, Jack. *Nashville's Grand Ole Opry*. New York: Abrams, 1975.

Hurst, Walter E. *The Music Industry Book*. Hollywood: Seven Arts Press, 1963.

International Research Associates. *A Guide for the Interpretation of the Musical Popularity Charts*. New York: Broadcast Music, Inc., 1956.

Jablonski, Edward. *The Encyclopedia of American Music*. New York: Doubleday, 1981.

Jacobs, Lewis. *The Rise of the American Film: A Critical History*. New York: Teachers College Press, 1939.

Jahn, Mike. *Rock: From Elvis Presley to the Rolling Stones*. New York: Times Books, 1973.

Jewell, Derek. *Duke: A Portrait of Duke Ellington*. New York: Norton, 1977.

John Edwards Memorial Foundation Quarterly. 1966–.

Johnson, E. F. Fenimore. *His Master's Voice Was Eldridge R. Johnson*. Milford, DE: State Media Press, 1974.

Johnson, James Weldon. *Along This Way: An Autobiography*. New York: Viking, 1933.

———. *Black Manhattan*. New York: Knopf, 1930.

Johnson, J. Rosamund. *Rolling Along in Song*. New York: Viking, 1934.

Jones, LeRoi. *Blues People: Negro Music in White America*. New York: Morrow, 1963.

Journal of Country Music. 1970–.

Journal of Jazz Studies. 1974–.

Jowett, Garth. *Film, the Democratic Art: A Social History of the American Film.* Boston: Little, Brown, 1976.

Kane, Henry. *How to Write a Song.* New York: Macmillan, 1962.

Karlin, Fred. *Edison Diamond Discs 50001–50651, 1912–1929.* Santa Monica, CA: Bona Fide, 1972.

Kaufmann, Helen L. *From Jehovah to Jazz: Music in America from Psalmody to the Present Day.* New York: Dodd, Mead, 1937.

Kaye, Sydney M. "A Blue-print for Broadcast Music, Inc." Prepared for the National Association of Broadcasters, September 8, 1939. Typescript.

Keil, Charles. *Urban Blues.* Chicago: University of Chicago Press, 1966.

Kenny, Nick. *How to Write, Sing and Sell Popular Songs.* New York: Hermitage Press, 1946.

Kimball, Robert, and William Bolcom. *Reminiscing with Sissle and Blake.* New York: Viking, 1973.

Kirschner, Roger. *The Music Machine.* Los Angeles: Nash Publications, 1971.

Kislan, Richard. *The Musical: A Look at the American Musical Theatre.* Englewood Cliffs, NJ: Prentice-Hall, 1980.

Knight, Arthur. *The Liveliest Art: A Panoramic History of the Movies.* New York: Macmillan, 1957.

Koenigsberg, Allan. *Edison Cylinder Records 1899–1912, with an Illustrated History of the Phonograph.* New York: Stellar Productions, 1969.

Kolodin, Irving. *The Musical Life.* New York: Knopf, 1958.

Krivine, John. *Jukebox Saturday Night.* Secaucus, NJ: Chartwell Books, 1977.

Krueger, Miles. *Showboat: The Story of a Classic American Musical.* New York: Oxford University Press, 1977.

Krummel, Donald W. "Counting Every Star, or Historical Statistics on Music Publishing in the United States." In *Interamerican Musical Research Yearbook 1974.*

Laing, Dave, ed. *The Electric Muse: The Story of Folk into Rock.* London: Methuen, 1975.

Lamb, Andrew. *Jerome Kern in Edwardian England.* Brooklyn: Institute for Studies in American Music, 1985.

Lambert, Dennis, with Ronald Zalkind. *Producing Hit Records.* New York: Schirmer, 1980.

Landau, Jon. *It's Too Late Now: A Rock 'n' Roll Journal.* San Francisco: Rolling Stone Books, 1972.

Landry, Robert J. *This Fascinating Radio Business.* Indianapolis: Bobbs-Merrill, 1956.

Lane, Burton. "Memorandum to House Special Subcommittee on Legislative Oversight Regarding Commercial Influences on the Selection and Promotion of Music for Radio and Television." October 29, 1959.

Lang, Paul Henry. *Music in Western Civilization.* New York: Norton, 1941.

Lang, Paul Henry, ed. *One Hundred Years of Music in America: A Centennial Publication on the Anniversary of G. Schirmer & Co.* New York: Grosset & Dunlap, 1960.

La Prade, Ernest. *Broadcasting Music.* New York: Stewart, 1942.

Larkin, Rochelle. *Soul Music.* New York: Lancer Books, 1970.

Larrabee, Eric, and Rolf Meyersohn, eds. *Mass Leisure.* Glencoe, IL: Free Press, 1958.

Laurie, Joe, Jr. *Vaudeville, from the Honky Tonks to the Palace*. New York: Henry Holt, 1953.

Lawless, Ray M. *Folksingers and Folksongs in America: A Handbook of Biography, Bibliography and Discography*. New York: Duell, Sloan and Pearce, 1965.

Lazarsfeld, Paul F., and Frank N. Stanton, eds. *Radio Research 1941*. New York: Duell, Sloan and Pearce, 1941.

———. *Radio Research 1942–43*. New York: Duell, Sloan and Pearce, 1944.

Ledbitter, Mike. *Delta Country Blues*. Bexhill-on-Sea: Blues Unlimited, 1969.

———. *From the Bayou: The Story of Goldband Records*. Bexhill-on-Sea: Blues Unlimited, 1969.

Ledbitter, Mike, ed. *Nothing but the Blues*. London: Hanover Books, 1971.

Lederman, Minna. *The Life and Death of a Small Magazine*. Brooklyn: Institute for Studies in American Music, 1983.

Lee, Edward. *Music of the People: A Study of Popular Music in Great Britain*. London: Barrie & Jenkins, 1970.

Leiter, Robert D. *The Musicians and Petrillo*. New York: Bookman Associates, 1953.

Levine, Faye. *The Culture Barons: An Analysis of Power and Money in the Arts*. New York: Crowell, 1976.

Levine, Lawrence W. *Black Culture and Black Consciousness: Afro-American Folk Thought from Slavery to Freedom*. New York: Oxford University Press, 1977.

Levy, Lester M. *Give Me Yesterday: American History in Music 1890–1920*. Norman: University of Oklahoma Press, 1975.

———. *Grace Notes in American History: Popular Sheet Music 1820–1900*. Norman: University of Oklahoma Press, 1967.

Lichter, Paul. *Elvis in Hollywood*. New York: Simon & Schuster, 1975.

Limbacher, James L., ed. *Film Music: From Violins to Video*. Metuchen, NJ: Scarecrow Press, 1974.

Lomax, Alan. *Mister Jelly Roll*. New York: Duell, Sloan and Pearce, 1950.

Lomax, John A. *Adventures of a Ballad Hunter*. New York: Macmillan, 1947.

Lombardo, Guy. *Auld Acquaintance*. Garden City, NY: Doubleday, 1976.

London, Kurt. *Film Music: A Summary of the Characteristics of Its History*. London: Faber & Faber, 1936.

Longley, Marjorie, Louis Silverstein, and Samuel A. Tower. *America's Taste 1851–1959: The Cultural Events of a Century Reported by Contemporary Observers in the Pages of the New York Times*. New York: Simon & Schuster, 1960.

Lopez, Vincent. *Lopez Speaking: My Life and How I Changed It*. Secaucus, NJ: Citadel Press, 1960.

Lowenthal, Daniel K. *Trends in the Licensing of Popular Song Hits 1940–1953*. New York: Bureau of Applied Social Research, Columbia University, 1953.

Luening, Otto. *The Odyssey of an American Composer*. New York: Scribner, 1980.

Lujack, Larry. *Super Jock: The Loud, Frantic, Nonstop World of a Radio DJ*. Chicago: Regnery, 1975.

Lydon, Michael. *Rock File: Portraits from the Rock 'n' Roll Musical Pantheon*. New York: Delta Books, Dell, 1973.

Lydon, Michael, and Ellen Mandel. *Boogie Lightning: How Music Becomes Electric*. New York: Dial, 1974.

Lyons, Eugene. *David Sarnoff: A Biography*. New York: Harper, 1966.

McCabe, John. *George M. Cohan, the Man Who Owned Broadway*. Garden City, NY: Doubleday, 1973.

McCabe, Peter, and Robert D. Schonfeld. *Apple to the Core: The Unmaking of the Beatles*. New York: Pocket Books, 1972.

McCarthy, Albert. *The Dance Band Era: The Dancing Decades from Ragtime to Swing*. London: Spring Books, 1974.

McCarthy, Albert, with Max Harrison. *Jazz on Record 1917–1967*. New York: Oak Publications, 1968.

McCarthy, Clifford, ed. *Film Composers in America: A Checklist of Their Work*. New York: Da Capo Press, 1972.

McCarthy, Todd, and Charles Flynn, eds. *Kings of the Bs: Working Within the Hollywood System: An Anthology of Film History and Criticism*. New York: Dutton, 1975.

MacDougald, Duncan, Jr. "The Popular Music Industry." In Lazarsfeld Paul F., and Frank N. Stanton, eds. *Radio Research 1941*. New York: Duell, Sloan & Pearce, 1941.

MacFarland, David Thomas. "The Development of the Top 40 Radio Format." 2 vols. Diss. University of Wisconsin, 1972.

MacKay, David R. "The National Association of Broadcasters: Its First 20 Years." Diss. Northwestern University, 1956.

MacLaughlin, M. C. "The Social World of American Popular Songs." Thesis. Cornell University, 1968.

McLean, Albert, F., Jr. *American Vaudeville as Ritual*. Lexington: University Press of Kentucky, 1965.

Malone, Bill C. *Country Music USA: A Fifty-Year History*. Austin: University of Texas Press, 1968.

———. Liner notes for the Smithsonian Collection of Classic Country Music. Washington, D.C.: Smithsonian Institution P8 #16450. 1983.

———. *Southern Music: American Music*. Lexington: University Press of Kentucky, 1979.

Malone, Bill C., with Judith McCulloh, eds. *The Stars of Country Music: Uncle Dave Macon to Johnny Rodriguez*. New York: Avon Books, 1976.

Manchester, William. *The Glory and the Dream: A Narrative History of America 1932–1972*. Boston: Little, Brown, 1972.

Mann, May. *Elvis and the Colonel*. New York: Pocket Books, 1976.

Marcus, Greil. *Mystery Train: Images of America in Rock 'n' Roll Music*. New York: Dutton, 1976.

———, ed. *Stranded: Rock and Roll for a Desert Island*. New York: Knopf, 1979.

Marks, J. *Rock and Other Four-Letter Words*. New York: Bantam, 1968.

Marston, William Moulton, and John Henry Fuller. *F. F. Proctor: Vaudeville Pioneer*. New York: Richard A. Smith, 1943.

Marx, Samuel, and Jan Clayton. *Rodgers & Hart: A Dual Biography*. New York: Putnam, 1976.

Matlaw, Myron, ed. *American Popular Entertainment: Papers and Proceedings of the Conference on the History of American Popular Entertainment*. Westport, CT: Greenwood Press, 1979.

Mattfield, Julius. *Variety Music Cavalcade: A Musical-Historical Review 1620–1969*. 3rd ed. Englewood Cliffs, NJ: Prentice-Hall, 1971.

Mellers, Wilfrid. *Music in a New Found Land: Themes and Developments in the History of American Music*. London: Barrie & Rocklif, 1964.

Melly, George. *Revolt Into Style: The Pop Arts*. New York: Anchor Books, Doubleday, 1971.

Metronome. 1871–1956.

Meyer, Hazel. *The Gold in Tin Pan Alley*. Philadelphia: Lippincott, 1958.

Michel, Trudi. *Inside Tin Pan Alley*. New York: Frederick Fell, 1948.

Middleton, Richard. *Pop Music and the Blues: A Study of the Relationship and Its Significance*. London: Gollancz, 1972.

Millar, Bill. *The Drifters*. New York: Collier Books, 1971.

Miller, Jim, ed. *The Rolling Stone Illustrated History of Rock 'n' Roll*. New York: Rolling Stone Press/Random House, 1976.

Miller, Manfred, Klaus Kuhnke, and Peter Schulze. *Geschichte der Pop-Musik Band 1 (Bis 1947)*. Bremen: Archiv für Populär Musik, 1976.

Mitchell, Loften. *Black Drama: The Story of the American Negro in the Theatre*. New York: Hawthorn Books, 1967.

Moogk, Edward B. *Roll Back the Years: A History of Canadian Recorded Sound and Its Legacy, Genesis to 1930*. Ottawa: National Library of Canada, 1975.

Moore, Grave. *You're Only Human Once*. New York: Doubleday, Doran, 1944.

Moore, Jerrold Northrup. *A Matter of Records: Fred Gaisberg and the Golden Era of the Gramophone*. New York: Taplinger, 1977.

Mordden, Ethan. *Better Foot Forward: The History of the American Musical Theatre*. New York: Grossman, 1976.

Morgenstern, Dan. Liner notes for "Souvenirs of Hot Chocolates." Smithsonian LP 14587.

Morris, Lloyd. *Not So Long Ago*. New York: Random House, 1949.

———. *Postscript to Yesterday*. New York: Random House, 1947.

Morse, Dave. *Motown and the Arrival of Black Music*. New York: Collier Books, 1972.

Mueller, John H. *The American Symphony Orchestra*. Bloomington: Indiana University Press, 1951.

Murray, Albert. *Stomping the Blues*. New York: McGraw-Hill, 1976.

Myrus, Donald. *Ballads, Blues and the Big Beat: Highlights of American Folk Singing from Leadbelly to Dylan*. New York: Macmillan, 1966.

Nanry, Charles, ed. *American Music: From Storyville to Woodstock*. New Brunswick, NJ: Transaction Books, 1972.

Nash, Roderick. *The Call of the Wild 1900–1916*. New York: Braziller, 1970.

National Association of Broadcasters (NAB). *Let's Stick to the Record*. Washington, D.C.: NAB, 1940.

———. *A Music Monopoly Is Reaching for Your Pocketbook*. Washington, D.C.: NAB, 1940.

———. *NAB Reports*. Washington, D.C.: NAB, 1933–.

———. *Portrait of a Protector*. Washington, D.C.: NAB, 1940.

National Music Publishers' Association Bulletin. 1970–.

Neale, A. D. *The Antitrust Laws of the United States of America: A Study of Competition Enforced by Law*. New York: Cambridge University Press, 1970.

Nelson, Ozzie. *Ozzie*. Englewood Cliffs, NJ: Prentice-Hall, 1973.

Nite, Norm K. *Rock On: The Illustrated Encyclopedia of Rock 'n' Roll*. New York: Crowell, 1974.

Noebel, David A. *Rhythm, Riots and Revolution*. Tulsa, OK: Christian Crusade Publications, 1966.

Oakley, Giles. *The Devil's Music: A History of the Blues*. New York: Harcourt Brace Jovanovich, 1978.

O'Connor, Eileen V. "Anti-ASCAP Legislation and Its Judicial Interpretation." *George Washington Law Review*, April 1941.

Offen, Carol. *Country Music: The Poetry*. New York: Ballantine, 1977.

Oliver, Paul. *Aspects of the Blues Tradition: A Fascinating Story of the Richest Vein of Black Music in America*. New York: Oak Publications, 1969.

———. *Blues Fell This Morning*. London: Cassell, 1960.

———. *Conversation with the Blues*. New York: Horizon Books, 1965.

———. *Savannah Syncopaters: African Retentions of the Blues*. London: Studio Books, 1970.

———. *Songsters and Saints: Vocal Tradition on Race Records*. New York: Cambridge University Press, 1984.

———. *The Story of the Blues*. New York: Chilton, 1969.

Olsson, Bengt. *Memphis Blues*. London: Studio Books, 1970.

Ord-Hume, Arthur W. J. G. *Player-Piano: The History of the Mechanical Piano*. New York: A. S. Barnes, 1970.

Ottley, Roi, and William J. Weatherby, eds. *The Negro in New York: An Informal Social History*. New York: New York Public Library, 1967.

Palmer, Robert. *Baby, That Was Rock & Roll: The Legendary Leiber & Stoller*. New York: Harcourt Brace Jovanovich, 1978.

———. *Deep Blues*. New York: Viking, 1981.

———. *The Rolling Stones*. New York: Rolling Stone Press, 1983.

Palmer, Tony. *All You Need Is Love: The Story of Popular Music*. New York: Grossman/Viking, 1976.

Parker, John W. "American Popular Music: An Emerging Field of Academic Study." Diss. University of Kentucky, 1962.

Passman, Arnold. *The Dee Jays*. New York: Macmillan, 1971.

Peacock, Alan, and Ronald Weir. *The Composer in the Market Place*. London: Faber & Faber, 1975.

Pearsall, Ronald. *Popular Music of the 20s*. London: David & Charles, 1976.

Peatman, John Gray. "Radio and Popular Music." In Lazarsfeld, Paul F., and Frank N. Stanton. *Radio Research 1942–43*. New York: Duell, Sloan and Pearce, 1944.

Perry, Dick. *Not Just a Sound: The Story of WLW*. Englewood Cliffs, NJ: Prentice-Hall, 1971.

Peterson, Lyman Ray. *Copyright in Historical Perspective*. Nashville: Vanderbilt University Press, 1968.

Peterson, Richard A. "Single-Industry Firm to Conglomerate Synergistics: Alternative Strategies for Selling Insurance and Country Music." In Blumstein, James, and Benjamin Walter, eds. *Growing Metropolis: Aspects of Development in Nashville*. Nashville: Vanderbilt University Press, 1975.

Pleasants, Henry. *The Great American Popular Singers*. New York: Simon & Schuster, 1974.

———. *Serious Music and All That Jazz.* New York: Simon & Schuster, 1969.

Pollack, Bruce. *In Their Own Words: Twenty Successful Songwriters Tell How They Write Their Songs.* New York: Collier Books, 1975.

Popular Music and Society. 1972–.

Porterfield, Nolan. *Jimmie Rodgers: The Life and Times of America's Blue Yodeler.* Urbana: University of Illinois Press, 1979.

Pratt, George C. *Spellbound in Darkness: A History of the Silent Film.* Greenwich, CT: New York Graphic Society, 1973.

Progriss, Jim. *The Language of Commercial Music.* New York: Hansen Books, 1976.

Radio Daily. 1936–1965.

Ramsaye, Terry. *A Million and One Nights: A History of the Motion Pictures Through 1925.* New York: Simon & Schuster, 1964.

Randle, William McK., Jr. "History of Radio Broadcasting and Its Social and Economic Effect on the Entertainment Industry. 3 vols. Diss. Western Reserve University, 1961.

Record Industry Association of America (RIAA). *Consumer Purchasing of Records and Prerecorded Tapes in the United States: A Three-Year Report 1979, 1980, 1981.* New York: RIAA, 1983.

———. *Issues Related to Market Expansion in the Recording Industry.* New York: RIAA, 1977.

———. *A Report to the Members of the RIAA.* New York: RIAA, 1953.

Record World. 1964–1982.

The Recorded Music Industry USA. New York: Frost & Sullivan, 1973.

Redd, Lawrence N. *Rock Is Rhythm and Blues.* East Lansing: Michigan State University Press, 1974.

Reed, Oliver, and Walter L. Welch. *From Tinfoil to Stereo: The Evolution of the Phonograph.* New York: Howard Sams/Bobbs-Merrill, 1959.

Ringer, Barbara. "Copyright in the 1980's." Sixth Annual Donald C. Brace Memorial Lecture. Washington, D.C.: 1976. Mimeo.

———. "Two Hundred Years of Copyright in America." Speech to Patent, Trademark, and Copyright Section of American Bar Association, August 10, 1976.

Roach, Hildred. *Black American Music.* Boston: Crescendo, 1973.

Roberts, John Storm. *Black Music of Two Worlds.* New York: Morrow, 1972.

———. *The Latin Tinge: The Impact of Latin American Music on the United States.* New York: Oxford University Press, 1979.

Rodgers, Mrs. Jimmie. *My Husband Jimmie Rodgers.* Nashville: Country Music Foundation Press, 1975. Reprint.

Rodgers, Richard. *Musical Stages: An Autobiography.* New York: Random House, 1975.

Rodnitzky, Jerome L. *Minstrels of the Dawn: The Folk-Protest Singer as a Cultural Hero.* Chicago: Nelson-Hall, 1976.

Roehl, Harvey. *Player Piano Treasury: The Scrapbook History of the Mechanical Piano in America.* Vestal, NY: Vestal Press, 1973.

Roffman, Frederick S. Liner notes for "Naughty Marietta." Smithsonian Collection, N 026, 1981.

Rogers, Eddy, as told to Mike Hennessey. *Tin Pan Alley.* London: Robert Hale, 1964.

Rolantz, Bob. *How to Get Your Song Recorded.* New York: Watson-Guptill, 1963.

Rolling Stone. *The Rolling Stone Interviews, Vol. 1.* New York: Warner Paperback Library, 1971.

———. *The Rolling Stone Interviews, Vol.* 2. New York: Warner Paperback Library, 1973.

———. *The Rolling Stone Library.* New York: Warner Paperbacks, 1974.

Romberg, Sigmund. " 'ASCAP', a Free and Open Discussion of Some of Its Difficulties, Together with a Few Remedies." *The Song Writers' Quarterly Bulletin,* April 1934.

Rosenberg, Bernard, and David Manning White, eds. *Mass Culture: The Popular Arts in America.* Glencoe, IL: Free Press, 1957.

Ross, Ted. *The Art of Music Engraving and Processing.* Miami: Hansen Books, 1970.

Roth, Ernst. *The Business of Music: Reflections of a Music Publisher.* London: Cassell, 1969.

Routt, Edd, James B. McGrath, and Frederic A. Weiss. *The Radio Format Conundrum.* New York: Hastings House, 1978.

Rowe, Mike. *Chicago Breakdown.* London: Eddison Press, 1973.

Roxon, Lillian. *Rock Encyclopedia.* New York: Grosset & Dunlap, 1969.

Russell, Ross. *Jazz Style in Kansas City and the Southwest.* Berkeley: University of California Press, 1971.

Russell, Tony. *Blacks, Whites and Blues.* London: Studio, 1970.

Rust, Brian. *The American Record Label Book: From the Nineteenth Century through 1942.* New Rochelle: Arlington House, 1979.

———. *Gramophone Records of the First World War.* North Pomfret, VT: David & Charles, 1975.

———. *The Victor Master Book.* Vol. 2. Highland Park, IL: Walter C. Allen, 1965.

Sampson, Henry T. *Blacks in Blackface: A Source Book on Early Black Musical Shows.* Metuchen, NJ: Scarecrow Press, 1980.

Sanjek, Russell. *From Print to Plastic: Publishing and Promoting America's Popular Music, 1900–1980.* Brooklyn: Institute for Studies in American Music, 1983.

———. "The War on Rock." *Downbeat Music '72.* Chicago: Maher Publications, 1972.

Sarlin, Robert. *Turn It Up (I Can't Hear the Words).* New York: Simon & Schuster, 1973.

Schicke, C. A. *Revolution in Sound: A Biography of the Recording Industry.* Boston: Little, Brown, 1974.

Schoener, Allon, ed. *Harlem on My Mind: Cultural Capitol of Black America 1900–1968.* New York: Random House, 1969.

Schuller, Gunther. *Early Jazz: Its Roots and Musical Development.* New York: Oxford University Press, 1968.

Seeger, Pete. *The Incomplete Folksinger.* New York: Simon & Schuster, 1972.

Seldes, Gilbert. *The Public Arts.* New York: Simon & Schuster, 1956.

———. *The Seven Lively Arts.* New York: Sagamore, 1957.

Shapiro, Nat, and Nat Hentoff. *Hear Me Talkin' to Ya.* New York: Rinehart, 1955.

———. *The Jazz Makers.* New York: Rinehart, 1957.

Shapiro, Nat, comp. *Popular Music: An Annotated Index of American Popular Songs, 1920–1969.* 6 vols. New York: Adrian Press, 1973.

Shaw, Arnold. *52nd Street: The Street That Never Slept*. New York: Coward, McCann, 1971.

———. *Honkers and Shouters: The Golden Years of Rhythm and Blues*. New York: Macmillan, 1978.

———. *The Lingo of Tin Pan Alley*. New York: Broadcast Music, Inc., 1960.

———. *The Rockin' 50s*. New York: Hawthorn Books, 1974.

———. *The Rock Revolution: What's Happening in Today's Music*. New York: Crowell-Collier, 1969.

———. *The World of Soul: Black America's Contribution to the Pop Music Scene*. New York: Cowles, 1970.

Shaw, Greg. *The Rolling Stone Illustrated History of Rock & Roll*. New York: Rolling Stone Press, and Random House, 1976.

Shemel, Sidney, and M. William Krasilovsky. *More About This Business of Music*. New York: Billboard Publications, 1978.

———. *This Business of Music*. New York: Billboard Publications, 1964, 1971, 1977, 1979.

Short, Bobby. *Black and White Baby*. New York: Dodd, Mead, 1971.

Silver, Abner, and Robert Bruce. *How to Write and Sell a Hit Song*. Englewood Cliffs, NJ: Prentice-Hall, 1939.

Simon, George. *Simon Says: The Sights and Sounds of the Swing Era*. New Rochelle, NY: Arlington House, 1955.

———. *The Big Bands*. New York: Collier Books, 1974.

———. *The Best of the Music Makers: From Acuff to Ellington to Presley to Sinatra to Zappa & 279 of the Most Popular Performers of the Last 50 Years*. New York: Doubleday, 1979.

Simpson, George Eaton. *Black Religion in the New World*. New York: Columbia University Press, 1978.

Sklar, Rick. *Rocking America: How the All-hit Radio Stations Took Over*. New York: St. Martin's Press, 1984.

Sklar, Robert. *The Plastic Age 1917–1930*. New York: Braziller, 1970.

Slate, Sam J., and Joe Cook. *It Sounds Impossible*. New York: Macmillan, 1963.

Smith, Cecil. *Musical Comedy in America*. New York: Theatre Arts Books, 1950.

Smith, Kate. *Living in a Great Big Way*. New York: Blue Ribbon Books, 1938.

Smith, Willie the Lion, and George Hoefer, *Music on My Mind*. New York: Doubleday, 1964.

Smith-Baxter, Derrick. *Ma Rainey and the Classic Blues Singers*. London: Studio Books, 1970.

Sobel, Bernard. *Pictorial History of Vaudeville*. New York: Citadel Press, 1961.

Somma, Robert, ed. *Nobody Waved Goodbye: A Casualty Report on Rock and Roll*. New York: Outerbridge & Dienstfrey, 1971.

Song Writers' Protective Association. *Song Writers' Protective Association Prospectus*. New York: SPA, 1947.

———. *The SPA Quarterly Bulletin*. 1933–.

———. *What Every Songwriter Should Know*. New York: The Songwriters' Protective Association, c. 1952.

Southern, Eileen, ed. *The Music of Black Americans*. New York: Norton, 1971. *Readings in Black American Music*. New York: Norton, 1971.

Spaeth, Sigmund. *The Facts of Life in Popular Music*. New York: Whittlesey House, 1934.

———. *Fifty Years With Music*. New York: Fleet, 1959.

———. *A History of Popular Music in America.* New York: Random House, 1943.

Speck, Samuel H. *The Song Writers' Guide: A Treatise on How Popular Songs Are Written and Made Popular.* Detroit: Remick Music, 1910.

Spiegel, Irwin O., and Jay L. Cooper. *Record and Music Publishing Forms of Agreement in Current Use.* New York: Law-Arts Books, 1971.

Spitz, Robert Stephen. *The Making of Superstars: Artists and Executives of the Rock Music World.* New York: Doubleday, 1978.

Stagg, Jerry. *The Brothers Shubert.* New York: Ballantine, 1969.

Stearns, Marshall and Jean: *Jazz Dance: The Story of American Vernacular Dance.* New York: Macmillan, 1969.

Steiner, Max. *The Real Tinsel.* New York: Macmillan, 1970.

Stokes, Geoffrey. *Starmaking Machinery: Inside the Business of Rock and Roll.* New York: Vintage, Random House, 1972.

Sublette, Richard H. "A History of the American Society of Composers, Authors and Publishers' Relationship with the Broadcasters," Thesis. University of Illinois. 1962.

Suchman, Edward A. "An Invitation to Music: A Study of the Creation of New Music Listeners by Radio." In Lazarsfeld, Paul F. and Frank N. Stanton, *Radio Research 1941.* New York: Duell, Sloan and Pearce, 1941.

Sullivan, Mark. *Our Times.* Vol. 1: *The Turn of the Century;* Vol. 2: *America Finding Herself;* Vol. 3: *Pre-War America.* New York: Scribner, 1926, 1927, 1930.

Summers, Harrison B., ed. *A Thirty-Year History of Programs Carried on National Radio Networks in the United States 1926–1956.* New York: Arno Press, 1971.

Sweeney, David. *Demystifying Compact Discs. A Guide to Digital Audio.* Blue Ridge: Tab Books, 1986.

Szwed, John. "Negro Music: Urban Renewal." In *Our Living Traditions,* edited by Tristram P. Coffin. New York: Basic Books, 1968.

The Talking Machine Journal. 1917–1956.

Tannenbaum, Frank. *Slave and Citizen: The Negro in America.* New York: Knopf, 1947.

Tanner, Louise. *All the Things We Were.* New York: Doubleday, 1968.

Taylor, John Russell. *The Hollywood Musical.* New York: McGraw-Hill, 1971.

Taylor, Theodore. *Jule: The Story of Composer Jule Styne.* New York: Random House, 1979.

Thomas, Bob. *King Cohn: The Life and Times of Harry Cohn.* New York: Putnam, 1967.

Thompson, Charles. *Bing: The Authorized Biography.* New York: David McKay, 1976.

Thorson, Theodore Winton. "A History of Music Publishing in Chicago." Diss. Northwestern University, 1961.

Thrasher, Fredric. *Okay for Sound: How the Screen Found Its Voice.* New York: Duell, Sloan and Pearce, 1946.

Tiomkin, Dimitri, and Prosper Buranelli. *Don't Hate Me: An Autobiography.* New York: Doubleday, 1959.

Titon, Jeff Dodd. *Early Downhome Blues: A Musical and Cultural Analysis.* Urbana: University of Illinois Press, 1977.

Toffler, Alvin. *The Culture Consumers: A Controversial Study in Culture and Affluence in America*. New York: St. Martin's Press, 1964.

Toll, Robert. *On with the Show! The First Century of Show Business in America*. New York: Oxford University Press, 1976.

Tosches, Nick. *Country: The Biggest Music in America*. New York: Stein & Day, 1977.

A Tour of the World's Record Markets 1967. London: EMI, 1967.

Townsend, Charles R. *San Antonio Rose: The Life and Music of Bob Wills*. Urbana: University of Illinois Press, 1976.

Traubner, Richard. *Operetta: A Theatrical History*. New York: Doubleday, 1983.

Tremlett, George. *The Osmond Story*. New York: Warner Books, 1975.

———. *The Rolling Stones*. New York: Warner Books, 1975.

———. *The Who*. New York: Warner Books, 1975.

Tucker, Sophie. *Some of These Days*. Garden City, NY: Doubleday, 1945.

Tuska, Jon. *The Filming of the West*. New York: Doubleday, 1976.

Vallee, Rudy. *Let the Chips Fall. . . .* New York: Stackpole Books, 1976.

Vallee, Rudy, and Gil McKean. *My Time Is Your Time*. New York: Oblensky, 1962.

Vance, Joel. *Fats Waller, His Life and Times*. Chicago: Contemporary Books, 1977.

Variety. 1905–.

Vassal, Jacques. *Electric Children*. New York: Taplinger, 1976.

Villiers, Douglas. "Jewish Influences in 20th Century Pop Music and Entertainment." In *Jerusalem: Portraits of the Jew in the Twentieth Century*. New York: Viking, 1976.

Wakeman, Frederic. *The Hucksters*. New York: Rinehart, 1946.

Walker, Leo. *The Wonderful World of the Great Dance Bands*. New York: Doubleday, 1972.

Walker, Stanley. *The Night Club Era*. New York: Frederick Stokes, 1933.

Waller, Maurice, and Anthony Calabrese. *Fats Waller*. New York: Schirmer, 1978.

Walley, David. *No Commercial Potential: The Saga of Frank Zappa*. New York: Outerbridge & Lazard, 1972.

Warner Communications, Inc. *Annual Report*. 1974, 1975.

———. *The Prerecorded Music Market: An Industry Survey*. New York: WCI, 1978.

Waters, Edward N. *Victor Herbert: A Life in Music*. New York: Macmillan, 1955.

Waters, Ethel, with Charles Samuels. *His Eye Is on the Sparrow*. New York: Doubleday, 1951.

Weinberg, Meyer. *TV in America: The Morality of Hard Cash*. New York: Ballantine, 1962.

Westin, Helen. *Introducing the Song Sheet: A Collector's Guide to Song Sheets*. Nashville: Thomas Nelson, 1976.

Whalen, Richard J. *The Founding Father: The Story of Joseph P. Kennedy*. New York: New American Library, 1964.

Whitcomb, Ian. *After the Ball*. London: Lane/Penguin, 1972.

———. *Rock Odyssey: A Musician's Chronicle of the 60s*. Garden City, NY: Dolphin Books, Doubleday, 1983.

———. *Tin Pan Alley 1919–1939: A Pictorial History*. New York: Two Continents, 1975.

White, Llewellyn. *The American Radio: A Report on the Broadcasting Industry in the United States from the Commission on Freedom of the Press*. Chicago: University of Chicago Press, 1947.

Whiteman, Paul. *Records for the Millions*. New York: Hermitage, 1948.

Whiteman, Paul, and Mary Margaret McBride. *Jazz*. New York: J. H. Sears, 1926.

Whittinghill, Dick, and Don Page. *Did You Whittinghill This Morning*. Chicago: Regnery, 1976.

Wickes, E. M. *Writing the Popular Song*. Springfield, MA: The Home Correspondence School, 1916.

Wilder, Alec. *American Popular Song: The Great Innovators, 1900–1950*. Edited by James T. Maher. New York: Oxford University Press, 1972.

Wilk, Max. *Memory Lane, 1890–1925: The Golden Age in American Popular Music*. New York: Ballantine, 1973.

———. *They're Playing Our Song: From Jerome Kern to Stephen Sondheim*. New York: Atheneum, 1973.

Williams, Hank. *Hank Williams Tells How to Write Folk and Western Music to Sell*. Nashville, 1951.

Williams, Hank, Jr., with Michael Bane. *Living Proof: An Autobiography*. New York: Putnam, 1979.

Williams, Richard. *Out of His Head: The Sound of Phil Spector*. New York: Outerbridge & Lazard, 1972.

Winkler, Max. *A Penny from Heaven: The Autobiography of a Man Who Entered the World of Music Through the Basement and Came Out as One of the World's Greatest Music Publishers*. New York: Appleton-Century, 1951.

Witmark, Isidore, with Isidore Goldberg. *From Ragtime to Swingtime: The House of Witmark*. New York: Furman, 1939.

Woolcott, Alec. *The Story of Irving Berlin*. New York: Putnam, 1925.

Yorke, Ritchie. *The History of Rock 'n' Roll*. Toronto: Methuen/Two Continents, 1976.

———. *The Led Zeppelin*. Toronto: Methuen, 1976.

Zeidman, Irving. *The American Burlesque Show: A History*. New York: Hawthorn Books, 1967.

Congressional Hearings on the Music Business

To Provide for Recordings in Coin-Operated Machines at a Fixed Royalty Rate. Hearing before Subcommittee No. 3 of the Committee on the Judiciary, House of Representatives, Eighty-second Congress, First Session, on H.R. 5473. Serial No. 11. Washington: GPO, 1951.

Rendition of Musical Compositions on Coin-Operated Machines. Hearings Before a Subcommittee of the Committee on the Judiciary, United States Senate, Eighty-third Congress, First Session, on S. 1106. Washington: GPO, 1954.

[The Cellar Hearings] *Monopoly Problems in Regulated Industries. Hearings before the Antitrust Subcommittee (Subcommittee No. 5) of the Committee on the Judiciary, House of Representatives, Eighty-fourth Congress, Second Session. Serial No.* 22. Washington: GPO, 1957.

Report of the Antitrust Subcommittee (Subcommittee No. 5) of the Committee on the Judiciary, House of Representatives, Eighty-fourth Congress, First Session, Pursuant to H. Res. 107 Authorizing the Committee on the Judiciary to Conduct Studies and Investigations Relating to Certain Matters within Its Jurisdiction on the Television Broadcasting Industry. Washington: GPO, 1957.

[The Smathers Bill] *Amendment to Communications Act of 1934 (Prohibiting Radio and Television Stations from Engaging in Music Publishing or Recording Business). Hearings before the Subcommittee on Communications of the Committee on Interstate and Foreign Commerce, United States Senate, Eighty-fifth Congress, Second Session, on S. 2834.* Washington: GPO, 1958.

[The Roosevelt Hearings] *Policies of American Society of Composers, Authors, and Publishers. Hearings before Subcommittee No. 5 of the Select Committee on Small Business, House of Representatives, Eighty-fifth Congress, Second Session, Pursuant to H. Res. 56.* Washington: GPO, 1958.

[The Payola Hearings] *Deceptive Practices in Radio and Television. Hearings before the Select Subcommittee on Legislative Oversight of the Committee on Interstate and Foreign Commerce, House of Representatives, Eighty-sixth Congress, Second Session.* Washington: GPO, 1960.

Copyright Law Revision. Studies Prepared for the Subcommittee on the Judiciary, United States Senate, Eighty-sixth Congress, Second Session, Pursuant to S. Res. 240. Studies 26–28. Washington: GPO, 1961.

Copyright Law Revision. Hearings Before the Subcommittee on Patents, Trademarks, and Copyrights of the Committee on the Judiciary, United States Senate, Ninetieth Congress, First Session, Pursuant to S. Res. 37 on S. 597. Washington: GPO, 1967.

Copyright Law Revision. Senate Report No. 94-473. 94th Congress, 1st Session. Calendar No. 460. 1975.

Copyright Law Revision. Hearings before the Subcommittee on Courts, Civil Liberties, and the Administration of Justice of the Committee on the Judiciary, House of Representatives, Ninety-fourth Congress, First Session, on H.R. 2223. Serial No. 36. Part 1, Part 2, Part 3. Washington: GPO, 1976.

Periodicals, with Pertinent Articles

Advertising Age.

Robertson, Bruce. "Broadcast Changes Face of Advertising, Selling." April 30, 1980.

American Mercury.

"Fortunes Made in Popular Songs." Oct. 1916.
"Putting Over Popular Songs." April 1917.
"Paul Whiteman Made Jazz Contagious." Jan. 1924.
"Report on the Musical Industry." March 1938.
Yark, Dane. "The Rise and Fall of the Phonograph." Sept. 1932.

American Musician.

"Irving Berlin Gives Nine Rules for Writing Popular Music." Oct. 1920.

American Opinion.

Allen, Gary. "That Music: There's More To It Than Meets the Ear." Feb. 1969.

Atlantic Monthly.

"Industry of Music Making." Jan. 1908.
Larner, Jeremy. "What Do They Get from Rock 'n' Roll?" Aug. 1964.

Barrons.

Pacey, Margaret. "Home Box Office Inc.: Pay Television Is Finally Making the Scene." May 19, 1975.
Kagan, Paul. "Big Broadcast: Radio and Television Command Handsome Premiums." Jan. 8, 1979.

Billboard.

"ASCAP Fiftieth Anniversary Issue." April 1, 1964.
"Twenty-five Years of Tape: A History of a Powerful Communications Tool." Nov. 11, 1972.
"Music to move to Discos." Nov. 1, 1975.
"CISAC: 50 Years of Protecting Intellectual Property Rights." Nov. 6, 1976.
"Black Music: A Genealogy of Sound." June 9, 1979.

Business Week.

"Tin Pan Alley Changes Tempo." April 16, 1930.
"Platter War: American Decca Starts Something." Nov. 10, 1934.
"Tempest in a Tune-Pot: Warner's Withdrawal from ASCAP Music Pool." Jan. 11, 1936.

"Platter Programs: Warner-ASCAP Dispute Involves Recordings for Radio." Feb. 8, 1936.
"Discord of the Air: National Association of Broadcasters Meets, but Doesn't Solve Copyright Problems." Oct. 18, 1936.
"Radio Raises Music War-Chest." Sept. 23, 1939.
"Battle Over Records." Feb. 24, 1940.
"Radio Out to Bust Music Trust: Drive to Break ASCAP's Current Tune Monopoly." March 9, 1940.
"Radio Music Battle Nears Showdown." July 6, 1940.
"ASCAP Defied." Nov. 16, 1940.
"Department of Justice Calls Tune: Big Radio Music Battle Becomes Trust Case." Jan. 4, 1941.
"ASCAP Deal Lags." May 3, 1941.
"Pax ASCAP: Final Armistice Awards Financial Victory to Radio." Oct. 18, 1941.
"Nine Year Peace." Nov. 8, 1941.
"Records Again? Senate Grilling May Force Petrillo to End Ban." Jan. 23, 1943.
"Score by Petrillo: Concludes Deal with Decca, Inc., for Extra Payments, and Musicians End 13-month Ban on Recordings." Sept. 25, 1943.
"Decca Cashes In: Meeting Petrillo's Terms Paid Dividends." March 11, 1944.
"Victory for ASCAP: Writer's Status Determines Whether Performing Rights Can Be Transferred." May 12, 1945.
"In the Groove: Sales Hit a Gold Mine." July 21, 1945.
"Petrillo Blows Hot: Disc-Making Ban Motivated by Sharp Decline in Live Jobs." Oct. 25, 1947.
"It's Music, Music, Music." July 22, 1950.
"St. Louis Station Smashes Records to End the Sway of Rock 'n' Roll." Jan. 25, 1958.
"The Record Industry Sounds a Note of Joy." Dec. 1, 1975.
"The Cable-TV Industry Gets Moving Again." Nov. 21, 1977.
"The Race to Dominate the Pay-TV Market." Oct. 2, 1978.
"Striking It Rich in Radio." Feb. 2, 1979.

Channels.

Traub, James. "Video Steps Out." June/July 1982.
Stokes, Geoffrey. "The Sound and Fury of MTV." Sept./Oct. 1982.

The Commonweal.

Hentoff, Nat. "They *Are* Playing Our Song: The Grinding Mediocre Level of Most Popular Music Today Is Due Essentially to What the Teen-agers Do Want." May 4, 1960.

Cue.

Taylor, Tim. "Disc Jockeys Rule the Airwaves." Nov. 5, 1955.

Current Literature.

"Popular Music—A Curse or a Blessing." Sept. 1912.

Current Opinion.

"Voice of the South in American Music." Sept. 1916.
"Black Music: And Its Future Transmutation into Real Art." July 1917.
"Jazz and Ragtime Are the Preludes to a Great American Music." May 1920.
"Public Music in America Today Is Petty Business." Aug. 1921.
"Secrets of Popular Songwriting." Jan. 1925.

The Economist.

"American Gramophone Records: Golden Oldies." Sept. 16, 1978.

Equity.

"The Facts of Vaudeville." Nov. 1923–March 1924.

The Etude.

"Will Ragtime Turn to Symphonic Poems." May 1920.
"What It Means to Put Over a Popular Song." March 1925.
Mills, E. C. "Protect Your Friends from This Monstrous Song Swindle." July 1925.
"Music, the Magic Carpet of Radio." Dec. 1935.
"Battle of Music." March 1941.
"Famous Composers Rally to ASCAP." March 1941.
"Publishing a Popular Song." Sept. 1946.

Forbes.

"$2 Billion Worth of Noise." July 15, 1968.
"Pay-TV: Is It a Viable Alternative?" May 1, 1978.
"The Gorillas Are Coming." July 10, 1978.
"Leisure: With home taping taking a deadly toll on the recorded music industry, retailers might end up switching to video accessories and allow records to bite the dust." Jan. 5, 1981.

Fortune.

"5,000,000 Songs Are the Commercial Heritage of ASCAP." Jan. 1933.
"Phonograph Records: From Fat to Lean and Halfway Back Again." Sept. 1939.
"Music for the Home." Oct. 1946.
"Phonograph Record Boom." Jan. 1950.
"The Money Makers of 'New Radio.'" Feb. 1958.
"Stereo Goes to Market." Aug. 1958.
"The Record Business: It's Murder." May 1961.
"The Motown Sound of Money." Sept. 1967.
"The Record Business: Rocking to the Big Money Beat." April 1979.

Forum.

Clark, Kenneth. "Why Our Popular Songs Don't Last." March 1934.

Harper's Magazine.

"The Gentle Art of Song Writing." Jan. 1910.
"What Petrillo's Up To: The Fight Against Canned Music, and a Possible Solution." Dec. 1942.
Asbell, Bernard. "Disk Jockeys and Baby Sitters." July 1957.
"Upheaval in Popular Music." May 1959.

High Fidelity.

Ackerman, Paul. "What Has Happened to Popular Music." June 1955.
Ramin, Jordan. "How to Launch a Hit Song." Aug. 1974.
Melanson, Jim. "Countdown to Monday: Charting the Top 100." May 1977.
Everett, Todd. "Automated Radio: The Future Is Upon Us." Sept. 1977.
———. "The Great American Radio Ratings Rat Race." Nov. 1977.
Mayer, Ira. "Record Distribution: The Big 6 Take Over." Oct. 1979.
Rea, Steven X. "Music and Recordings in 1981: Bottom-Line Blues." Jan. 1982.
Sutherland, Sam. "The Indies Are Coming." Aug. 1982.

John Edwards Memorial Foundation Quarterly.

Green, Archie. "Graphics #39: Vernacular Music Albums." Winter 1981.

Journal of American Culture.

Slezak, Mary. "The History of Charlton Press, Inc. and Its Song Lyric Publications." Spring 1980.

Journal of American Folklore.

"The Hillbilly Issue." July/Sept. 1965.
Green, Archie. "Hillbilly Music: Source and Symbol." July/Sept. 1965.
Hellmann, John M., Jr. " 'I'm a Monkey': The Influence of Black American Blues Argot on the Rolling Stones." Oct./Dec. 1973.

Journal of Applied Psychology.

Wiebe, G. D. "A Comparison of Various Ratings Used in Judging the Merits of Popular Songs." Feb. 1939.

Journal of Geography.

Ford, Larry. "Geographic Factors in the Origin, Evolution and Diffusion of Rock and Roll Music." Nov. 1971.

The Journal of Popular Culture.

Austin, Mary. "Petrillo's War." Summer 1978.
Miller, Douglas T. "Popular Religion in the 1950s." IX, 1975.
Crawford, David. "Gospel Songs in Court: From Rural Music to Urban Industry." Fall 1977.

Journal of the Royal Society of Arts, London.

Wood, L. G. "The Growth and Development of the Recording Industry." Sept. 1971.

Literary Digest.

"Ethics of Ragtime." Aug. 10, 1912.
"To Censor Popular Songs." May 24, 1913.
"Our $600,000,000 Music Bill." June 28, 1913.
"Sources of Our Popular Song and Dance." Aug. 30, 1913.
"How to Tell Good Songs from Bad." Dec. 6, 1913.
"Canning Negro Melodies." May 27, 1916.
"Fortune in a Popular Song." July 1, 1916.
"Birth of Our Popular Songs." Oct. 7, 1916.
"There's Millions in the Pop of Popular Songs." March 3, 1918.
"Can Popular Songs Be Stamped Out." Aug. 14, 1920.
"Organize, Not Sell, Music." June 9, 1928.
"Radio, Friend or Foe." Nov. 3, 1928.
"Natural History of a Song." May 3, 1930.

"Broadcasters and Composers in a Clinch." Aug. 20, 1932.
"How Music Is Murdered." Aug. 5, 1933.
"Strident Song Writers Fight Duffy Copyright Bill." March 14, 1936.

Look.

"Great Rock 'n' Roll Controversy." June 26, 1956.
Schickel, Richard. "The Big Revolution in Records." April 15, 1958.
"Teen-agers Scream, Stomp and Shake for the Big Band Beat of Pop Rock." June 15, 1965.

Modern Music.

Copland, Aaron. "The Composers Get Wise." 4, 1940.

Music Business.

"The Story of BMI." Sept.–Nov. 1946.

Music Educators Journal.

Schwartz, Elliott. "Directions in American Composition Since the Second World War." Part I. Feb. 1975.
Childs, Barney. "Directions in American Composition Since the Second World War." Part II. March 1975.
Gary, Charles L. "A Closer Look at the New Copyright Law." Nov. 1977.

Music Trade Review.

Lux, Peter F. "When All's Said and Done, It's the Same Old Popular Song, but Its Dress Changed Occasionally." March 1931.

The Nation.

"Music and Monopoly: ASCAP and the Radio Industry." Dec. 4, 1940.

Nation's Business.

"Music Industry Plays Billion $ Tune." Sept. 1954.
"Country Music Makes the Bottom Line Boom." Feb. 1979.

Newsweek.

"Why a Bandleader Can't Broadcast a Song He Wrote." Jan. 11, 1936.
"Warner Publishers Give 36,000 Songs Back to Crooners." Aug. 15, 1936.
"America's Jukebox Craze: Coin Phonographs Reap Harvest." June 3, 1940.
"Radio's Battle of Music." Nov. 18, 1940.
"Dialers Face Tune Blackout as ASCAP-Radio Feud Deepens." Dec. 23, 1940.
"Peace on the Air." Nov. 10, 1941.
"Lac Bug vs. Jitterbugs." April 27, 1942.
"Czar of Platters: Petrillo Breaks Promise: Continues Ban on Musical Recordings." Oct. 23, 1944.
"Covers Up: What Packaging Has Done for the Record Industry." Dec. 25, 1944.
"Great Record Boom." Dec. 16, 1946.
"War on Wax: Behind Petrillo's Fight with the Disc Makers." Nov. 3, 1947.
"Caesar in Reverse." Dec. 8, 1947.
"Petrillo Peace." Feb. 9, 1948.
"And the Ban Played On." Dec. 27, 1948.
"White Council vs. Rock 'n' Roll." April 23, 1956.
"Rocking and Rolling." June 18, 1956.
"Golden Days: Profits on Platters." Dec. 2, 1957.
"The Folk and the Rock." Sept. 20, 1965.
"Mick Jagger and the Future of Rock." Jan. 4, 1971.
"Stars of the Cathode Church." Feb. 4, 1980.
"Record Companies Turn for the Better." Oct. 12, 1980.
"Is Rock on the Rocks?" April 19, 1982.
"The New Boom in Laser Discs." Jan. 21, 1983.
"Rocking Video: Suddenly Rock and Roll Is Here to See." April 18, 1983.
"The New Sound of Music." June 11, 1983.
"Not the Sound of Silence." Nov. 14, 1983.
"Britain Rocks in America—Again." Jan. 23, 1984.
"Michael Jackson, Inc." Feb. 27, 1984.
"Motown's 25 Years of Soul." May 23, 1984.

New York Daily News.

Stearn, Jess. "The Big Payola: New Tune in Tin Pan Alley." March 27–30, 1956.
———. "Rock 'n' Roll Rolls into Trouble." April 11–12, 1956.
Marsh, Dave. "The Rock 'n' Roll Recession." Sept. 30, 1979.

New York Journal-American.

Kilgallen, James L. "Irving Berlin: Fifty Years of Songs." August 5–9, 1957.
Baer, Atra. "The War of Songs." Oct. 13–16, 1957.
Horan, James, Dom Frasca, and John Mitchell. "The Fabulous 'Juke Box' Empire." Oct. 19–24, 1958.

New York Magazine.

Egan, Jack. "Breaking Records in the Record Business." March 26, 1979.
———. "Hollywood vs. Sony: Betamax on Trial." June 11, 1979.
———. "Pop Records Go Boom: CBS and MTV Are the Leaders." Oct. 31, 1983.

New York Post.

Greenberg, Charles, with Peter McElroy and Bernard Schiff. "The Rock and Roll Story." Oct. 2–12, 1958.
Schwartz, Bernard. "The Real Payola: What the Harris Committee Pigeonholed." Dec. 14–19, 1959.
Scaduto, Anthony. "The World of Rock." June 16–21, 1960.
Carr, William H. A., and Gene Grove. "Inside the Record Business." Nov. 12–26, 1962.

The New York Times.

"Lure of Viennese Waltz Wins Wealth for Composers." July 14, 1910.
"How Popular Song Factories Manufacture a Hit." Sept. 10, 1910.
"The Music Trust That Reigns Over Italian Opera." Jan. 8, 1911.
"Demand Royalties for French Music: SACEM Is Reestablished in New York." Jan. 12, 1911.
"New York Pays About $7,000,000 Yearly for Its Music." March 19, 1911.
"Trust for Control of Music Business: ASCAP Organized at Meeting Here." Feb. 14, 1914.
"Music Industries Chamber of Commerce Formed." Feb. 17, 1916.
"Supreme Court Rules Against Vanderbilt Hotel and Shanley's Restaurant for Playing Music by Victor Herbert." Jan. 23, 1917.
"Vaudeville Heads Accused by Board: Federal Commission Charges Managers Have Formed an Illegal Combination." May 15, 1918.
"Reciprocal Agreement for Copyrights of Musical Compositions Reached by France and U.S." May 29, 1918.
"National Association of Sheet Music Dealers Adopts Resolution to Bar German Titles and Reduce Size of Sheet Music." June 11, 1918.
"Music Publishers Sued as Trust: Irving Berlin, Leo Feist and Others Charged with Breaking Sherman Act." Aug. 4, 1920.
"Songwriters Plan Formation of Protective Association." Nov. 19, 1920.
"Music Publishers' Protective Association in Dispute with Lyric Writers and Composers Protective League Over Song Royalties." May 4, 1921.
"Chicago Songwriters Go on Strike Over Royalties." May 28, 1921.
"Tin Pan Alley: Where Popular Songs Are Manufactured." Feb. 18, 1923.
"ASCAP's J. C. Rosenthal Alleges Infringement of Copyright Law When Stations Broadcast Society's Works Without Payment of Royalties." March 22, 1923.
"50 Broadcasting Stations Begin to Negotiate with ASCAP." March 23, 1923.
"E. C. Mills Suggests Reduction in Broadcasting Stations." March 25, 1923.

"Radio Broadcasting Society Will Contest ASCAP Stand." April 12, 1923.

"ASCAP Demands Licenses from Every Commercial Station Broadcasting Copyrighted Music." April 14, 1923.

"Motion Picture Producers and Distributors of America Ready for Fight to Finish with Composers Over Copyright Fees." May 23, 1923.

"E. A. Wealti Causes Arrest for Playing 'Yes, We Have No Bananas' on Phonograph All Day." Aug. 7, 1923.

"Representatives of Broadcasting Stations Meet in Chicago: Organize National Association of Broadcasters to Press Fight to Broadcast Copyrighted Music." April 26, 1924.

"Classical Music Publishers Adopt Music Publishers' Association's Plan to Offer Radio Their Music Without Charge." May 14, 1924.

"E. C. Mills of MPPA on His Work to Censor Songs." July 6, 1924.

"ASCAP Wins Suit Requiring Movie Houses to Pay License Fees." July 18, 1924.

"Motion Picture Theater Owners Attack ASCAP 10¢ per Seat Charge for Music." Aug. 19, 1924.

"John McCormack and Lucrezia Bori Broadcast Over Eight-Station Network: Victor Talking Machine Company Produces Program." Jan. 2, 1925.

"Theatrical Managers Protective Association Will Refuse Contracts with Composers Who Do Not Retain Copyright Control." Jan. 4, 1925.

"E. C. Mills, Chairman of MPPA, Says Broadcasting Hurts Sheet Music Sales." Feb. 15, 1925.

"Passing of 'the Ragtime Queen' Causes Drop in Sheet Music Sales." Feb. 15, 1925.

Article on Methods employed by Tin Pan Alley to make songs popular. Feb. 15, 1925.

"NAB States That Radio Stations Have Concluded They Should Not Pay for Copyrighted Music." April 5, 1925.

"National Association of Broadcasters Discusses ASCAP Situation at Convention: Plans to Ask Congress to Fix Royalty Rates for Music." Sept. 17, 1925.

"E. C. Mills of MPPA and ASCAP Comments on Recent U.S. Supreme Court Decision Upholding Appeals Court Ruling That Copyright Law Applies to Broadcast Programs." Oct. 25, 1925.

"Justice Department Ends Two-Year Inquiry with Decision That ASCAP Is Not Violating Antitrust Law." Aug. 6, 1926.

"E. C. Mills Describes ASCAP's Use of Musical Scouts to Detect Radio Infringements." Sept. 5, 1926.

"E. C. Mills Believes Controversy Between Broadcasters and ASCAP Is Near Peaceful Settlement." Oct. 17, 1926.

"ASCAP Warns Members of American Hotel Association That They Must Have Licenses to Play Copyrighted Music." March 26, 1927.

"National Association of Music Merchants Urges Censorship of Suggestive Popular Songs." May 29, 1927.

"ASCAP Moves to Curtail Too Frequent Radio Play of Popular Songs." May 29, 1927.

"MPPA's Mills Signs Contract with Electric Research Products Inc., Licensors of Vitaphone and Movietone, on Behalf of Its 63 Members." Dec. 21, 1927.

"House Committee on Patents Holds Hearing on Bill to Legalize Artists' Bargaining for Compensation for Mechanical Reproduction of Works." April 4, 1928.

Article on trying out songs in Tin Pan Alley. April 8, 1928.

Article on hits in musical comedies and revues. Oct. 21, 1928.

Article on Tin Pan Alley's new methods of marketing songs. June 8, 1929.

"Bankrupt Music Publishers Blame Radio and Sound Films for Failure." Aug. 30, 1929.

"ASCAP Has Five-Year Contract with Majority of Members." Oct. 13, 1929.

"NBC-RCA Forms Own Publishing Arm." Dec. 5, 1929. Further stories on Dec. 8, 15, 1929.

"Publishers Discuss Use of Radio and Records to Stimulate Sales of Sheet Music." June 6, 1930.

Article on Tin Pan Alley's new methods of marketing. June 8, 1930.

"Some Radio Announcers Use 'Played by Permission of the Copyright Owners' to Protect Stations." March 22, 1931.

"U.S. Supreme Court Rules in Jewell-LaSalle Realty Co.: Hotels Must Pay for Rebroadcasting Music to Rooms." April 14, 1931.

"ASCAP Announces New License Terms for Radio." Nov. 8, 1931.

"ASCAP President Gene Buck Denies Charges of Racketeering." Feb. 27, 1932.

Feature article on Tin Pan Alley's rise to respectability. March 27, 1932.

"ASCAP Favors 300% Increase in Radio License Fees." April 17, 1932.

"Broadcasters Hold Out for Flat Fee While ASCAP Asks Percentage for Music Use." Aug. 21, 1932.

"NAB Agrees to Pay Percentage of Annual Receipts to ASCAP." Aug. 25, 1932.

"Richmond-Mayer Music Corp. of New York and Chicago Sues Music Dealers Service for Alleged Conspiracy in Publishing and Distribution Trade." Oct. 1, 1932.

Feature article on royalties for publishers and composers. Oct. 2, 1932.

"Composers Seek Higher Fees: Broadcasters Plan Formation of Radio Program Foundation to Establish Independent Music Catalogue." July 20, 1933.

"ASCAP Lays Drop in Sales of Sheet Music and Records to Radio." July 19, 20, 1933.

"Rudy Vallee Says Radio Does Not Pay Fair Share for Use of Popular Songs." Sept. 22, 1933.

"NAB Song Reservoir Cited." Oct. 8, 1933.

"ASCAP Sued by Government on Popular Music." Aug. 31, 1934.

"Seeking to Void Music License System, Justice Department Opens Suit Charging ASCAP Violation of Antitrust Act in Popular Music Business." June 12, 1935.

"NAB Favors Per-Piece Payment for Use of Copyrighted Music." June 23, July 9, 1935.

"Warner Brothers Music Subsidiary Plans to Secede from ASCAP Because of Broadcast Royalties." Nov. 22, 1935.

"Eleven Music Companies Withdraw from ASCAP." Nov. 27, 1935.

"Expiration of Warner Brothers ASCAP Contract Forces Popular Song Hits from Air." Jan. 1, 1936.

"Independent Stations Accept Three-Month Warner Contract. Networks Not Offered Deal." Jan. 3, 1936.

"Broadcasters Protest High ASCAP Rates Because 25% of Its Music Was Withdrawn by Warner." Jan. 11, 1936.
"ASCAP May Cause Warner Brothers to Establish Its Own Nationwide Radio Chain." March 7, 1936.
"Warner Music Houses Rejoin ASCAP: All Damage Suits Dropped." Aug. 4, 1936.
Review of Warner-ASCAP fight. Aug. 9, 1936.
"ASCAP Lists Hit Songs of 1935 According to Frequency of Airplay on CBS and NBC." Sept. 2, 1936.
"ASCAP Head Gene Buck Says Radio Popularity Kills Life of Songs. Suggests Limit on Airplay of New Music." Sept. 9, 1936.
"FTC Publishes Code for Popular Music Business." March 30, 1937.
"SAP Discusses Validity of Publishers' Claim to Copyright Renewals." March 30, 1938.
Fox Sylvan. "Disks Today: New Sounds and Technology." Aug. 20, 1967.
Gelb, Arthur. "Record Companies Taking Major Role as Theatre Angels." Sept. 25, 1967.
Shepard, Richard F. "Hunt for Talent for Pop Disks Goes On." Jan. 30, 1968.
Ferretti, Fred. "Witness Details Workings of the Recording Industry." June 8, 1973.
Lichtenstein, Grace. "Some Find Rock 'n' Roll Now Rock 'n' Recession." Jan. 30, 1975.
Rockwell, John. "Latin 'Salsa' Music Gains Popularity and Recognition." May 5, 1975.
———. "The Volatile Pop Field Is Bubbling." Aug. 29, 1976.
———. "A Mixed Bag of Treats for a Mixed Audience." Aug. 28, 1977.
Briggs, Kenneth. "Religious Broadcasting: The Fourth Network." Jan. 29, 1978.
Brown, Les. "The Networks Cry Havoc: All the Way to the Bank." Feb. 12, 1978.
Ditlea, Steve. "Rock Sings an International Tune." June 11, 1978.
McDowell, Edward. "Record Pirates: Industry Sings the Blues." June 30, 1978.
———. "Religious Networks Blossom." July 23, 1978.
Sheppard, Nathaniel, Jr. "Cities, Once Deaf to Rock, Turn On to Concert Revenues." Feb. 6, 1979.
Kirkeby, Marc. "Changing Face of Record Distribution: Cash Security Lures the Independents." Feb. 18, 1979.
Kornbluth, Jesse. "Merchandising Disco for the Masses." Feb. 18, 1979.
———. "High Court Says Single Permit Fee for TV Music Is Not Price Fixing." April 18, 1979.
Rockwell, John. "Decade-Old TV Music Question Still Open." April 18, 1979.
———. "Digital Recording Techniques Are Already Being Widely Used in Pop Music." July 3, 1979.
White, Timothy. "The Life and Times of Reggae." July 22, 1979.
Rockwell, John. "Industry's Sales Slowing After 25 Years of Steady Growth." Aug. 8, 1979.
Holsendolph, Ernest. "Religious Broadcasts Bring Rising Revenues and Create Rivalries." Dec. 7, 1979.
Rockwell, John. "The Music Craze Was All Disco." Dec. 23, 1979.
Hollie, Pamela. "Record Industry: Big Changes." Jan. 12, 1980.
Palmer, Robert. "Rock No Longer 'Devil's Music.'" Sept. 16, 1980.

Kerr, Peter. "Music Video's Uncertain Payoff." Aug. 29, 1981.
Palmer, Robert. "New Bands on Small Labels Are the Innovators of the 80's." Sept. 6, 1981.
"Battle Looms on Decontrol of Cable TV: Copyright Fees Could Rise Sharply." March 3, 1982.
McDowell, Edwin. "A New Copyright Law Is Authors' Target." April 30, 1982.
Wills, Kendall J. "Gospel Music on the Ascent." June 27, 1982.
Holden, Stephen. "RCA Gambling on Kenny Rogers." July 28, 1982.
Palmer, Robert. "Pop Music's Heyday Said To Be Waning Amid Falling Sales." Aug. 14, 1982.
DiNardo, Robert. "Musicians vs. Tape: A Survival Battle." Aug. 22, 1982.
"Music Charges for Local TV Are Ruled Illegal: Court Bars Music Licensing System." Sept. 23, 1982.
Yarrow, Peter. "AM Stations, Hurt by FM, Are Going Stereo." Oct. 14, 1982.
Palmer, Robert. "The Pop Record Industry Is Under Electronic Siege." Oct. 28, 1982.
Palmer, Robert. "In Hard Times, Pop Music Surges with Fresh Energy." Dec. 26, 1982.
Maslin, Janet. "A Song Is No Longer Strictly a Song: Now It's a 'Video.' " Jan. 23, 1983.
"Audiodisk: Record of Future?" March 18, 1983.
Bedell, Sally. "How Copyright Fees Affect Cable TV." March 19, 1983.
Salmans, Sandra. "Sales of Records on the Rise: Hit Albums Aid Upturn." March 29, 1983.
Holland, Bernard. "Digital Compact Disks: Replacement for LPs?" March 31, 1983.
Levine, Ed. "TV Rocks with Music." May 3, 1983.
Harmetz, Aljean. "Cassettes Are Changing Movie-Audience Habits." July 11, 1983.
Rockwell, John. "Country Music Is No Small-town Affair." July 17, 1983.
O'Connor, John J. "MTV: A Success Story with a Curious Shortcoming." July 24, 1983.
Palmer, Robert. "Will Video Clips Kill Radio as a Maker of Rock's Top 10." Aug. 1, 1983.
Lacayo, Richard. "The Rock Competition Steps Up a Beat." Aug. 7, 1983.
Lohr, Steve. "Hard Hit Sony Girds for a Fight in the American Electronics Market." Aug. 14, 1983.
Levine, Ed. "Music Video Turns on the Industry." Aug. 21, 1983.
"The Woes of the Little Guy." Aug. 21, 1983.
Palmer, Robert. "The Pop Record Business Shows Signs of Recovery." Aug. 31, 1983.
Specter, Michael J. "Rock Puts on a Three-Piece Suit." Oct. 2, 1983.
Pareles, Jon. "Copyrights, Tapes and Royalties Issue." Nov. 2, 1983.
Holden, Stephen. "Pop Music Surges Along New and Unexpected Paths." Nov. 11, 1983.
Salmans, Sandra. "What's New in Cable?" Nov. 27, 1983.
Pareles, Jon. "Pop Record Business Shows Signs of Recovery." Nov. 28, 1983.
Palmer, Robert. "Energy and Creativity Added Up to Exciting Pop." Dec. 25, 1983.
Palmer, Robert. "Songwriters Express Copyright Law Concern." Jan. 16, 1984.

Greenhouse, Linda. "Television Taping at Home Is Upheld by Supreme Court." Jan. 18, 1984.

Fantel, Hans. "50 Years Ago: The Birth of Tape." Feb. 12, 1984.

Welles, Merida. "Tower's Costly Gamble: But Will It Pay Off?" Aug. 5, 1984.

Pareles, Jon. "Rock Video, All Day and All Night." Aug. 21, 1984.

"Audio Disk Players Coming of Age." Nov. 26, 1984.

Palmer, Robert. "Pop Music Makes a Comeback and Video Helped It Out." Dec. 30, 1984.

Fantel, Hans. "Digital Sound Began To Live Up To Its Potential." Dec. 30, 1984.

Smith, Sally Bedell. "HBO Altering Its Plan After Year of Bad News." Jan. 22, 1985.

The New York Times Magazine.

Wolf, S. J. "What Makes a Song: A Talk with Irving Berlin." July 28, 1940.

Miller, Mitch. "June, Moon, Swoon and KoKoMo." April 24, 1955.

Samuels, Gertrude. "Why They Rock 'n' Roll and Should They." Jan. 12, 1958.

"Global Report on Rock 'n' Roll." July 27, 1958.

Wilson, John S. "How No-talent Singers Get Talent." June 21, 1959.

Levin, Phylis Lee. "The Sound of Music." March 14, 1965.

Greenfield, Jeff. "They Changed Rock, Which Changed the Culture, Which Changed Us." Feb. 16, 1975.

Slater, Jack. "A Sense of Wonder: To Be Young, Gifted and Blind." Feb. 23, 1975.

Rockwell, John. "Rock Lives." Feb. 27, 1977.

Harmetz, Aljean. "Hollywood's Video Gamble." March 28, 1982.

Lindsay, Robert. "Home Box Office Moves in on Hollywood." June 12, 1983.

New York World-Telegram & Sun.

Fischer, Muriel. "The Music Makers." Dec. 30, 1957.

The New Yorker.

Johnston, Alva. "Czar of Song: Gene Buck." Dec. 17, 24, 1932.

"Pulse of the Public: Career of Jack Kapp." Aug. 24, 1940.

Notes of the Music Library Association.

Schultz, Lucia A. "Performing Rights Societies of the United States." March 1979.

Option.

Kemp, Mark. “Name That Tune.” May/June 1989.
Unterberger, Richie. “Changes in the Independent Sector, Part I.” May/June 1988.

Outlook.

Arnoth, D. G. “At the Popular Music Publishers.” May 26, 1920.

The Player Monthly.

Kitchener, Frederick. “Popular Music and Its Popularity: A Study.” No. 1, 1910.

Printers Ink.

Peterson, Eldridge. “Radio Quarrel with Composers Now Approaching Climax.” June 28, 1940.

Publishers Weekly.

Wagner, Susan. “New Copyright Law Primer.” Dec. 26, 1977; Jan. 30, 1978.

Radio Daily.

“BMII: A Story of Free Enterprise.” April 17, 1950.

Record World.

“A Special Tribute to Sam Goody: 35 Years of Creative Record Retailing.” March 2, 1974.

Redbook.

Herndon, Booton. “The Battle Over the Music You Hear.” Dec. 1957.

The Reporter.

Mannes, Marya. “Who Decides What Songs Are Hits?” Jan. 10, 1957.

Rolling Stone.

"Record Quality: A Pressing Problem Remains Unsolved." March 18, 1982.
"New Music Thriving on Small Labels." April 29, 1982.
"Record Executives Sing the Blues at NARM." May 13, 1982.
"Record Rental Stores Booming in US." Sept. 2, 1982.
"Record Industry Nervous as Sales Drop Fifty Percent." Sept. 30, 1982.
Letters on home taping. Oct. 28, 1982.
Connelly, Christopher. "Rock Radio: A Case of Racism?" Dec. 9, 1982.
"Video Royalty Wars?" July 7, 1983.
Goodman, Fred. "Record Industry Preparing to Bury the LP." March 10, 1988.
Wilkinson, Peter. "What is DAT and Why Are the Record Companies Trying To Keep It from You?" September 10, 1987.

Sales Management.

"The Rocking Rolling Record Industry: The Tuff Generation Turned Us On." Dec. 15, 1966.

The Saturday Evening Post.

"Trouble in Tin Pan Alley." Oct. 19, 1935.
Grevatt, Ren, and Merrill Pollack. "It All Started with Elvis." Sept. 26, 1959.
Portis, Charles. "That New Sound from Nashville." Feb. 2, 1966.
Aaronowitz, Alfred A. "Pop Music: The Most? Or Just a Mess?" July 15, 1967.

Saturday Review.

Diamond, Morris. "Following the Ban: Trust Agreement Providing Opportunities for the Playing of Live Music." Jan. 29, 1949.
Hammerstein, Oscar, II. "Some Dissonances in Tin Pan Alley: ASCAP vs. BMI." Feb. 23, 1957.
Haverlin, Carl. "Answer to Oscar Hammerstein." March 2, 1957.
"ASCAP vs. BMI II." March 3, 1957.
"Twenty Years of Recordings." Aug. 26, 1967.
Shayon, Robert Louis. "The Copyright Dilemma." Nov. 11, 1967.
Heinsheimer, Hans W. "Music from the Conglomerates." Feb. 22, 1969.
"The Phonograph Celebrates a Birthday." July 2, 1977.
Mariani, John. "Television Evangelism: Milking the Flock." Feb. 3, 1979.

Sponsor.

"Is Radio Playing the Wrong Music?" June 27, 1955.

Television.

Land, Herman. "The Storz Bombshell." May 1957.

Time.

"U.S. vs. ASCAP." July 1, 1935.
"Merchants of Music." Aug. 10, 1935.
"Millworkers: Hollywood Songwriters." March 23, 1936.
"Phonograph Boom." Sept. 4, 1939.
"Record Price Cut: Columbia Recording Corporation." Aug. 19, 1940.
"Arnold to the Music World." Jan. 6, 1941.
"ASCAP's First Blow." Jan. 13, 1941.
"ASCAP Returns." May 19, 1941.
"Peace On Air." Aug. 11, 1941.
"Tunes Back." Nov. 10, 1941.
"Music's Moneybags: Money-Making Records." Sept. 21, 1942.
"One with the Dough: Petrillo's Contract with the Record Companies." Oct. 11, 1943.
"Platter for the Lion: MGM in the Phonograph Record Industry." Feb. 24, 1947.
"Petrillo's Resolve." Dec. 29, 1947.
"Record Mixup." Dec. 27, 1948.
"Top Jock." Feb. 14, 1955.
"Yeh-heh-heh-heh, Baby: Rock 'n' Roll." June 18, 1956.
"Rock 'n' Roll." July 23, 1956.
"The Voice and Payola." Sept. 9, 1957.
"Rock Is Solid." Nov. 4, 1957.
"The Singing Land." Dec. 23, 1957.
"Rock 'n' Roll: The Sound of the Sixties." May 21, 1965.
"Rock 'n' Roll: The Return of the Big Beat." Aug. 15, 1969.
"Pop Records: Moguls, Money & Monsters." Feb. 17, 1973.

Tower Records Pulse.

Santoro, Gene. "Shanachie's World-Beat Spectrum." March 1988.

TV Guide.

Gunther, Max. "The Beat Doesn't Go On: Rock Musicians Are Earning Fortunes—But Not from Television." July 22, 29, 1978.

USA Today.

Baker, Rob. "Music Videos Inspire Fast, Flashy Flicks." June 6, 1983.
White, Miles. "Megahits Revive Record Industry." Dec. 14, 1983.

Heller, Karen. "Rock Channel Grabs Us Overtime." Dec. 14, 1983.
An Interview with Les Garland: "MTV's World Turns Around Rock Culture." Feb. 20, 1984.
Wishnik, Debra. "Registers Ringing at Record Stores." March 5, 1984.

Village Voice.

Kopkind, Andrew. "The Dialectics of Disco: Gay Music Goes Straight." Feb. 12, 1979.
Smith, Howard, and Cathy Cox. "What's Wrong with the Record Business? 12 Presidents Speak Out." Feb. 26, 1979.
Fergusson, Isaac. "So Much to Say: The Journey of Bob Marley." May 18, 1982.
Christgau, Robert. "Rock 'n' Roller Coaster: The Music Biz on a Joyride." Feb. 7, 1984.
Reynolds, Steve. "Rate That Tune: The PTA Would Like the Recording Industry to Label Records Containing Vulgar or Violent Lyrics." Dec. 4, 1985.
Christgau, Robert. "Significance and Its Discontents in the Year of the Blip." March 1, 1988.

Village Voice Rock & Roll Quarterly.

Aaron, Charles. "Gettin' Paid. Is Sampling Higher Education or Grand Theft Audio?" Vol. 2, No. 3, Fall 1989.

Wall Street Journal.

Penn, Stanley. "Slipping Discs: Teen-agers Cut Buying of 'Pop' Records, Slow Industry's Total Gain." Aug. 31, 1959.
"Broadcast Music, Inc. Accused of Monopoly in Popular Music Field." Dec. 11, 1964.
O'Connor, John J. "Pop Music Explosion: Rock and Roll and Bankroll." April 9, 1969.
Gottschalk, Earl. "The Sound of Money: Rock Records Spawn Fortunes and Attract Growth-Minded Firms." Jan. 13, 1970.
Schmedel, Scott R. "The Trend Buckers: Record Firms Spin to Sweetest Music They Ever Heard—The Sound of Money." April 16, 1970.
Revzin, Philip. "The Rocky Road: Record Talent Scout Studies Highs and Lows for the Bottom Line." June 15, 1976.
Wysocki, Bernard Jr. "Higher Fi?: Computerized System May Free Recordings of Distortion Problems." Jan. 3, 1979.
Drinkhall, Jim. "Are Mafia Mobsters Acquiring a Taste for Sound of Rick?" Jan. 29, 1979.
Grover, Stephen. "Record Industry May Be in Groove Again After One of Worst Slumps in Its History." July 12, 1979.
"TV Stations Win Ruling on Payments for Rights to Music." Aug. 23, 1982.

Landro, Laura. "Record Industry Finding Financial Revival in Promoting Artists on Video Music Shows." Nov. 19, 1982.

———. "Video-Game Firms Face Tough Christmas As Industry Approaches a Major Shakeout." Sept. 2, 1983.

———. "Movie Studios' Cut in Videocassette Prices Stirs Battle with Retailers on Video Rentals." Sept. 23, 1983.

Kneale, Kevin. "Weirder Is Better in the Red-Hot Land of Rock Videos." Oct. 17, 1983.

Western World (Brussels).

Sinatra, Frank. "The Diplomacy of Music." Nov. 1957.

Index